ESSENTIAL CALCULUS

with applications in business, biology and behavioral sciences

Margaret L. Lial
Charles D. Miller

American River College
Sacramento, California

Scott, Foresman and Company   Glenview, Illinois
Dallas, Tex.   Oakland, N.J.   Palo Alto, Cal.   Tucker, Ga.

Cover: "Circus Wagon" by Joseph A. Burlini
Title page: Federal Reserve Bank of Minneapolis

Library of Congress Catalog Card Number: 74-82771
ISBN: 0-673-07959-7
AMS 1970 Subject Classification: 98A20

Copyright © 1975 Scott, Foresman and Company.
Philippines Copyright 1975 Scott, Foresman and Company.
All Rights Reserved.
Printed in the United States of America.

Preface

This book is designed for a one-semester or one- or two-quarter course in calculus for students of business, social science, or biology. The only prerequisite is a course in algebra.

Chapter 1, Fundamentals of Algebra, reviews and reinforces the ideas from algebra that will be needed in this text. A diagnostic pretest at the beginning of the chapter can be used to find those areas where student review will be helpful.

Functions and analytic geometry are discussed in Chapter 2. Chapter 3 introduces the first of the ideas from calculus—the limit. One of our goals in writing this book was to introduce calculus ideas early because students in a calculus course want to see some calculus as soon as possible.

A pretest is included before each chapter. The pretest for Chapter 1 is diagnostic, as mentioned above. The other pretests review those ideas from algebra, and previous chapters, needed for success in each new chapter.

Each chapter has one or more cases that show the ideas from that chapter being used in a practical situation. These cases add interest and variety to the course, and help answer that age-old question, "Yes, but what is this course really good for?" We would like very much to receive favorite examples of cases for future editions of the book.

We have tried to write a book that can really be covered in one semester or in one or two quarters—we used preliminary versions nicely in one semester. Optional sections are marked with an asterisk, and can be used as time permits. All of Chapter 8, on differential equations, is optional. Chapter 9, on calculus of more than one variable, has been designed so that the basic material is in the first half of the chapter. The other sections can be covered as time permits.

Many people suggested ways to improve the preliminary drafts of the manuscript. These reviewers were frank, open, and honest in their attempt to help us make this book one that will be successful in the classroom. The reviewers, teaching in many different types of colleges and universities, included Judith Clark, Larew Collister, Vern Heeren, Edgar Kelly, Robert Mosher, Lyman Peck, Roland Sink, Virginia Skinner, and Michael Williams. We are grateful for their ideas and advice.

Scott, Foresman seems to have more than its share of really talented and creative people. Pamela Conaghan and Robert Runck work very hard to insure utility and clarity in the books they edit, and are experts at finding reviewers who can really help authors with the writing and rewriting of a book.

Margaret L. Lial
Charles D. Miller

Contents

Pretest 1

1   Fundamentals of Algebra   2

1.1   The Real Numbers   2
1.2   Linear Equations and Inequalities   7
1.3   Quadratic Equations and Inequalities   11
1.4   Rational Expressions   16
1.5   Exponentials   20
        Case 1   Warehouse Location—The FMC Corporation   24

Pretest 28

2   Functions   29

2.1   Definition of a Function   29
2.2   Linear Functions   35
2.3   Some Nonlinear Functions   43
2.4   Polynomial and Rational Functions   52
2.5   Operations on Functions   59
        Case 2   Estimating Oil Tanker Construction Costs   62

Pretest 64

3   Limits   65

3.1   The Limit of a Function   65
3.2   Properties of Limits   75
3.3   Continuity   80
3.4   Limits to Infinity   86
        Case 3   Limit of a Sequence   89

Pretest 95

4   The Derivative   96

4.1   Definition of the Derivative   96
4.2   Techniques of Differentiation   103
4.3   Applications of the Derivative   108
4.4   Derivatives of Products and Quotients   116

4.5 The Chain Rule *120*
4.6 Implicit Differentiation* *123*
     Case 4 *Marginal Cost—Booz, Allen and Hamilton* *129*

Pretest 132

5 Further Applications of the Derivative 133

5.1 Optimization Theory—The First Derivative Test *133*
5.2 The Second Derivative Test *142*
5.3 Applications of the Theory of Extrema *146*
5.4 Curve Sketching *153*
5.5 Newton's Method* *161*
5.6 L'Hopital's Rule* *165*
     Case 5 *Minimizing Manufacturing Costs* *168*
     Case 6 *Minimum Warehouse Cost* *170*
     Case 7 *Pricing an Airliner—The Boeing Company* *172*

Pretest 175

6 Exponential and Logarithmic Functions 176

6.1 Exponential Functions *176*
6.2 Logarithmic Functions *182*
6.3 Applications of Exponential and Logarithmic Functions *189*
6.4 Derivatives of Exponential and Logarithmic Functions *195*
     Appendix *Common Logarithms* *199*
     Case 8 *Compound Interest* *203*
     Case 9 *Allocating Catalog Advertising—Montgomery Ward* *206*

Pretest 209

7 Integration 210

7.1 The Antiderivative *210*
7.2 Area and the Definite Integral *214*
7.3 The Fundamental Theorem of Calculus *218*
7.4 Some Applications of Integrals *221*
7.5 The Area Between Two Curves *226*
7.6 Additional Techniques of Integration *235*
7.7 Tables of Integrals *241*
7.8 Integration by Parts* *244*
7.9 Numerical Integration* *248*
     Case 10 *Estimating Depletion Dates for Minerals* *251*
     Case 11 *How Much Does a Warranty Cost?* *254*

*Optional section.

Pretest 257

8   Differential Equations   258

8.1   General and Particular Solutions   *258*
8.2   Separation of Variables   *263*
8.3   Applications of Differential Equations   *266*
       Case 12   *Differential Equations in Ecology*   *271*

Pretest   275

9   Multivariate Functions   276

9.1   Functions of Several Variables   *276*
9.2   Graphing in Three Dimensions   *279*
9.3   Partial Derivatives   *287*
9.4   Maxima and Minima   *293*
9.5   Lagrange Multipliers   *300*
9.6   An Application—The Least Squares Line   *304*
       Case 13   *Lagrange Multipliers for a Predator*   *309*

Appendix   313

Answers   320

Bibliography   342

Index   344

Chapter 1 Pretest

Evaluate. **[1.1]***

1. $|-3|$ 2. $|4-6|$ 3. $|-2|+|-8|$

Graph on the number line. **[1.1]**

4. $\{x \mid x \text{ is an integer}, -2 \leq x < 3\}$ 5. $\{x \mid x \geq 2\}$

Solve. **[1.2]**

6. $2(x+3) - 4x = x + 5$
7. $\dfrac{2}{x-1} + \dfrac{4}{2(x-1)} = \dfrac{4}{3}$
8. $3a - 4 > 2a$
9. $m - (3 - m) < 4m + 6$

Factor. **[1.3]**

10. $6x^2 - 7x - 3$
11. $9y^2 - 4$
12. Solve $m^2 - 3m - 10 = 0$ by factoring. **[1.3]**
13. Solve $-3x^2 - 2x = -1$ by factoring. **[1.3]**
14. Solve $m^2 - 2m - 3 > 0$. **[1.3]**

Perform the indicated operations. **[1.4]**

15. $\dfrac{m^2 + m}{4m^2} \cdot \dfrac{8}{m^2 + 2m + 1}$
16. $\dfrac{3}{p-1} - \dfrac{p}{p+1}$
17. $\left(\dfrac{a^2}{a+b} + \dfrac{ab}{a^2 - b^2} \right) \dfrac{1}{a}$

Evaluate the following and write all answers without exponents. **[1.5]**

18. 5^{-2} 19. $(2/3)^{-1}$ 20. $(16)^{1/2}$ 21. $(-8)^{2/3}$
22. $2^{-1} - 3^{-2}$

Simplify the following and write answers with only positive exponents. **[1.5]**

23. $\dfrac{x^5}{x^2}$ 24. $\dfrac{(3m^2)^{-2}}{m^{-3}}$ 25. $5^{-1/2} \cdot 5^{5/2}$ 26. $\dfrac{y^{2/3} \cdot y^{1/3}}{y^{-4/3}}$

*The bracketed bold-faced number refers to the section in the text which discusses the following exercises.

1 Fundamentals of Algebra

This chapter includes the essential algebra topics needed for the remainder of the text. Since this chapter is intended as a review, the topics are presented briefly without any attempt at a rigorous presentation. The emphasis is on drill problems so that students will be able to brush up on the required algebraic techniques. By taking the pretest for Chapter 1, students can determine which sections, if any, should be reviewed. Chapter 1 can also be used as a reference in later chapters, as indicated in their pretests.

1.1 The Real Numbers

The concept of *set* is perhaps the one mathematical idea most commonly associated in the public's mind with the "new math" of the late 1950's and early 1960's. A brief treatment of sets is beneficial for the following reasons: (1) Sets can help to clarify and classify some of the ideas we wish to discuss. (2) Sets make some mathematical ideas easier to understand, in particular, the important concept of a function, which we consider in Chapter 2.

A **set** is a collection of **elements** or **members**. Thus, we can speak of the set of all products manufactured by General Motors, or the set of all species of bacteria in a given culture, or the set of all families in a clan. Sets are written using **set braces**, { }. Thus, the set whose elements are the numbers 1, 2, 3, and 4 can be written

$$\{1, 2, 3, 4\}.$$

The order in which the elements are listed is unimportant, so that we could write {1, 2, 4, 3} or any other arrangement of the four numbers. It is customary to use capital letters to denote sets. Thus, we might call the set described above S, so that

$$S = \{1, 2, 3, 4\}.$$

It is sometimes more useful to describe a set by a *general rule* or *common property*. For example, it is perhaps difficult to see a common property of the elements in the set

{lizards, tortoises, finches},

but if we write the same set as

{$x \mid x$ is a species found by Charles Darwin on the Galápagos Islands},

then a common property of the elements is clear. This last set is read "The set of all elements x such that x is a species found by Charles Darwin on the Galápagos Islands." A set expressed in this way is said to be written in **set-builder notation**. The set

$$F = \{x \mid x \text{ is a number between 0 and 1}\}$$

cannot be described by listing its elements. Set F is an example of an *infinite set*, one which has an unending number of elements. On the other hand, a *finite set* is one which has a limited number of elements. Some infinite sets, unlike F, can be described by a listing process. For example, the set of numbers used for counting, called the **natural numbers** or **counting numbers**, can be written as

$$N = \{1, 2, 3, 4, \ldots\}.$$

A set whose elements are all elements of a second set is said to be a **subset** of that second set. Each element of the set $S = \{1, 2, 3, 4\}$ described above is an element of N, and so S is a subset of N, which is written as $S \subset N$. In this case, N is not a subset of S, written $N \not\subset S$, since there are elements in N which are not in S (the elements 5 and 6, for example).

The set P of all presidents of corporations is a subset of the set E of all executives of corporations. However, $E \not\subset P$. Two sets A and B are said to be **equal** whenever $A \subset B$ and $B \subset A$. Thus, $A = B$ means that the two sets contain exactly the same elements. The set of all bacteria of diameter greater than two centimeters contains no elements. Such a set is called the **empty set**, or **null set**, and symbolized $\varnothing$.

Relationships between numbers and sets of numbers are often clearer and easier to understand with the aid of a diagram called a **number line**. To construct a number line, choose any point on a horizontal line and label it 0. Then choose any point to the right of 0 and label it 1. The distance between 0 and 1 gives a unit measure that can be used repeatedly to locate points to the right of 1, which we label 2, 3, 4, and so on, and similarly, to locate points to the left of 0, labeled $-1, -2, -3$, and so on. Any number which can be associated with a point on the number line is called a **real number**.* All the numbers used in this text are real numbers. Since we frequently need to refer to some subsets of the set of real numbers, we identify them by name as follows:

---

*An example of a number that cannot be associated with a point on the number line is $\sqrt{-1}$.

4   Fundamentals of Algebra

Counting Numbers $\{1, 2, 3, \ldots\}$
Whole Numbers $\{0, 1, 2, 3, \ldots\}$
Integers $\{\ldots, -2, -1, 0, 1, 2, \ldots\}$
Rational Numbers $\left\{\frac{p}{q} \mid p, q \text{ are integers}, q \neq 0\right\}$
Irrational Numbers $\{x \mid x \text{ is a real number that is not rational}\}$

The relationships among these sets of numbers are shown in Figure 1.1.

Figure 1.1

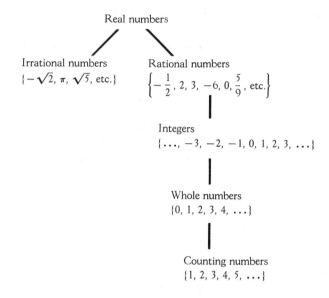

An important use of numbers is in measurement. Whole numbers are used for counting. Measurements are usually given as rational numbers, often in decimal form, such as 31.2 centimeters, 6.5 (or 13/2) grams, 1.0 liters, and so on. Some numbers, which cannot be expressed exactly as rational numbers, are irrational numbers. One example of an irrational number is π, the ratio of the circumference of a circle to its diameter. The number π can be approximated by writing $\pi \approx 3.14159$, or $\pi \approx 22/7$ ($\approx$ means "approximately equal to"), but there is no rational number that is exactly equal to π. Another irrational number can be found by constructing a triangle having a 90° angle, with the two shortest sides each of length 1, as shown in Figure 1.2. The third side can be shown to have a length which is irrational (and is denoted $\sqrt{2}$). Many whole numbers have square roots which are irrational numbers.

Figure 1.2

Given a subset of the set of real numbers, we can graph its elements on a number line by placing solid circles at the points on the number line associated with the elements of the set. The number line then becomes the graph of the set. For example, we can graph the set $\{-4, -2, 0, \sqrt{2}, 9/4, \pi\}$ as shown in Figure 1.3. (To locate π and $\sqrt{2}$ on the number line, we use the approximations $\pi \approx 3.142$ and $\sqrt{2} \approx 1.414$. Table 2 in the appendix gives approximate values of square roots.)

Figure 1.3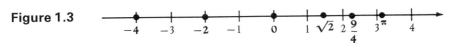

Example 1 Graph $\{x \mid x$ is an integer between 1 and 5$\}$.

The numbers 2, 3, and 4 are the only elements of this set. The graph is as shown in Figure 1.4.

Figure 1.4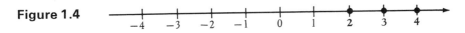

The set given in Example 1, $\{x \mid x$ is an integer between 1 and 5$\}$, could also have been written

$$\{x \mid x \text{ is an integer, } 1 < x < 5\},$$

where $1 < x < 5$ is read "1 is less than x and x is less than 5," or "x is between 1 and 5." (The word "between" implies that the numbers 1 and 5 themselves are excluded.)

Example 2 Graph $\{x \mid -2 < x < 3\}$.

To graph this set, which includes all the real numbers between -2 and 3 (not just integers), we draw a heavy, solid line from -2 to 3 on the number line, as shown in Figure 1.5. Open circles are drawn at -2 and at 3 to show that neither of these points belongs to the set.

Figure 1.5

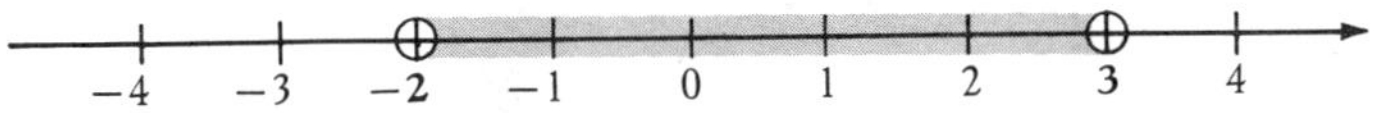

Example 3 Graph $\{x \mid x \geq -2\}$.

The symbol $\geq$ is read "greater than or equal to." Since the number -2 satisfies the condition $x \geq -2$, a solid circle is drawn on the number line at -2 and the graph is completed with a heavy line to the right of -2, as shown in Figure 1.6.

Figure 1.6

The distance between two places is always given as a positive number. For example, on the number line the distance from 0 to -2 is 2, the same as the distance from 0 to 2. We say that 2 is the absolute value of both numbers, 2 and -2. In general, the **absolute value** of a number a is the distance on the number line between a and 0. To express the absolute value of a, we write $|a|$, so that

$$|a| = \begin{cases} a & \text{if } a \geq 0 \\ -a & \text{if } a < 0. \end{cases}$$

Thus, $|5| = 5$, $|-5| = 5$, and $-|-5| = -5$.

Since absolute value represents the distance between two points on the number line, we can represent the distance between the numbers a and b by writing $|a - b|$, where

$$|a - b| = \begin{cases} a - b & \text{if } a - b \geq 0 \\ -(a - b) & \text{if } a - b < 0. \end{cases}$$

Using this definition, the distance on the number line between 4 and 6 can be expressed as either $|4 - 6| = 2$ or $|6 - 4| = 2$.

1.1 Exercises

Label each of the following true *or* false.

1. Every integer is a rational number.
2. Every integer is a whole number.
3. Some rational numbers are not integers.
4. Some whole numbers are not natural numbers.
5. Some whole numbers are natural numbers.
6. Some rational numbers are natural numbers.
7. No natural number is a rational number.
8. If a number is not a whole number, then it is not a rational number.

Graph each of the following on the number line.

9. $\{x \mid x \text{ is an integer}, -5 < x < 5\}$
10. $\{x \mid x \text{ is an integer}, -4 < x \leq -2\}$
11. $\{x \mid x \text{ is a whole number}, x \leq 3\}$
12. $\{x \mid x \text{ is a whole number}, 1 < x < 8\}$
13. $\{x \mid x \text{ is a natural number}, x \leq 2\}$
14. $\{x \mid x \text{ is a natural number}, -1 < x < 5\}$
15. $\{x \mid x < 4\}$
16. $\{x \mid x > 8\}$
17. $\{x \mid x > 5\}$
18. $\{x \mid x > -2\}$
19. $\{x \mid 6 \leq x\}$
20. $\{x \mid -4 < x\}$
21. $\{x \mid -5 < x < -4\}$
22. $\{x \mid 3 \leq x \leq 5\}$
23. $\{x \mid -3 \leq x \leq 6\}$
24. $\{x \mid 8 \leq x \leq 14\}$
25. $\{x \mid 1 < x < 6\}$
26. $\{x \mid -4 < x < 3\}$

Complete the following statements, using = whenever it applies, otherwise either ≤ or ≥.

27. $|5|$ ___ $|-5|$
28. $|3|$ ___ $|-3|$
29. $-|7|$ ___ $|7|$
30. $-|-4|$ ___ $|4|$
31. $|10 - 3|$ ___ $|3 - 10|$
32. $|6 - (-4)|$ ___ $|-4 - 6|$
33. $|1 - 4|$ ___ $|4 - 1|$
34. $|10 - 8|$ ___ $|8 - 10|$
35. $|-2 + 8|$ ___ $|2 - 8|$
36. $|3 + 1|$ ___ $|-3 - 1|$
37. $|3| \cdot |-5|$ ___ $|3(-5)|$
38. $|3| \cdot |2|$ ___ $|3(2)|$
39. $|4 + 3|$ ___ $|4| + |3|$
40. $|5 - 1|$ ___ $|5| + |-1|$
41. $|-2 + 3|$ ___ $|-2| + |3|$
42. $|5 - 1|$ ___ $|5| - |1|$
43. $|3 - 2|$ ___ $|3| - |2|$
44. $|3 - 4|$ ___ $|3| - |4|$
45. $\dfrac{|7|}{|-1|}$ ___ $\left|\dfrac{7}{-1}\right|$
46. $\dfrac{|-1|}{|-3|}$ ___ $\left|\dfrac{-1}{-3}\right|$

47. In general, if a and b are any real numbers having the same sign (both negative or both positive), is it always true that

$$|a + b| = |a| + |b|?$$

48. If a and b are any real numbers, is it always true that

$$|a + b| = |a| + |b|?$$

1.2 Linear Equations and Inequalities

An equation or inequality which approximately describes a real-life situation is an example of a **mathematical model**. Real-world problems can often be solved by writing appropriate equations or inequalities and finding their solutions. In this section and the next, we discuss methods of solving some of the most common types of equations and inequalities.

If the number 9 is substituted for x in the equation $2x + 1 = 19$, the result, $2(9) + 1 = 19$, is true. We say 9 **satisfies** the equation $2x + 1 = 19$ or that 9 is a solution of $2x + 1 = 19$. Any value of x other than 9 will make the equation $2x + 1 = 19$ false. Thus, 9 is the only solution to the equation. In the same way, the equation $|x| = 4$ has two solutions, 4 and -4, and the equation $x^2 = 9$ has the two solutions -3 and 3. In this section, we shall discuss methods for solving linear equations.

A **linear equation** in standard form is an equation of the form

$$ax + b = c,$$

where a, b, and c are real numbers, with $a \neq 0$. The equations $6x + 1 = 13$, $8x = 32$, and $x = 4$ are all examples of linear equations. The following three properties are used to solve linear equations.

Distributive Property For any real numbers a, b, and c,

$$a(b + c) = ab + ac \quad \text{and} \quad (b + c)a = ba + ca.$$

Addition Property of Equality If P, Q, and R represent any mathematical expressions, then the equations $P = Q$ and $P + R = Q + R$ each have the same solution. (If the same expression is added to both sides of an equation, the solution is unchanged.)

Multiplication Property of Equality If P, Q, and R ($R \neq 0$) are any mathematical expressions, then the equations $P = Q$ and $PR = QR$ each have the same solution. (If both sides of an equation are multiplied by the same nonzero expression, the solution is unchanged.)

For example, to solve the linear equation

$$5x - 3 = 12,$$

we first use the addition property of equality and add 3 to both sides of the given equation, obtaining

$$5x - 3 + 3 = 12 + 3$$
$$5x = 15.$$

We now multiply both sides of this last equation by 1/5, getting

$$\frac{1}{5}(5) = \frac{1}{5}(15)$$

$$x = 3.$$

Thus, the solution is 3, which can be checked by substituting 3 for x in the original equation.

Example 4 Find the solution of the equation

$$2k + 3(k - 4) = 2(k - 3).$$

First, use the distributive property to simplify $3(k - 4)$ and $2(k - 3)$. Doing this yields

$$2k + 3k - 12 = 2k - 6.$$

Using the distributive property again, we have $2k + 3k = (2 + 3)k = 5k$. This means we can write

$$5k - 12 = 2k - 6.$$

Now, using the addition property of equality, add $-2k + 12$ to both sides of the equation to get

$$3k = 6,$$

from which the equation $k = 2$ follows from the multiplication property of equality. Thus, the solution is 2.

1.2 Linear Equations and Inequalities

Example 5 Solve the equation $\frac{r}{10} - \frac{2}{15} = \frac{4}{5} \cdot r - \frac{3}{20}$.

To solve this equation, first eliminate all denominators by multiplying both sides of the equation by a *common denominator*, a number which can be divided evenly by each denominator in the equation. Here, 60 is such a number. Multiplying both sides of the equation by 60 and using the distributive property gives

$$60\left(\frac{r}{10} - \frac{2}{15}\right) = 60\left(\frac{4}{5} \cdot r - \frac{3}{20}\right)$$

$$(60)\frac{r}{10} - (60)\frac{2}{15} = (60)\frac{4}{5} \cdot r - (60)\frac{3}{20}$$

$$6r - 8 = 48r - 9.$$

The equation can now be solved by using the addition and multiplication properties of equality.

$$1 = 42r$$
$$r = 1/42.$$

Example 6 Solve $\frac{4}{k+2} - \frac{1}{3(k+2)} = \frac{11}{9}$.

Fractions can be cleared by multiplying both sides of the equation by $9(k+2)$, an expression which can be divided evenly by each denominator in the equation, as follows.

$$9(k+2)\left(\frac{4}{k+2}\right) - 9(k+2)\left(\frac{1}{3(k+2)}\right) = 9(k+2)\frac{11}{9}$$

$$36 - 3 = 11(k+2)$$
$$33 = 11k + 22$$
$$11 = 11k$$
$$k = 1.$$

If both sides of an equation are multiplied by an expression containing a variable, then there is a chance that extraneous solutions (solutions that do not satisfy the original equation) may be introduced. Extraneous solutions occur whenever the solution to an equation includes a value which makes any denominator of the equation zero. To avoid extraneous solutions, we must be certain that the proposed solution does not make any denominator zero. This is illustrated in the next example.

Example 7 Solve $\frac{x}{x-2} = \frac{2}{x-2} + 2$.

We note that $x - 2$ cannot equal 0 so that x cannot equal 2. The common denominator is $x - 2$. The equation is solved as follows:

$$(x-2)\left(\frac{x}{x-2}\right) = (x-2)\left(\frac{2}{x-2}\right) + (x-2)(2)$$

$$x = 2 + 2x - 4$$
$$x = -2 + 2x$$
$$x = 2.$$

The solution 2 makes both denominators zero, and must be rejected. Thus the equation has no solution. Recall that the multiplication property of equality requires that the quantity R which multiplies both sides of the equation be nonzero.

To solve a linear inequality, such as $3x - 5 < 7$, we use the following properties of inequalities, which are similar to the properties of equality.

Addition Property of Inequality* If P, Q, and R represent any mathematical expressions, then the inequalities $P < Q$ and $P + R < Q + R$ have the same solution.

Multiplication Property of Inequality* If P, Q, and R are any mathematical expressions, then
(a) when R is positive, the inequalities $P < Q$ and $PR < QR$ have the same solution.
(b) when R is negative, the inequalities $P < Q$ and $PR > QR$ have the same solution.

It is important to note that multiplication by a negative quantity reverses an inequality from $<$ to $>$. The multiplication axiom excludes the case where $R = 0$ since the two products would then be equal.

To use these properties to solve the inequality $3x - 5 < 7$, first add 5 to both sides, as follows.

$$3x - 5 + 5 < 7 + 5$$
$$3x < 12$$

Then multiply both sides by the positive number 1/3, to get

$$x < 4.$$

The solution set can be expressed in set-builder notation as $\{x \mid x < 4\}$. In this text, we shall write simply $x < 4$ as the solution.

Example 8 Solve $7 + 2y \geq 4 - 3y$.

First, use the addition property of inequality:

$$7 + 2y \geq 4 - 3y$$
$$3 \geq -5y.$$

Next, use part (b) of the multiplication property of inequality and multiply both sides by the negative number $-1/5$. The result is

$$-\frac{3}{5} \leq y \quad \text{or} \quad y \geq -\frac{3}{5}.$$

---

*Both the addition and the multiplication properties of inequality can also be stated for $>$, $\geq$, and $\leq$.

1.2 Exercises

Solve each of the following equations.

1. $4x - 1 = 11$
2. $3x + 5 = 23$
3. $-2k + 1 = 19$
4. $-5y - 6 = 14$
5. $4p - 11 + 3p = 2p - 1$
6. $8y - 5y + 4 = 2y - 9$
7. $9x - 2(x - 6) = 10x + 3$
8. $5r - 6(r + 4) = -5r - 4$
9. $2(k - 5) + 4k - 6 = 2k$
10. $2(z - 4) = 7z + 2 - 2z$
11. $4(1 - k) - 2(k + 3) = -6$
12. $3(2r + 1) - 2(r - 2) = 5$
13. $\dfrac{2f}{5} - \dfrac{f-3}{5} = 2$
14. $\dfrac{3w}{4} - \dfrac{2w}{3} = \dfrac{w-6}{3}$
15. $\dfrac{2}{r} + \dfrac{3}{2r} = \dfrac{7}{6}$
16. $\dfrac{2}{q} - \dfrac{3}{4q} = \dfrac{-5}{12}$
17. $\dfrac{1}{m-1} + \dfrac{2}{3(m-1)} = \dfrac{-5}{12}$
18. $\dfrac{4}{a+2} - \dfrac{1}{3(a+2)} = \dfrac{11}{9}$
19. $\dfrac{3}{4(z-2)} - \dfrac{2}{3(z-2)} = \dfrac{1}{36}$
20. $\dfrac{2}{3x+7} - \dfrac{5}{2(3x+7)} = \dfrac{-1}{56}$

Solve the following inequalities.

21. $2x + 1 \leq 9$
22. $3y - 2 < 10$
23. $-3p - 2 \leq 1$
24. $-5t + 3 \leq -2$
25. $6k - 4 < 3k - 1$
26. $2a - 2 > 4a + 2$
27. $m - (4 + 2m) + 3 < 2m + 2$
28. $2p - (3 - p) \leq -7p - 2$
29. $-2(3y - 8) > 5(4y - 2)$
30. $5r - (r + 2) \geq 3(r - 1) + 5$
31. $3p - 1 < 6p + 2(p - 1)$
32. $x + 5(x + 1) > 4(2 - x) + x$

1.3 Quadratic Equations and Inequalities

In this section we discuss the solution of quadratic equations and inequalities. An equation of the form

$$ax^2 + bx + c = 0;$$

where $a \neq 0$, is called a **quadratic equation** in standard form. Examples of quadratic equations include $x^2 - 4x - 5 = 0$, $16x^2 - 25 = 0$, and $4x^2 = 25$. A simple method of solving quadratic equations is by **factoring**—that is, writing the expression $ax^2 + bx + c$ as a product.

The simplest type of factoring is to use the distributive property to "factor out" any greatest common factor in the terms of the polynomial. For example, the polynomial $3x^2 + 9x + 15$ can be written as

$$3x^2 + 9x + 15 = 3(x^2) + 3(3x) + 3(5).$$

Then, by the distributive property, we have

$$3x^2 + 9x + 15 = 3(x^2 + 3x + 5).$$

Example 9 Factor out the greatest common factor in each of the following.
(a) $6x^2 + 18x = 6x(x) + 6x(3) = 6x(x + 3)$.
(b) $4y + 8y^2 + 12y^3 = 4y(1 + 2y + 3y^2)$.

When there is no factor (other than 1) common to all the terms of a quadratic expression, other methods of factoring can sometimes be used. For example, to factor the expression $m^2 + 2m - 24$, we need to find two numbers a and b such that

$$(m + a)(m + b) = m^2 + 2m - 24.$$

Since

$$(m + a)(m + b) = m^2 + (a + b)m + ab,$$

we must have $a + b = 2$ and $ab = -24$. That is, we need two numbers whose sum is 2 and whose product is -24. The numbers -4 and 6 satisfy these requirements, so we have

$$m^2 + 2m - 24 = (m - 4)(m + 6).$$

Example 10 Factor $x^2 - 9$.
This expression can be written as

$$x^2 - 9 = x^2 + 0x - 9 = (x + a)(x + b).$$

We want two numbers a and b so that $a + b = 0$ and $ab = -9$. The numbers that satisfy these conditions are 3 and -3, so that

$$x^2 - 9 = (x + 3)(x - 3).$$

In general, for the *difference between two squares*, $a^2 - b^2$, we have

$$a^2 - b^2 = (a + b)(a - b).$$

Example 11 Factor $6p^2 - 7p - 5$.
Here we use the method of trial and error to find numbers a, b, c, and d so that

$$6p^2 - 7p - 5 = (ap + b)(cp + d).$$

By inspection we find that

$$6p^2 - 7p - 5 = (3p - 5)(2p + 1).$$

Now we can discuss the solution of quadratic equations by factoring. The method depends on the following property.

Zero-Factor Property If a and b are real numbers with $ab = 0$, then $a = 0$ or $b = 0$.

1.3 Quadratic Equations and Inequalities

We can use the zero-factor property to solve the quadratic equation $x^2 + 5x + 6 = 0$. First, factor $x^2 + 5x + 6$:

$$x^2 + 5x + 6 = 0$$
$$(x + 2)(x + 3) = 0.$$

We have a product of two factors, $(x + 2)$ and $(x + 3)$, equal to 0. Thus, by the zero-factor property,

$$x + 2 = 0 \quad \text{or} \quad x + 3 = 0$$
$$x = -2 \quad \text{or} \quad x = -3.$$

The numbers -2 and -3 are both solutions of the equation $x^2 + 5x + 6 = 0$.

Example 12 Solve $6x^2 + 7x = 3$.

To use the zero-factor property, first write the equation so one side is zero, and then factor.

$$6x^2 + 7x = 3$$
$$6x^2 + 7x - 3 = 0$$
$$(3x - 1)(2x + 3) = 0$$

$$3x - 1 = 0 \quad \text{or} \quad 2x + 3 = 0$$
$$x = \frac{1}{3} \quad \text{or} \quad x = \frac{-3}{2}.$$

Not all quadratic equations can be solved by factoring. Hence we need a more general method. The solution of any quadratic equation can be found by using the quadratic formula given below.

Quadratic Formula The solutions of the quadratic equation, $ax^2 + bx + c = 0$, $a \neq 0$, are given by

$$x = \frac{-b \pm \sqrt{b^2 - 4ac}}{2a}.$$

Example 13 Solve $x^2 - 4x - 5 = 0$ by using the quadratic formula.

First, identify the values of a, b, and c that are required by the quadratic formula. Here $a = 1$, $b = -4$, and $c = -5$. Then, by the quadratic formula, we have

$$x = \frac{-(-4) \pm \sqrt{(-4)^2 - 4(1)(-5)}}{2(1)}$$

$$= \frac{4 \pm \sqrt{16 + 20}}{2}$$

$$= \frac{4 \pm 6}{2}.$$

14 Fundamentals of Algebra

The two solutions are

$$x = \frac{4+6}{2} = 5 \quad \text{and} \quad x = \frac{4-6}{2} = -1.$$

Example 14 Solve $x^2 + 1 = 4x$ by using the quadratic formula.

To use the quadratic formula, the equation must be written in the form $ax^2 + bx + c = 0$. We can rewrite the given equation in that form by adding $-4x$ to both sides, to get

$$x^2 - 4x + 1 = 0.$$

Now we find $a = 1$, $b = -4$, and $c = 1$. Thus,

$$x = \frac{4 \pm \sqrt{16 - 4}}{2}$$

$$= \frac{4 \pm \sqrt{12}}{2}.$$

To write the solutions in the simplest form, simplify $\sqrt{12}$ by writing $\sqrt{12} = \sqrt{4 \cdot 3} = \sqrt{4} \cdot \sqrt{3} = 2\sqrt{3}$. Substituting $2\sqrt{3}$ for $\sqrt{12}$ gives

$$x = \frac{4 \pm 2\sqrt{3}}{2}$$

$$= \frac{2(2 \pm \sqrt{3})}{2}$$

$$= 2 \pm \sqrt{3}.$$

The two solutions are $2 + \sqrt{3}$ and $2 - \sqrt{3}$.

To solve a *quadratic inequality*, we use the following procedure, which is illustrated by solving the inequality $x^2 - x - 12 < 0$.

Step 1. Find the solution of the corresponding quadratic equation.

$$x^2 - x - 12 = 0$$
$$(x - 4)(x + 3) = 0$$
$$x - 4 = 0 \quad \text{or} \quad x + 3 = 0$$
$$x = 4 \quad \text{or} \quad x = -3.$$

Step 2. Locate the solutions to the quadratic equation on a number line. These points divide the number line into regions. In our example, there are three regions, A, B, and C, as indicated in Figure 1.7.

Figure 1.7

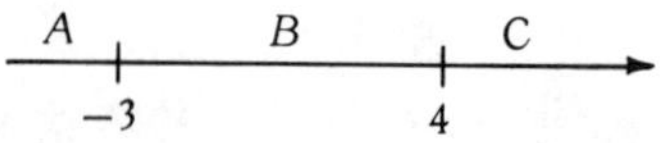

```
        A       B       C
    ────+───────────+──────►
        -3          4
```

Step 3. For each region, select any point in that region and use it to determine whether $x^2 - x - 12$ is positive or negative in that region. Let us select the values -4 from region A, 0 from region B, and 5 from region C. Testing these values in the inequality, we have

$$(-4)^2 - (-4) - 12 = 16 + 4 - 12 \quad \text{(positive)}$$
$$0^2 - 0 - 12 = -12 \quad \text{(negative)}$$
$$5^2 - 5 - 12 = 25 - 5 - 12 \quad \text{(positive)}$$

This means that the inequality $x^2 - x - 12 < 0$ is satisfied only by numbers in region B. Thus, the solution set, which includes all points in region B, is $\{x \mid -3 < x < 4\}$.

Example 15 Solve $x^2 + 3x > 4$.
First rewrite the inequality so that one side is zero.

$$x^2 + 3x > 4$$
$$x^2 + 3x - 4 > 0.$$

Now follow the steps mentioned above.

Step 1. Solve $x^2 + 3x - 4 = 0$:

$$(x - 1)(x + 4) = 0$$
$$x - 1 = 0 \quad \text{or} \quad x + 4 = 0$$
$$x = 1 \quad \text{or} \quad x = -4.$$

Step 2. Determine the required regions. See Figure 1.8.

Figure 1.8

Step 3. Select any number from each region and determine the value of $x^2 + 3x - 4$ for that number. Let us choose -5 from region A, 0 from B, and 2 from C. Then

$$(-5)^2 + 3(-5) - 4 = 25 - 15 - 4 \quad \text{(positive)}$$
$$0^2 + 3(0) - 4 = 0 + 0 - 4 \quad \text{(negative)}$$
$$2^2 + 3(2) - 4 = 4 + 6 - 4 \quad \text{(positive)}$$

Since any number in region A or any number in region C makes $x^2 + 3x - 4$ positive, the solution set can be written as $\{x \mid x < -4 \text{ or } x > 1\}$.

1.3 Exercises

Factor the following expressions.

1. $12m^2 + 8m$
2. $3p^2 - 18p$
3. $6z^2 - 8z^3 - 12z$
4. $9m^2 - 18m^3 + 27m^4$

16 Fundamentals of Algebra

5. $4p^2 + 3p - 1$
6. $6x^2 + 7x - 3$
7. $2x^2 + 7x - 30$
8. $4z^2 + z - 3$
9. $6q^2 - q - 12$
10. $10b^2 - 19b - 15$
11. $12r^2 + 24r - 15$
12. $12p^2 + p - 20$
13. $18r^2 - 3r - 10$
14. $12m^2 + 16m - 35$
15. $m^2 - n^2$
16. $z^2 - 4$
17. $1 - a^2$
18. $9b^2 - c^2$
19. $25x^2 - 4y^2$
20. $9 - 16p^2$

Solve the following equations by factoring.

21. $m^2 - 2m - 15 = 0$
22. $p^2 - p - 6 = 0$
23. $x^2 - 3x - 4 = 0$
24. $m^2 + 5m + 6 = 0$
25. $k^2 = 24 - 5k$
26. $r^2 - 16 = 0$
27. $x^2 - 9 = 0$
28. $2z^2 + 5z - 3 = 0$
29. $3k^2 + 7k - 6 = 0$
30. $2p^2 + 3p - 20 = 0$
31. $2k^2 - k = 0$
32. $6r^2 - 4r = 0$

Solve the following equations by the quadratic formula.

33. $2r^2 - 7r + 5 = 0$
34. $8x^2 - 8x = -3$
35. $y^2 - 6y + 8 = 0$
36. $p^2 - 15 = 2p$
37. $-r^2 - 3r + 3 = 0$
38. $2z^2 = 2z + 3$
39. $6k^2 - 11k + 4 = 0$
40. $8m^2 - 10m + 3 = 0$
41. $x^2 + 3x = 10$
42. $p^2 = 30 - p$
43. $r^2 + r = 12$
44. $y^2 - 2y = 3$

Solve the following inequalities.

45. $x^2 - 9 < 0$
46. $p^2 > 16$
47. $y^2 - 10y + 25 < 25$
48. $m^2 + 6m + 9 < 9$
49. $r^2 + 4r + 4 \geq 9$
50. $z^2 + 6z + 9 \leq 1$
51. $x^2 - x \leq 6$
52. $r^2 + r < 12$
53. $2k^2 - 9k > -4$
54. $3n^2 < 10 - 13n$

1.4 Rational Expressions

In later chapters, we shall need to simplify **rational expressions**—fractions which include algebraic expressions—such as

$$\frac{2x + 5}{3x^3}, \quad \frac{4y + \dfrac{1}{y}}{y + 1}, \quad \text{and} \quad \frac{2 + x}{\dfrac{1}{x}}.$$

The rules for simplifying rational expressions are summarized below.

> **Properties of Rational Expressions** For all mathematical expressions, P, $Q \neq 0$, R, and $S \neq 0$,
>
> (a) $\dfrac{P}{Q} = \dfrac{R}{S}$ if and only if $PS = QR$.
> (Equality test for rational expressions.)

(b) $\dfrac{P}{Q} = \dfrac{P \cdot S}{Q \cdot S}$.

(Fundamental property of rational expressions.)

(c) $\dfrac{P}{Q} \cdot \dfrac{R}{S} = \dfrac{PR}{QS}$.

(Multiplication of rational expressions.)

(d) $\dfrac{P}{Q} + \dfrac{R}{Q} = \dfrac{P + R}{Q}$.

(Addition of rational expressions.)

(e) $\dfrac{P}{Q} - \dfrac{R}{Q} = \dfrac{P - R}{Q}$.

(Subtraction of rational expressions.)

(f) $\dfrac{P}{Q} \div \dfrac{R}{S} = \dfrac{P \cdot S}{Q \cdot R}$, $R \neq 0$.

(Division of rational expressions.)

Example 16 Simplify $\dfrac{x^2 + 2x + 1}{x^2 - 1}$.

Here we can use the fundamental property of rational expressions by first factoring the numerator (top) and the denominator (bottom) of the fraction.

$$\dfrac{x^2 + 2x + 1}{x^2 - 1} = \dfrac{(x + 1)(x + 1)}{(x + 1)(x - 1)}$$

$$= \dfrac{x + 1}{x - 1}.$$

When working with rational expressions, it is important to realize that any number which makes the denominator zero cannot be used to replace the variable. Thus, in Example 16, we have the implicit restrictions $x \neq 1$ and $x \neq -1$. When solving equations which involve rational expressions, these restrictions must be considered. From now on, we shall not state (unless specifically required) these restrictions on the variable. But, keep in mind that no statement involving rational expressions is meaningful for values of the variable which make any denominator zero.

Example 17 Simplify $\dfrac{3m^2 - 2m - 8}{3m^2 + 14m + 8} \cdot \dfrac{3m + 2}{3m + 4}$.

We use part (c) above, multiplication of rational expressions. Again, we must first factor.

$$\dfrac{3m^2 - 2m - 8}{3m^2 + 14m + 8} \cdot \dfrac{3m + 2}{3m + 4} = \dfrac{(m - 2)(3m + 4)}{(m + 4)(3m + 2)} \cdot \dfrac{3m + 2}{3m + 4}$$

$$= \dfrac{(m - 2)(3m + 4)(3m + 2)}{(m + 4)(3m + 2)(3m + 4)}$$

$$= \dfrac{m - 2}{m + 4}.$$

18   Fundamentals of Algebra

Example 18 $\dfrac{x^2 - x}{x^2 - 1} \div \dfrac{x^2}{x + 1} = \dfrac{x(x - 1)}{(x + 1)(x - 1)} \cdot \dfrac{x + 1}{x^2}$ (By (f) above)

$$= \frac{x(x - 1)(x + 1)}{x^2(x - 1)(x + 1)}$$

$$= \frac{1}{x}.$$

Example 19   Simplify $\dfrac{y + 2}{y - 1} + \dfrac{y}{y + 1}$.

To use part (d), addition of rational expressions, both rational expressions must have the same denominator. A common denominator here is $(y - 1)(y + 1)$. We can use the fundamental property to change both rational expressions to fractions with the same denominator.

$$\frac{y + 2}{y - 1} + \frac{y}{y + 1} = \frac{(y + 2)(y + 1)}{(y - 1)(y + 1)} + \frac{y(y - 1)}{(y + 1)(y - 1)}$$

$$= \frac{y^2 + 3y + 2}{(y - 1)(y + 1)} + \frac{y^2 - y}{(y - 1)(y + 1)}$$

$$= \frac{y^2 + 3y + 2 + y^2 - y}{(y - 1)(y + 1)}$$

$$= \frac{2y^2 + 2y + 2}{(y - 1)(y + 1)}$$

$$= \frac{2(y^2 + y + 1)}{(y - 1)(y + 1)}.$$

It is customary to leave the denominator in factored form.

Example 20   Simplify $\dfrac{1}{a + b} - \dfrac{1}{a}$.

First, choose a common denominator. Here we can use $a(a + b)$. Next, change both fractions to fractions with that denominator.

$$\frac{1}{a + b} - \frac{1}{a} = \frac{a}{a(a + b)} - \frac{(a + b)}{a(a + b)}$$

$$= \frac{a - (a + b)}{a(a + b)}$$

$$= \frac{a - a - b}{a(a + b)}$$

$$= \frac{-b}{a(a + b)}.$$

1.4 Exercises

Perform the indicated operations to simplify each of the following expressions.

1. $\dfrac{2x-2}{3} \cdot \dfrac{6x-6}{(x-1)^3}$

2. $\dfrac{3m-15}{4m-20} \cdot \dfrac{m^2-10m+25}{12m-60}$

3. $\dfrac{a^2-a-6}{a+2} \div \dfrac{a-3}{a+2}$

4. $\dfrac{m^2+11m+30}{m+6} \div \dfrac{m+5}{m+6}$

5. $\dfrac{p^2-p-12}{p^2-2p-15} \cdot \dfrac{p^2-9p+20}{p^2-8p+16}$

6. $\dfrac{x^2+2x-15}{x^2+11x+30} \cdot \dfrac{x^2+2x-24}{x^2-8x+15}$

7. $\dfrac{2n^2-5n-12}{2n^2+5n-12} \div \dfrac{2n^2+9n+9}{2n^2+3n-9}$

8. $\dfrac{3z^2+z-2}{4z^2-z-5} \div \dfrac{3z^2+11z+6}{4z^2+7z-15}$

9. $\dfrac{x^2-y^2}{(x-y)^2} \div \dfrac{x^2-xy+y^2}{x^2-2xy+y^2}$

10. $\dfrac{4y^2-25}{4y^2-20y+25} \cdot \dfrac{2y-5}{y}$

11. $\dfrac{m}{m+n} + \dfrac{n}{m+n}$

12. $\dfrac{3}{y} + \dfrac{4}{y}$

13. $\dfrac{1}{y} + \dfrac{1}{y+1}$

14. $\dfrac{2}{3(x-1)} + \dfrac{1}{4(x-1)}$

15. $\dfrac{2}{a+b} - \dfrac{1}{2(a+b)}$

16. $\dfrac{3}{m} - \dfrac{1}{m-1}$

17. $\dfrac{1}{a+1} - \dfrac{1}{a-1}$

18. $\dfrac{1}{x+z} + \dfrac{1}{x-z}$

19. $\dfrac{m+1}{m-1} + \dfrac{m-1}{m+1}$

20. $\dfrac{2}{x-1} + \dfrac{1}{1-x}$

    (Hint: $1-x = -(x-1)$)

21. $\dfrac{3}{a-2} - \dfrac{1}{2-a}$

22. $\dfrac{3}{m-n} - \dfrac{m}{m+n}$

23. $\dfrac{1}{a^2-5a+6} - \dfrac{1}{a^2-4}$

24. $\dfrac{-3}{m^2-m-2} - \dfrac{1}{m^2+3m+2}$

25. $\dfrac{1}{x^2+x-12} + \dfrac{1}{x^2-7x+12}$

26. $\dfrac{2}{2p^2-9p-5} + \dfrac{p}{3p^2-17p+10}$

27. $\left(\dfrac{3}{p-1} - \dfrac{2}{p+1}\right)\left(\dfrac{p-1}{p}\right)$

28. $\left(\dfrac{y}{y^2-1} - \dfrac{y}{y^2-2y+1}\right)\left(\dfrac{y-1}{y+1}\right)$

29. $\dfrac{\dfrac{1}{x+h} - \dfrac{1}{x}}{h}$

30. $\dfrac{1}{h}\left(\dfrac{1}{(x+h)^2+9} - \dfrac{1}{x^2+9}\right)$

(Hint: first simplify the numerator and denominator)

31. $\dfrac{1+\dfrac{1}{x}}{1-\dfrac{1}{x}}$

32. $\dfrac{2-\dfrac{2}{y}}{2+\dfrac{2}{y}}$

33. $\dfrac{1 + \dfrac{1}{1-b}}{1 - \dfrac{1}{1+b}}$

34. $m - \dfrac{m}{m + \dfrac{1}{2}}$

1.5 Exponentials

The expression a^m is called an **exponential**. The number a is called the *base* and m is the *exponent* or *power*. If m and n are nonnegative integers and if a is any real number, then the exponentials a^m and a^n satisfy the following properties.

> **Properties of Exponentials** For any real numbers a and b, and any nonnegative integers m and n,
>
> (a) $a^m = a \cdot a \cdot a \cdots a$, where a appears as a factor m times $(m \neq 0)$
>
> (b) $a^m \cdot a^n = a^{m+n}$
>
> (c) $\dfrac{a^m}{a^n} = \begin{cases} a^{m-n} & \text{if } m > n \\ 1 & \text{if } m = n \\ \dfrac{1}{a^{n-m}} & \text{if } m < n \end{cases}$ $(a \neq 0)$
>
> (d) $(a^m)^n = a^{mn}$
>
> (e) $(ab)^m = a^m b^m$
>
> (f) $a^0 = 1$ $(a \neq 0)$
>
> (g) $\left(\dfrac{a}{b}\right)^n = \dfrac{a^n}{b^n}$ $(b \neq 0)$

Using these properties for nonnegative integer exponents, we have, for example,

$$\dfrac{a^5}{a^2} = a^{5-2} = a^3,$$

and

$$\dfrac{a^2}{a^5} = \dfrac{1}{a^{5-2}} = \dfrac{1}{a^3},$$

if $a \neq 0$. If we try to work this second example by subtracting exponents directly, as we did with the first example above, we would have

$$\dfrac{a^2}{a^3} = a^{2-5} = a^{-3},$$

an expression of the form a^n, where n is a *negative* integer. By using two different methods to work the problem a^2/a^5, we have obtained two different answers,

1.5 Exponentials

$1/a^3$ and a^{-3}. In mathematics, if a problem is worked in two different ways, we would hope to get the same answer. We can guarantee that by agreeing that

$$a^{-3} = \frac{1}{a^3}$$

and, in general, if $a \neq 0$ and n is a positive integer,

$$a^{-n} = \frac{1}{a^n}.$$

With this definition of a^{-n}, it can be proved that the properties of exponentials are valid for all integer values of m and n, and not only positive values.

The quotient a^m/a^n can now be stated in more compact form as

$$\frac{a^m}{a^n} = a^{m-n}, \quad a \neq 0.$$

Example 21 (a) $2^{-5} = \dfrac{1}{2^5} = \dfrac{1}{32}$

(b) $4^{-1} = \dfrac{1}{4}$

(c) $\dfrac{4^3}{4^8} = 4^{3-8} = 4^{-5} = \dfrac{1}{4^5}$

(d) $\dfrac{2^{-3} \cdot 2^5}{2^4 \cdot 2^{-7}} = \dfrac{2^2}{2^{-3}} = 2^{2-(-3)} = 2^5 = 32$

(e) $2^{-1} + 3^{-1} = \dfrac{1}{2} + \dfrac{1}{3} = \dfrac{5}{6}$

(f) $(4x^{-3})^2 = (4^2)(x^{-6}) = 16x^{-6} = \dfrac{16}{x^6}$

We have discussed and assigned meaning to exponentials of the form a^m for all nonzero real numbers a and all *integer* values of m, both positive and negative. Now we need to define a^m for *rational* values of m. Let us first consider an exponential of the form $a^{1/n}$, where n is a positive integer. We want any meaning that we might assign to $a^{1/n}$ to be consistent with the properties of exponentials listed above. For example, we know that for any real number a, and integers m and n, $(a^m)^n = a^{mn}$. If this property is to hold for the expression $a^{1/n}$, we must have

$$(a^{1/n})^n = a^{(1/n)n} = a^1 = a,$$

or

$$(a^{1/n})^n = a.$$

Thus, the nth power of $a^{1/n}$ must be a. For this reason, $a^{1/n}$ is called an **nth root** of a. For example, $a^{1/2}$ denotes a second root, or **square root** of a; and $a^{1/3}$ denotes a third root, or **cube root** of a. (We will soon associate $a^{1/2}$ with the more familiar $\sqrt{a}$.)

22 Fundamentals of Algebra

There are two possible square roots of 16, namely 4 and -4. Also, there are two possible fourth roots of 81, that is, 3 and -3. In all such cases, we reserve the symbol $a^{1/n}$ for the *positive* root, and write

$$16^{1/2} = 4 \quad \text{and} \quad 81^{1/4} = 3,$$

and so forth.

Example 22 (In evaluating the following roots, Tables 1 and 2 in the Appendix may be helpful.)

(a) $121^{1/2} = 11$, since 11 is positive and $11^2 = 121$.
(b) $625^{1/4} = 5$, since $5^4 = 625$.
(c) $-81^{1/2} = -(81^{1/2}) = -9$.
(d) $-64^{1/6} = -2$.

There is no real number x such that $x^2 = -16$. Therefore, $(-16)^{1/2}$ is not a real number. (Contrast this with $-16^{1/2}$, which equals -4.) In general, if $a < 0$, and n is an *even* integer then $a^{1/n}$ is not a real number.

Since $2^3 = 8$, we can write $8^{1/3} = 2$. Since $(-2)^3 = -8$, we can write $(-8)^{1/3} = -2$. In general, if a is any real number, and n is an *odd* integer, then there is exactly one real number which is equal to $a^{1/n}$.

Example 23 (a) $27^{1/3} = 3$
(b) $(-32)^{1/5} = -2$
(c) $-32^{1/5} = -(32^{1/5}) = -2$
(d) $(-49)^{1/2}$ is not a real number.

We have now defined a^m for all integer values of m and for all exponents of the form $1/n$, where n is a positive integer. To extend the definition of a^m to include all rational values of m, we use the following property.

For all real numbers a, for all positive integer values of n such that $a^{1/n}$ is a real number, and for all integer values of m,

$$(a^{1/n})^m = (a^m)^{1/n}.$$

We can now define $a^{m/n}$ as follows: for all values of a and all positive integer values of n such that $a^{1/n}$ is a real number, and for all integer values of m,

$$a^{m/n} = (a^{1/n})^m, \quad \text{or equivalently,} \quad a^{m/n} = (a^m)^{1/n}.$$

Example 24 (a) Using the definition of $a^{m/n}$, we have

$$27^{2/3} = (27^{1/3})^2 = 3^2 = 9.$$

(b) We could also evaluate $27^{2/3}$ as follows:

$$27^{2/3} = (27^2)^{1/3} = 729^{1/3} = 9.$$

In this case, as in most practical examples, the first method involved less computation and was easier.

Example 25 (a) $32^{2/5} = (32^{1/5})^2 = 2^2 = 4$.
(b) $64^{4/3} = (64^{1/3})^4 = 4^4 = 256$.

It is common to express $a^{1/2}$ as $\sqrt{a}$, where the symbol $\sqrt{}$ is called a **radical sign**. In general, if n is an integer greater than 2, the symbol $a^{1/n}$ can be denoted $\sqrt[n]{a}$. Using radical signs, and the work above, we see that for any integer m, and any positive integer n,

$$a^{m/n} = \sqrt[n]{a^m} = (\sqrt[n]{a})^m,$$

whenever all these roots exist. Using these results, we can summarize the properties of exponentials as follows.

Properties of Exponentials For all rational numbers m and n, and all real numbers a and b for which all of the following roots exist, it is true that

(a) $a^m \cdot a^n = a^{m+n}$

(b) $\dfrac{a^m}{a^n} = a^{m-n}$ $(a \neq 0)$

(c) $(a^m)^n = a^{mn}$

(d) $(ab)^m = a^m b^m$

(e) $\left(\dfrac{a}{b}\right)^n = \dfrac{a^n}{b^n}$ $(b \neq 0)$

We shall discuss real number exponents in Section 6.1.

1.5 Exercises

Evaluate each of the following. Write all answers without exponents.

1. 3^4
2. 4^2
3. 4^{-3}
4. 2^{-6}
5. 8^{-1}
6. 1^{-5}
7. $\left(\dfrac{3}{2}\right)^{-2}$
8. $\left(\dfrac{11}{10}\right)^{-2}$
9. $64^{1/2}$
10. $121^{1/2}$
11. $216^{1/3}$
12. $125^{1/3}$
13. $216^{2/3}$
14. $125^{2/3}$
15. $(-128)^{1/7}$
16. $(-343)^{1/3}$
17. $(-32)^{3/5}$
18. $(-64)^{4/3}$
19. $625^{-3/4}$
20. $16^{-5/4}$
21. $2^{-1} + 4^{-1}$
22. $2^{-2} + 3^{-2}$
23. $16^{1/2} + 25^{1/2}$
24. $36^{1/2} - 49^{1/2}$

Simplify each of the following. Write all answers using only positive exponents.

25. $\dfrac{3^8}{3^2}$
26. $\dfrac{4^9}{4^7}$
27. $\dfrac{3^{-5}}{3^{-2}}$

28. $\dfrac{6^{-1}}{6}$

29. $4^{-2} \cdot 4^2$

30. $\dfrac{1}{5^{-3}} \cdot 5^{-3}$

31. $\left(\dfrac{5^{-6} \cdot 5^3}{5^{-2}}\right)^{-1}$

32. $\left(\dfrac{8^{-3} \cdot 8^4}{8^{-2}}\right)^{-2}$

33. $2^{1/2} \cdot 2^{3/2}$

34. $5^{3/8} \cdot 5^{5/8}$

35. $27^{2/3} \cdot 27^{-1/3}$

36. $9^{-3/4} \cdot 9^{1/4}$

37. $\dfrac{4^{2/3} \cdot 4^{5/3}}{4^2}$

38. $\dfrac{3^{-5/2} \cdot 3^{3/2}}{3^{7/2} \cdot 3^{-9/2}}$

39. The supply of a certain item is given by

$$S = 2x^{1/2} + 3x^{3/4},$$

where S is the supply at a price of x. Find the supply at a price of
   (a) $x = 1$ (b) $x = 16$

40. The demand for a certain commodity is given by

$$D = 1000 - 200x^{-2/3}, \quad x \neq 0$$

where x is the price of the commodity. Find the demand at a price of
   (a) $x = 27$ (b) $x = 64$

(In this example, the demand goes up as the price goes up. While this seems strange, there are certain products where this can be true—cosmetics and dog food are two examples.)

• Case 1   Warehouse Location—The FMC Corporation*

The FMC Corporation is a large diversified producer of machinery, chemicals, films, and fibers such as nylon. The company has annual sales which place it in the top hundred corporations in the nation. The study presented in this case was done for FMC's Link-Belt Products Division, manufacturers of a broad line of industrial equipment. The study was done by FMC's own consultants, people who are available to work with any of the company's divisions.

A few years before the beginning of this study, Link-Belt management began to feel that perhaps it should reduce the number of its warehouses. This feeling was based on several factors, including the decrease in transportation time necessary to reach customers, the lower cost of communication services, higher labor costs, and improvements in techniques of automating warehouses.

The company had warehouses in Philadelphia; Atlanta; Columbus, Ohio; Chicago; Kansas City; Dallas; Reno; Seattle; Houston; and Portland, Oregon. The question presented by the Link-Belt management to the consultants: Should any of the current warehouses be closed, and, in general, what possible configuration of warehouse sites would provide the lowest possible cost while still providing good service to customers?

Dollar amounts reflecting total warehouse sales, tonnages handled, and total operating costs are considered confidential by FMC. However, in the most

---

*Case supplied by Manher D. Naik, Economic and Decision Analysis Coordinator, FMC Corporation.

recent year for which an analysis of figures could be made, the percentage breakdown for Link-Belt's warehouse operating costs were as follows: 19% for freight, 42% for inventory investment, and 39% for operating expenses.

To begin the analysis, 17 additional cities were selected as potential warehouse sites. Since construction and land costs vary from city to city, it was necessary to develop for each city an equation which represented the local costs of construction. In developing this equation, we shall use the following variables.

A = warehouse floor area (in thousands of square feet)
C_c = cost of labor and materials to build a warehouse (in thousands of dollars)
L = amount of land needed for a warehouse (in acres)
C_1 = cost of land (in thousands of dollars)
I = inventory in a warehouse at a given time (in pounds)
T = total quantity of merchandise going through a warehouse in a year (in thousands of pounds)

For example, if A represents the warehouse area in thousands of square feet, and C_c represents the cost of labor and materials in thousands of dollars, then

$$C_c = 12.5 + 3.75A$$

was found to provide a good approximation to the cost for labor and materials in Atlanta, while

$$C_c = 18.75 + 5.6A$$

is a similar equation for Chicago. These equations were obtained by studying construction costs in the cities in question.

Land prices also vary from city to city. Again using information obtained about each of the cities in question, it was estimated that the amount of land, L, in acres, needed for a warehouse of area A, in thousands of square feet, is given by

$$L = 0.875 + 0.0315A.$$

For Chicago, the cost of this land, C_1, in thousands of dollars, is given by

$$C_1 = 30.6 + 1.10A,$$

while the cost equation for Atlanta is

$$C_1 = 14.8 + 0.94A.$$

Based on past records, the company knows that one square foot of warehouse area can store about 70 pounds of merchandise, or, if I represents inventory measured in pounds in a warehouse at a given time, then

$$I = 70A.$$

The inventory at a given time, again from experience, is also given by

$$I = 180 + 0.1435T,$$

where T is the total weight of merchandise in thousands of pounds that go through the warehouse in a year.

Using the above equations, we can find the cost of land, labor, and materials for a new warehouse in Chicago in terms of T—that is, we can find the cost in terms of the quantity of merchandise going through the warehouse in a year. To find the cost for labor and materials, we begin with

$$C_c = 18.75 + 5.6A,$$

and since $I = 70A$, or $A = I/70$, we get

$$C_c = 18.75 + 5.6\left(\frac{I}{70}\right).$$

We also know that $I = 180 + 0.1435T$; thus

$$C_c = 18.75 + \frac{5.6}{70}(180 + 0.1435T),$$

which simplifies to

$$C_c = 33.15 + 0.0115T.$$

To find the cost of land, go through the same steps to obtain

$$C_1 = 33.47 + 0.00225T$$

for the equation which gives the cost of land for a warehouse in Chicago.

Using these equations, the analysts prepared the following chart.

Cost of a Warehouse in Chicago	Fixed cost	Variable cost (dollars per 1000 pounds)
Labor, materials	$33,000	$11.50
Land	$33,500	$2.25
Total	$66,500	$13.75

The numbers in this chart were obtained as follows. We know that the cost of land in Chicago is given by $C_1 = 33.47 + 0.00225T$. The fixed cost is found by letting $T = 0$: $C_1 = 33.47 + 0.00225(0) = 33.47$, which represents a fixed cost of about $33,500. The variable cost is given by 0.00225 thousands of dollars, which is about $2.25 per thousand pounds of merchandise.

Charts similar to the one above could be made for each of the other cities under discussion. Using all these results, and a process called linear programming, the analysts recommended the following consolidation of warehouse sites. All warehousing should be centralized in five warehouses, located in Philadelphia, Atlanta, Indianapolis, Dallas, and San Francisco. Operating from these five cities will save $660,000 annually, with an additional $730,000 to be realized from selling the warehouses which would be closed. The analysts estimated that service to customers would be as follows: it would be possible to reach eighty-seven percent of the market from these five warehouses in two days or less

Case 1   Warehouse Location   27

(compared to current delivery times of one day or less for 89% of all customers), with the remaining 13% reached in three days. About half the market will have delivery times of one day or less.

Case 1   Exercises

1. Complete each of the following steps.
   (a) Cost of labor and materials in Atlanta:
   $$C_c = \underline{\hspace{4cm}}.$$
   (b) Since $A = I/70$ and $I = 180 + 0.1435T$, we have
   $$C_c = 12.5 + 3.75(\underline{\hspace{2cm}})$$
   $$= 12.5 + \frac{3.75}{70}(\underline{\hspace{2cm}})$$
   $$= \underline{\hspace{4cm}}.$$
   (c) The equation for the cost of land in Atlanta is
   $$C_1 = \underline{\hspace{4cm}}.$$
   (d) We have
   $$C_1 = 14.8 + 0.94(\underline{\hspace{2cm}})$$
   $$= 14.8 + \frac{0.94}{70}(\underline{\hspace{2cm}})$$
   $$= \underline{\hspace{4cm}}.$$

2. Complete the following chart.

Cost of a Warehouse in Atlanta	Fixed Cost	Variable Cost
Labor, materials		
Land		
Total		

3. Suppose the cost equations for a Sacramento warehouse can be given by
   $$C_c = 11.4 + 4.20A,$$
   $$C_1 = 12.9 + 0.90A.$$
   (a) Obtain C_c and C_1 in terms of T. (Hint: Go through the steps of Exercise 1 above.)
   (b) Complete a table, similar to the one of Exercise 2, for a warehouse in Sacramento.

4. Show that warehouse area A, in thousands of square feet, needed for a certain annual total quantity of goods, T, in thousands of pounds, is given by
   $$A = 2.57 + 0.00205T,$$
   or
   $$T = 487A - 1250.$$

Chapter 2 Pretest

1. Solve $|x| = 4$. [1.1]
2. Solve for y: [1.2]

   (a) $2x + y = 3$ (b) $\frac{2}{3}x - 4y = 10$

3. Solve for x: $y = \frac{2x + 3}{x - 1}$. [1.2]
4. Find y if $x = -2$, given $8 - 2x = 3y$. [1.2]
5. Find $y = \frac{5}{3 - 2x}$ if $x = 0; 1; 10; 100$. [1.2]
6. Find the value of x which makes $3x - 4$ and $\frac{x}{2} + 7$ equal. [1.2]
7. Solve: [1.3]

   (a) $2x^2 - 3x = 0$ (b) $x^2 + 5x - 2 = 0$
8. Given $y = 2x^2 - 3x$, find y for $x = 0; -1; 2$. [1.3]
9. Given $y = \sqrt{2x^2 - 4}$, find y if $x = 2; 3; 4$. [1.5]
10. Given $y = -\sqrt{x + 3}$, find y if $x = 0; 1; 6$. [1.5]
11. Find $y = 2x^3 - 3x^2 + x - 5$ if $x = -1$. [1.5]

2 Functions

Functions are used to describe the connection between two related quantities. The function concept is particularly useful for expressing relationships in many practical situations. In this chapter, we discuss the equations of several useful functions.

2.1 Definition of a Function

A common problem in many real-life situations is to describe relationships between quantities. For example, assuming that the number of hours a student studies each day is related to the grade received in a course, how can the relationship be expressed? One way is to set up pairs of symbols representing hours of study and the corresponding grades. For example, we could use

$$(3, A), (2.5, B), (2, C), (1, D), (0, F)$$

to represent the relationship that three hours of study would result in an A, 2.5 hours of study would result in a B, and so on.

In general, (a, b) is called an **ordered pair**; a is called the **first component** of the ordered pair, and b is called the **second component**. A **relation** is defined as any set of ordered pairs. Thus the set

$$\{(3, A), (2.5, B), (2, C), (1, D), (0, F)\}$$

is a relation.

In a more complex relationship, a formula of some sort can often be used to describe how one quantity changes with respect to the other. For example, if the area of a room is to be 120 square feet, we can write length times width equals 120, or the relation $LW = 120$, to describe the relationship between the possible lengths and widths of the room. A few of the infinite number of ordered pairs belonging to this relation are

$$(10, 12), (6, 20), \text{ and } (8, 15),$$

where width is given first, length second.

Another example of a relation in which a formula or rule shows how the second element of the ordered pairs of the relation is obtained from the first is the set of ordered pairs

$$\{(x, y) \mid y = x + 3; x = 1, 2, 3, \text{ or } 4\}.$$

In this relation, the second component is obtained by adding 3 to the first component, which can be either 1, 2, 3, or 4. Thus (1, 4), (2, 5), (3, 6), and (4, 7) are the ordered pairs that belong to the relation. The set of first components $\{1, 2, 3, 4\}$ is called the **domain** of the relation, while the set of second components $\{4, 5, 6, 7\}$ is called the **range** of the relation.

Example 1 Find the ordered pairs belonging to the relation

$$R = \{(x, y) \mid y \leq 2x + 1; x = -1 \text{ or } 0; y = -2, -1, 0, \text{ or } 1\},$$

and give the domain and range of R.

By inspection, we find that R contains the following ordered pairs:

$$R = \{(-1, -2), (-1, -1), (0, -2), (0, -1), (0, 0), (0, 1)\}.$$

The domain is $\{-1, 0\}$; the range is $\{-2, -1, 0, 1\}$.

A **function** is a special type of relation, where to each domain element x, there corresponds exactly one range element y. We call x the **independent variable** and y the **dependent variable**. For example,

$$\{(-1, 2), (2, 3), (3, 3)\}$$

is a function, since to each domain element (the domain is $\{-1, 2, 3\}$), we assign exactly one range element (the range is $\{2, 3\}$). On the other hand,

$$\{(-1, 2), (-1, 3), (4, 7)\}$$

is not a function. Here we assign *two* range elements, 2 and 3, to the single domain element -1.

It is often useful to graph a function. To do this, we use the perpendicular crossed number lines of a **Cartesian coordinate system**, as shown in Figure 2.1. The horizontal number line, or **x-axis**, represents the first component of the ordered pairs of the function, while the vertical or **y-axis** represents the second component. The point where the number lines cross is the zero point on both lines; this point is called the **origin**.

To locate a point corresponding to an ordered pair such as $(-2, 4)$, start at the origin, and count two units to the left on the horizontal axis, and then 4 units up, parallel to the vertical axis. This point is labeled in Figure 2.1, as are several other sample points. The numbers -2 and 4 are called the **x-** and **y-coordinates** of the point $(-2, 4)$. As shown in Figure 2.1, all points above the horizontal axis and to the right of the vertical axis are in **Quadrant I**. The other three quadrants are named as indicated in Figure 2.1. Points on the two axes belong to no quadrant.

Figure 2.1

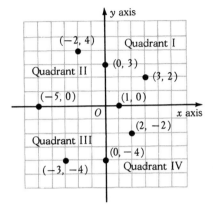

Example 2 Graph the following functions:

(a) $\{(-1, 2), (2, 3), (3, 3)\}$
(b) $\{(-3, -1), (-2, 0), (-1, 3), (1, 1), (2, 2), (3, 0)\}$.

The graphs are shown in Figure 2.2.

Figure 2.2

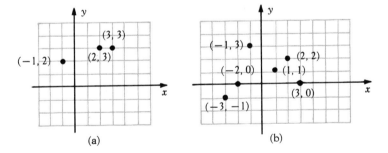

By definition, we must be able to assign exactly one value of y to each value of x in the domain of a function. Figure 2.3 shows the graph of a relation. Note that for the value $x = x_1$, the graph gives the two y-values, y_1 and y_2. Since to the x-value x_1 there correspond two y-values, y_1 and y_2, the relation is not a function. Generalizing, we have the vertical line test for a function:

> **Vertical Line Test** If any vertical line intersects the graph of a relation in more than one point, then the relation does not represent a function.

Figure 2.3

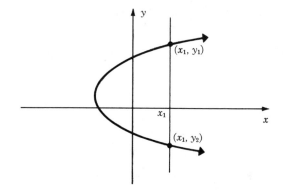

Example 3 Use the vertical line test to determine which of the graphs in Figure 2.4 represent functions.

Figure 2.4

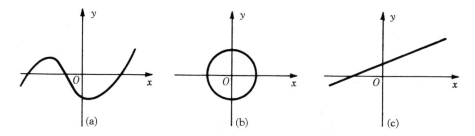

Since no vertical line would intersect them in more than one point, the graphs in Figure 2.4 (a) and (c) are graphs of functions. There are many vertical lines which would intersect the graph in Figure 2.4 (b) in two places. Therefore, the graph in Figure 2.4 (b) is not the graph of a function.

Lower-case letters, such as f, g, and h, are often used to name functions. Thus, we can write

$$f = \{(x, y) \mid y = 2x + 3\},$$

where f is used as a name for this function. When the domain is unspecified, as it is here, we shall assume it to be the set of all real numbers. Verify that the range of f is also the set of all real numbers. If we select any real number x, the equation $y = 2x + 3$ can be used to determine exactly one value of y from the range. For example, if we choose the value $x = 4$ from the domain, then the range element is $y = (2)(4) + 3 = 11$, so that $(4, 11)$ belongs to the function. In the same way, $(-3, -3)$, $(-1, 1)$, and $(0, 3)$ also belong to the function.

We saw above that $(4, 11)$ belongs to the function f. This relationship can also be expressed by writing

$$f(4) = 11,$$

(read "f of 4 equals 11"). In the same way, $f(-3) = -3$, $f(-1) = 1$, and $f(0) = 3$.

We shall often abbreviate

$$f = \{(x, y) \mid y = 2x + 3\}$$

by writing

$$f(x) = 2x + 3,$$

or even

$$y = 2x + 3.$$

This last way of writing the function f is especially useful, since it shows that x is the independent variable and y is the dependent variable. Another way of saying this is to say that **y is a function of x**.

Example 4 Let $f(x) = \dfrac{5 + x}{4 - x}$. Find $f(3)$, $f(0)$, and $f(a)$.

We have

$$f(3) = \frac{5 + 3}{4 - 3} = \frac{8}{1} = 8$$

$$f(0) = \frac{5 + 0}{4 - 0} = \frac{5}{4}$$

$$f(a) = \frac{5 + a}{4 - a} \quad (a \neq 4).$$

Example 5 Suppose the sales of a company are given by

$$S(x) = 100 + 80x,$$

where $S(x)$ represents the total sales in thousands of dollars in year x, with $x = 0$ representing 1974. Find the sales in 1974 ($x = 0$) and in 1977 ($x = 3$).

We have

$$S(0) = 100 + 80(0) = 100,$$

so that sales totaled $100,000 during 1974. Also,

$$S(3) = 100 + 80(3) = 340,$$

which shows that sales will be $340,000 during 1977.

2.1 Exercises

List the ordered pairs belonging to each of the following relations. Assume the domain of x in each exercise is $\{-2, -1, 0, 1, 2, 3\}$. Graph each relation. Tell which relations are functions.

1. $x = y + 1$
2. $2x - 1 = y$
3. $y = x^2$
4. $y(x + 3) = 1$

5. $y = x(x + 1)$
6. $y = |x|$
7. $y = |2x - 1|$
8. $y = 2$

Identify any of the following graphs that represent functions.

9.
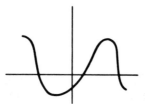

11.

10.
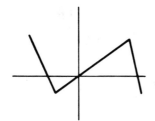

12.
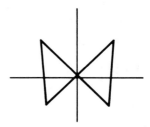

Use the definition of function to determine which of the following relations are also functions.

13. $y = 2x + 1$
14. $y - 1 = 4x$
15. $x = |y|$
16. $y = x^2$
17. $x = 4$
18. $y = -1$
19. $y = \dfrac{1}{x}, \quad x \neq 0$
20. $xy = 2$

For each of the following functions, find

(a) $f(4)$ (b) $f(-3)$ (c) $f(0)$ (d) $f(a)$ (e) $f(a + 2)$.

21. $f(x) = -6x + 2$
22. $f(x) = -x^2 + 4x - 6$
23. $y = 4x + 5$
24. $y = -3x + 4$

Let $f = \{(x, y) \mid y = 2x - 3\}$. Find each of the following.

25. $f(0)$
26. $f(-1)$
27. $f(a)$
28. $f(-r)$
29. $f(a + 2)$
30. $f(x + 2)$

31. In a biology experiment, a student found that the growth in millimeters of a plant, $g(x)$, was related to x, the number of hours of light, by the formula

$$g(x) = x(x - 1).$$

Find the growth which corresponds to the following amounts of light. (a) 0 hours, (b) 1 hour, (c) 2 hours, (d) 5 hours, (e) 8 hours. (f) Graph the resulting ordered pairs.

32. Suppose the sales of a company in year t are given by $S(t)$, where

$$S(t) = 1000 + 50(t + 1),$$

where t is time in years, with $t = 0$ representing the year 1975. Find each of the following. (a) $S(0)$, (b) $S(9)$, (c) the sales in 1976, (d) the sales in 1979.

33. Assume that the post office charges 10¢ per ounce, or fraction of an ounce, to mail a letter. Let us use $f(x)$ to represent the cost in cents of mailing a letter weighing x ounces.
    (a) Graph $f(x)$.
    (b) Find the domain and range of the function.

34. A taxi company charges a flat rate of 50¢ plus 10¢ per mile or fraction of a mile. Let $C(x)$ represent the cost of a trip of x miles by taxi.
    (a) Graph $C(x)$.
    (b) Find the domain and range of the function.

2.2 Linear Functions

Any function of the form

$$f(x) = mx + b,$$

where m and b are real numbers, is called a **linear function**. For example, $f(x) = -5x$, $y = 2x - 3$, $y = -x + 8$, $y = 3$, and $y = x$ are all linear functions. To graph a linear function, we first find several ordered pairs that satisfy the function. For example, the function $y = x + 1$ is satisfied by the ordered pairs

$$(2, 3), (0, 1), (4, 5), (-2, -1), (-5, -4), \text{ and } (-3, -2),$$

among others, as we can verify by substitution. Next, plot the ordered pairs. The graph of the ordered pairs above is shown in Figure 2.5(a). These points all lie on the same straight line which is drawn through the points in Figure 2.5(b). The vertical line test confirms that this is the graph of a function.

Figure 2.5a, 2.5b

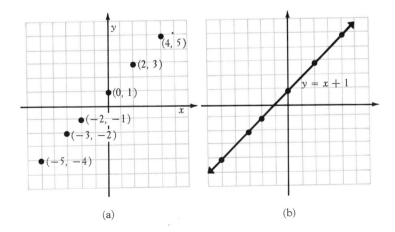

(a)   (b)

It can be shown that the graph of any linear function is a straight line. Since a straight line is completely determined if we know two distinct points that it passes through, it is only necessary to locate two different points of the graph. Two points that are often useful for this purpose are the points where the graph crosses the axes. The **x-intercept** is the x-value (if any) at which the graph of the function crosses the x-axis, while the **y-intercept** (if any) is the y-value at which the graph crosses the y-axis. Where the graph crosses the y-axis, $x = 0$. (See Figure 2.6.) Thus, to find the y-intercept of a line such as $y = -2x + 5$, we let $x = 0$ and solve for y. Hence, the y-intercept is

$$y = -2(0) + 5 = 5.$$

In the same way, we find the x-intercept by letting $y = 0$ and solving for x:

$$0 = -2x + 5$$
$$2x = 5$$
$$x = 5/2.$$

Thus, the graph of $y = -2x + 5$ passes through $(0, 5)$ and $(5/2, 0)$. Using these two points, we can draw the graph shown in Figure 2.6.

Figure 2.6

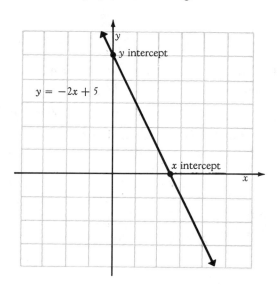

In the discussion of intercepts above, we added the phrase "if any" with reference to the x-value at which a graph crosses the x-axis. To see that we can sometimes fail to have an x-intercept, consider the example below.

Example 6 Graph $y = -3$.

Using $y = -3$ or, equivalently, $y = 0x - 3$, we always get the same y-value, -3, for any value of x. Therefore there is no value of x that will make $y = 0$, so that the graph has no x-intercept. By definition, $y = -3$ is a linear function, with a straight line graph, and since the graph cannot cross the x-axis, the line must be parallel to the x-axis. For any value of x, we have $y = -3$.

Therefore, the graph is the horizontal line which is parallel to the x-axis and has y-intercept -3, as shown in Figure 2.7. In general, the graph of $y = k$, with k a real number, is the horizontal line having y-intercept k. Although the equation $x = k$, with k a real number, does not describe a function (why?), its graph is also a straight line, which is vertical, has x-intercept k, and no y-intercept.

Figure 2.7

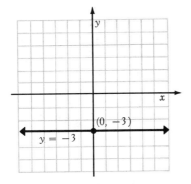

As mentioned earlier, a straight line is completely determined if we know two distinct points that the line passes through. A line is also completely determined if we know one point that the line passes through, along with some measure of the "steepness" of the line. To get a numerical measure which represents the "steepness" of a line, we define the *slope* of a line in the next paragraph.

Figure 2.8 shows a line going through the two points $(x_1, y_1) = (-3, 5)$ and $(x_2, y_2) = (2, -4)$. The difference in the x-values, $x_2 - x_1$, is called the **change in x**, and is denoted Δx (read "delta x") while the difference in the y-values, $y_2 - y_1$, called the **change in y**, is denoted Δy. The slope of the line through the two distinct points (x_1, y_1) and (x_2, y_2) is defined as

$$\text{slope} = \frac{\text{change in } y}{\text{change in } x} = \frac{\Delta y}{\Delta x} = \frac{y_2 - y_1}{x_2 - x_1}.$$

Figure 2.8

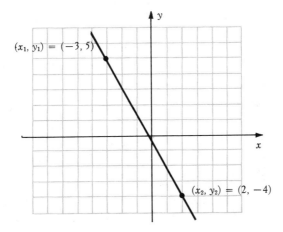

38 Functions

The slope of the line in Figure 2.8 is thus given by

$$\frac{\Delta y}{\Delta x} = \frac{-4 - 5}{2 - (-3)} = \frac{-9}{5}.$$

It can be shown that the slope of a particular line is independent of the choice of points used to find the slope. That is, the same slope will be obtained no matter which pair of distinct points on the line is selected. Also, the slope of the line will be the same whether we calculate Δx and Δy as we move from (x_1, y_1) to (x_2, y_2) or from (x_2, y_2) to (x_1, y_1). For example, we found above that the slope of the line through $(-3, 5)$ and $(2, -4)$ is $-9/5$. This same result is obtained if we subtract the coordinates in the reverse order from the way we did above:

$$\text{slope} = \frac{\Delta y}{\Delta x} = \frac{5 - (-4)}{-3 - 2} = \frac{9}{-5} = -\frac{9}{5}.$$

Example 7 Find the slope of the line $3x - 4y = 12$.

First find any two distinct points that lie on the line. Two such points can be found from the intercepts: $(0, -3)$ and $(4, 0)$. The change in y is $\Delta y = 0 - (-3) = 3$ and the change in x is $\Delta x = 4 - 0 = 4$. The slope, therefore, is $\Delta y / \Delta x = 3/4$.

In Example 7, we found that the slope of the line $3x - 4y = 12$ is $3/4$. This same result can be obtained by first solving $3x - 4y = 12$ for y. Doing this, we have

$$3x - 4y = 12$$
$$-4y = -3x + 12$$
$$y = \frac{3}{4}x - 3.$$

Note that after solving for y, the slope is the same as the coefficient of x. (Also, verify that -3 is the y-intercept of $3x - 4y = 12$.) This is always the case, as stated below.

> When a linear function is written in the form $y = mx + b$, then m is the slope of the line and b is the y-intercept.

Example 8 Find the slope and y-intercept of $3x - 2y = 2$. Graph the line.

First write $3x - 2y = 2$ in the form $y = mx + b$. (In other words, solve for y.) Here we have

$$y = \frac{3}{2}x - 1.$$

From this form of the equation of the line, we see that the slope is $m = 3/2$ and the y-intercept is $b = -1$. To draw the graph, first locate the y-intercept (see Figure 2.9). By the definition of slope, if m represents the slope then

$$m = \frac{\Delta y}{\Delta x} = \frac{\text{change in } y}{\text{change in } x}.$$

2.2 Linear Functions

Here we know $m = 3/2$. Thus, if x changes 2 units, y will change 3 units. We can use this fact to find a second point of the graph, as shown in Figure 2.9. Using this point and the y-intercept, we can complete the graph.

Figure 2.9

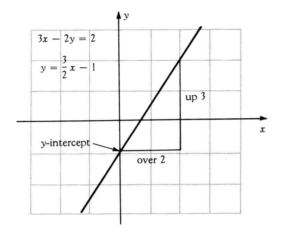

Because the slope and y-intercept can be read quickly from $y = mx + b$, this form is often called the **slope-intercept form** of the equation of a line. Sometimes, however, we know the slope of a line, together with one point on the line. In such a case, to find the equation of the line, it is often helpful to use the **point-slope form** of the equation of a line, as given below.

> If a line has slope m and passes through the point (x_1, y_1), then the equation of the line is given by
> $$y - y_1 = m(x - x_1).$$

Example 9 Write the equation of the line going through $(-4, 1)$ with slope $m = -3$.

Here $x_1 = -4$, $y_1 = 1$, and $m = -3$. Using the point-slope form of the equation of a line, we have

$$y - 1 = -3[x - (-4)]$$
$$y - 1 = -3(x + 4)$$
$$y = -3x - 11.$$

Example 10 Find the equation of the line through the points $(5, 4)$ and $(-10, -2)$.

Before we can use the point-slope form of the equation of a line, we must find the slope of the line. To do this we use the definition of slope:

$$\text{slope} = \frac{\Delta y}{\Delta x} = \frac{-2 - 4}{-10 - 5} = \frac{-6}{-15} = \frac{2}{5}.$$

40 Functions

We can now use the point-slope form, with $m = 2/5$ and (x_1, y_1) equal to either $(5, 4)$ or $(-10, -2)$. If we choose $(-10, -2)$, we have

$$y - (-2) = \frac{2}{5}[x - (-10)]$$

$$y + 2 = \frac{2}{5}(x + 10)$$

$$5(y + 2) = 2(x + 10)$$

$$5y - 2x = 10.$$

Verify that we would have obtained the same result if we had used $(5, 4)$ for (x_1, y_1).

Linear functions can be used in solving problems involving supply and demand for a commodity. Both the supply and the demand of a commodity are related to the price of the commodity. That is, if x is the price, then the supply S and the demand D can be expressed as functions of x. For example, suppose the demand, $D(x)$, for a certain item is given by the linear function

$$D(x) = -\frac{4}{3}x + 80,$$

where x represents the price of the item in some units. At a price of $x = 0$ units, the demand is given by $D(0) = 80$ units, and at a price of $x = 40$ units, the demand is given by $D(40) = 80/3$ units. Using these two points, $(0, 80)$ and $(40, 80/3)$, and the fact that $D(x)$ is a linear function, we can draw the graph shown in Figure 2.10. Only the part of the graph in the first quadrant is shown, since problems involving supply and demand are meaningful only for non-negative price and quantity.

Figure 2.10

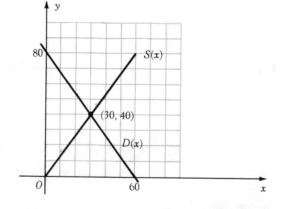

Suppose the supply function for the commodity shown is given by

$$S(x) = \frac{4}{3}x,$$

2.2 Linear Functions

where x represents the price and $S(x)$ represents supply. Verify that the supply function goes through the two points $(0, 0)$ and $(60, 80)$. Using these two points, we can draw the graph which is also shown in Figure 2.10.

As shown on the graph of Figure 2.10, both the supply and the demand functions pass through the point $(30, 40)$. If the price of the commodity is more than $x = 30$, the supply will exceed the demand. At a price less than $x = 30$, demand will exceed supply. Only at a price of $x = 30$ will demand and supply be equal. The price at which supply and demand are equal is called the **equilibrium price**. Thus $x = 30$ is the equilibrium price, while $D(30) = S(30) = 40$ is called the **equilibrium demand** or **equilibrium supply**.

While the equilibrium price was found above by an inspection of the graphs of $D(x)$ and $S(x)$, the price can also be found using algebra. Since we want supply and demand to be equal, we write

$$S(x) = D(x)$$

$$\frac{4}{3}x = -\frac{4}{3}x + 80$$

$$8x = 240$$

$$x = 30.$$

2.2 Exercises

Use the x- and y-intercepts to graph each of the following linear functions.

1. $y = 2x + 6$
2. $y = 3x - 9$
3. $y = 4x + 8$
4. $y = x + 5$
5. $3y + 4x = 12$
6. $y + 5x = 10$
7. $6x + y = 12$
8. $x + y = 9$
9. $2y + 5x = 20$
10. $2x + 7y = 14$
11. $3y + 8x = 24$
12. $9y = 4x - 36$

Find the slopes of the lines going through the given points.

13. $(-8, 6), (2, 4)$
14. $(-3, 2), (5, 9)$
15. $(-1, 4), (2, 6)$
16. $(3, -8), (4, 1)$
17. The origin and $(-4, 6)$
18. The origin and $(8, -2)$
19. $(-2, 9), (-2, 11)$
20. $(7, 4), (7, 12)$
21. What is the slope of any vertical line?
22. $(3, -6), (-5, -6)$
23. $(5, -11), (-9, -11)$
24. What is the slope of any horizontal line?

Find the slope of the following lines.

25. $y = 3x + 4$
26. $y = -3x + 2$
27. $y = -4x + 8$
28. $y - x = 3$
29. $3x + 4y = 5$
30. $2x - 5y = 8$
31. $y = -4$
32. $x = 3$

Functions

Graph each of the following lines.

33. Through $(-4, 2)$, $m = 2/3$
34. Through $(3, -2)$, $m = -3/4$
35. Through $(-5, -3)$, $m = -2$
36. Through $(-1, 4)$, $m = 2$
37. Through $(8, 2)$, $m = 0$
38. Through $(2, -4)$, no slope
39. Through $(0, -2)$, $m = -3/4$
40. Through $(0, -3)$, $m = 3/5$

Find equations for the lines having the following y-intercepts and slopes.

41. 4, $m = -2/3$
42. -3, $m = 3/4$
43. -2, $m = -1/2$
44. $3/2$, $m = 1/4$
45. $5/4$, $m = 3/2$
46. $-3/8$, $m = 3/4$

Find an equation for each of the following lines.

47. Through $(-4, 1)$, $m = 2$
48. Through $(5, 1)$, $m = -1$
49. Through $(3, 2)$, $m = 1/4$
50. Through $(0, 3)$, $m = -3$
51. Through $(-2, 3)$, $m = 3/2$
52. Through $(0, 1)$, $m = -2/3$
53. Through $(-1, 1)$ and $(2, 5)$
54. Through $(4, -2)$ and $(6, 8)$
55. Through $(9, -6)$ and $(12, -8)$
56. Through $(-5, 2)$ and $(7, 5)$
57. Through $(-8, 4)$ and $(-8, 6)$
58. Through $(2, -5)$ and $(4, -5)$

59. Suppose the demand for a certain item is given by

$$D(x) = \frac{40 - 4x}{5},$$

where $D(x)$ is the demand in units at a price x.
(a) Find the demand at a price of 0.
(b) Find the demand at a price of 10.
(c) Graph $D(x)$.

60. Suppose the supply for the item of Exercise 59 is given by

$$S(x) = \frac{4}{5}x,$$

where $S(x)$ is the supply at a price of x.
(a) Find the supply at a price of 0.
(b) Find the supply at a price of 10.
(c) Graph $S(x)$ on the same graph used for $D(x)$ in Exercise 59.
(d) From the graph, estimate the equilibrium price. Verify the answer using algebra.

61. Let the supply and demand of a certain type of candy be given by

$$S(x) = \frac{2x}{3} \quad \text{and} \quad D(x) = \frac{300 - 4x}{3},$$

where $S(x)$ represents supply, $D(x)$ demand, and x price.
(a) Graph $S(x)$ and $D(x)$ on the same graph.
(b) Find the equilibrium price.
(c) Find the equilibrium demand.

62. Let the supply and demand for ice cream be given by

$$S(x) = \frac{5x}{2} \quad \text{and} \quad D(x) = \frac{500 - 5x}{2},$$

where $S(x)$ represents supply, $D(x)$ demand, and x price.
   (a) Graph $S(x)$ and $D(x)$ on the same graph.
   (b) Find the equilibrium price.
   (c) Find the equilibrium supply.
   (d) Find the equilibrium demand.

2.3 Some Nonlinear Functions

Many of the functions which we shall use in later work are nonlinear. Thus, the equations defining these functions *cannot* be written in the form $f(x) = mx + b$. In this section we shall discuss several nonlinear functions.

The Absolute Value Function To graph the **absolute value function**, $y = f(x) = |x|$, recall that $|x| \geq 0$ for all values of x. As long as $x \geq 0$, then $|x| = x$. On the other hand, if $x < 0$, then $|x| = -x$. Since we have $y = |x|$, we can write

$$\begin{aligned} y &= x & \text{if } x \geq 0, \\ y &= -x & \text{if } x < 0. \end{aligned}$$

Using these facts, and selecting some ordered pairs belonging to the function, we obtain the graph shown in Figure 2.11. The domain is the set of real numbers, while the range is restricted to the set of all nonnegative numbers.

Figure 2.11

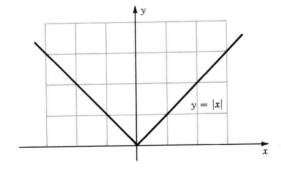

Example 11 Graph $y = |4x - 8|$.
By the definition of absolute value,

$$y = |4x - 8| = \begin{cases} 4x - 8 & \text{if } 4x - 8 \geq 0 \\ -(4x - 8) & \text{if } 4x - 8 < 0. \end{cases}$$

The values of x which make $4x - 8 \geq 0$ can be found by solving the inequality
$$4x - 8 \geq 0$$
$$4x \geq 8$$
$$x \geq 2.$$
Thus,
$$y = |4x - 8| = 4x - 8 \quad \text{if } x \geq 2,$$
while
$$y = |4x - 8| = -(4x - 8) = -4x + 8 \quad \text{if } x < 2.$$
Graphing these two lines gives the graph shown in Figure 2.12.

Figure 2.12

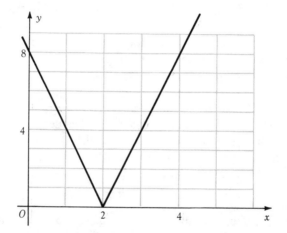

Quadratic Functions A function of the form
$$f(x) = ax^2 + bx + c,$$
a, b, and c real numbers, $a \neq 0$, is called a **quadratic function**. Without the restriction $a \neq 0$, we could have $y = bx + c$, which is a linear function. In general, the domain of a quadratic function includes all real numbers, but the range is more restricted, as we shall see.

The simplest quadratic function is $y = x^2$, where $a = 1$, $b = 0$, and $c = 0$. To graph this function, choose some values for x and then find the corresponding values for y, as is done in the chart of Figure 2.13. The domain of a quadratic function, as mentioned above, is all real numbers, while the range of this quadratic function, $y = x^2$, is the set of all nonnegative real numbers. The graph shown in Figure 2.13 is called a **parabola**. It can be shown that every quadratic function has a graph which is a parabola. The lowest point of the graph in Figure 2.13, (0, 0), is called the **vertex** of the parabola.

Parabolas have properties which make the parabolic shape particularly useful in many real-life applications. The cross-sections of radar reflectors, spotlights, and telescope reflectors form parabolas. The circular disks visible on the sidelines at football games have parabolic cross-sections, and are used by

Figure 2.13

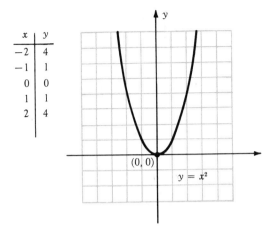

television networks to pick up the shouted signals of the quarterback. Paths of projectiles approximate parabolas.

The following examples illustrate some other quadratic functions and compare their graphs to that of the quadratic function $y = x^2$ which can be used as a kind of standard.

Example 12 Graph $y = -x^2$.

The y-values of the ordered pairs of this parabola are all nonpositive. The graph (in Figure 2.14) is "upside down" when compared to the graph of $y = x^2$. Here, the vertex $(0, 0)$ is the *highest* point on the graph.

Figure 2.14

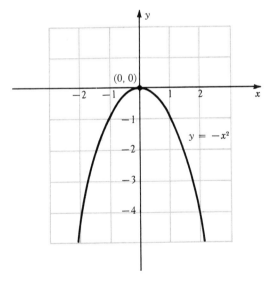

Example 13 Graph $y = 2x^2$.

The coefficient of 2 causes the value of y to increase more rapidly than in the graph of $y = x^2$, so that the graph of this example (Figure 2.15) is "thinner" than the graph of $y = x^2$.

Figure 2.15

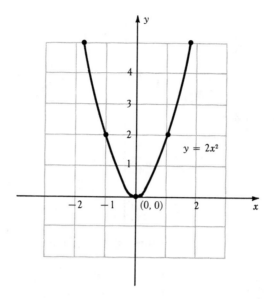

In general, the graph of the quadratic function $f(x) = ax^2 + bx + c$ ($a \neq 0$) will open upward if $a > 0$, and open downward if $a < 0$. If $0 < |a| < 1$, the parabola is "fatter" than the graph of $y = x^2$, while if $|a| > 1$, the parabola is "thinner" than the graph of $y = x^2$.

Example 14 Graph $y = x^2 - 2x + 3$.

Verify that the ordered pairs shown in the chart of Figure 2.16 satisfy the equation. Since the coefficient of x^2 is 1, the graph should have the same shape as that of $y = x^2$ and should open upward. The graph drawn through the given ordered pairs is shown in Figure 2.16. Note that the range of this parabola is the set of all real numbers greater than or equal to 2. At this point, we cannot always determine the vertex exactly for a parabola of this type; however, in Chapter 5, with the use of calculus, we present a method for determining the vertex of any parabola from its equation.

Figure 2.16

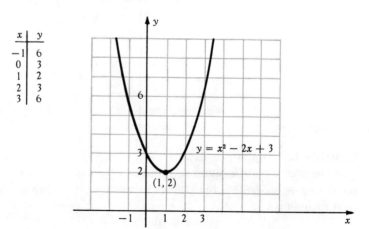

2.3 Some Nonlinear Functions

Example 15 Elmyra makes Aunt Elmyra's Blueberry Pies. Always wanting to maximize her profits, she hires a consulting firm to analyze her operations. The consultant tells Elmyra that her profits are given by

$$P(x) = 120x - x^2,$$

where x is the number of units of pies that she makes. How many units of pies should Elmyra make in order to produce maximum profit? What is her maximum profit?

Elmyra (or the consultant) can sketch the graph of that part of her profit function which is in Quadrant I. (See Figure 2.17.) Note that the maximum profit, which occurs at the vertex, is halfway between the two x-values where the profit is 0. To find these points, set $P(x) = 0$ and solve the equation

$$0 = -x^2 + 120x.$$

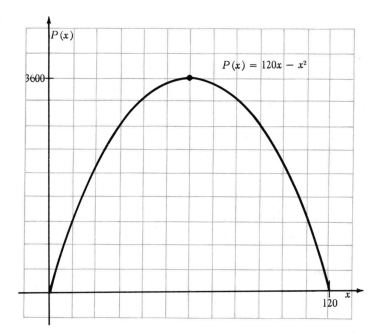

Figure 2.17

Verify that $P(0) = 0$ and $P(120) = 0$. Therefore, the maximum profit, which occurs at $x = 60$, is $P(60) = 3600$. The maximum profit is obtained when she makes 60 units of pies, with her profit at that point equal to $3600. The graph shows that profit increases as more and more pies are made, up to a certain point, and then decreases as more are made past this point.

Root Functions The parabolas we have graphed so far have all been vertical, with graphs which opened upward or downward. Horizontal parabolas, whose graphs open either to the right or to the left, have equations of the form

$$x = ay^2 + by + c,$$

where a, b, and c are real numbers, $a \neq 0$. Note the interchange in the roles played by x and y. Because of it, these parabolas are not the graphs of functions, as can be verified by using the vertical line test on the graph in Figure 2.18.

Figure 2.18

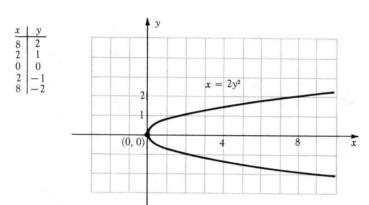

Example 16 Graph $x = 2y^2$.

Using methods similar to those discussed above for vertical parabolas, we get the graph shown in Figure 2.18. The vertex $(0, 0)$ is now the extreme left point and the parabola opens to the right.

If an equation of the form $x = ay^2 + by + c$ is solved for y (by using the quadratic formula), the result is two expressions of the form

$$y = k + \sqrt{dx + e} \quad \text{or} \quad y = k - \sqrt{dx + e},$$

where k, d, and e are real numbers, $d \neq 0$. Each of these two square-root equations defines a function: the graph of the first one is the upper half of the parabola $x = ay^2 + by + c$, while the graph of the second one is the lower half.

Example 17 Graph $y = \sqrt{1 - x}$.

If we square both sides of the given equation, we have $y^2 = 1 - x$ or $x = -y^2 + 1$, which represents a horizontal parabola with vertex $(1, 0)$. The

Figure 2.19

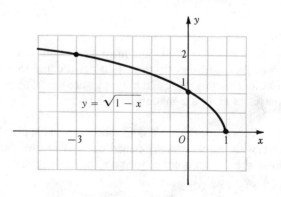

negative y^2 term indicates that this parabola should open to the left (the negative x direction). The range of the equation we want to graph, $y = \sqrt{1-x}$, is the set of nonnegative numbers (the *positive* square root is given), so the graph will be only the upper half of the parabola as shown in Figure 2.19. Note that the result is the graph of a function.

Circles A **circle** is the set of all points in a plane which lie a fixed distance from a fixed point. The fixed point is called the **center**, and the fixed distance is called the **radius**. For a circle with center at the origin, the points (x, y) on the circle must satisfy the relation $x^2 + y^2 = r^2$, where r is the length of the radius.

Example 18 Find an equation for the circle which has center at the origin and radius 5.

Substituting $r = 5$ into the equation of a circle given above, we have

$$x^2 + y^2 = 25.$$

The graph of this circle is shown in Figure 2.20.

Figure 2.20

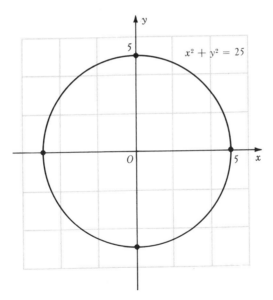

Example 19 Graph $y = -\sqrt{36 - x^2}$.

Again, we begin by squaring both sides of the equation. The result is $y^2 = 36 - x^2$ or $x^2 + y^2 = 36$, which represents a circle of radius 6, with center at (0, 0). However, $-\sqrt{36 - x^2}$ is nonpositive, so that $y \leq 0$. Hence, the graph is a semicircle, as shown in Figure 2.21. Although $x^2 + y^2 = 36$ is not a function, $y = -\sqrt{36 - x^2}$ is a function, another root function.

Figure 2.21

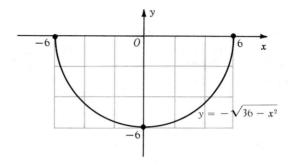

Example 20 Suppose the demand for a certain item, $D(x)$, in thousands of units, is given by

$$D(x) = \sqrt{9 - x^2},$$

while the supply of the item, $S(x)$, in thousands, is given by

$$S(x) = \sqrt{7 + x},$$

where x is the price of the item in dollars. Graph the supply and demand curves and find the equilibrium price.

The graph of the demand function in the first quadrant is the quarter-circle shown in Figure 2.22. The supply function has a graph which is the positive half of a horizontal parabola. Figure 2.22 shows the part of the graph which lies in the first quadrant.

Figure 2.22

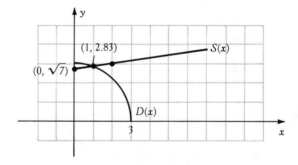

To find the coordinates of the point at which supply equals demand, we write

$$D(x) = S(x)$$
$$\sqrt{9 - x^2} = \sqrt{7 + x}.$$

To solve an equation with square roots, square both sides and be careful to check all solutions in the original equation. Here, squaring gives

$$9 - x^2 = 7 + x,$$

which can be solved either by factoring or by using the quadratic formula with the result $x = 1$ or $x = -2$. We cannot have a price of $x = -2$, and so the equilibrium price is $x = 1$. Note that this solution does satisfy the equation $D(x) = S(x)$. At the price of $1, the supply and demand functions both equal $\sqrt{8}$ which is approximately 2.83. Thus the equilibrium supply and demand occur at about 2830 items.

2.3 Exercises

Graph each of the following functions.

1. $y = |x + 1|$
2. $y = |x - 2|$
3. $y = |x - 3|$
4. $y = |x + 4|$
5. $y = |2x - 1|$
6. $y = |3x + 2|$
7. $y = 3x^2$
8. $y = \frac{1}{2}x^2$
9. $y = -2x^2$
10. $y = x^2 + 1$
11. $y = -x^2 + 1$
12. $y = (x + 1)^2$
13. $y = -(x + 3)^2$
14. $y = -2(x - 3)^2$
15. $y = x^2 - 2x - 2$
16. $y = x^2 - 4x + 6$
17. $y = x^2 + 6x + 3$
18. $y = -x^2 + 6x + 4$
19. $y = 2x^2 + 4x + 1$
20. $y = 3x^2 + 6x + 5$
21. $y = \sqrt{2x}$
22. $y = \sqrt{1 - x}$
23. $y = -\sqrt{x - 9}$
24. $y = -\sqrt{x + 4}$
25. $y = \sqrt{x - 1}$
26. $y = \sqrt{x + 5}$
27. $y = \sqrt{16 - x^2}$, $x \geq 0$
28. $y = \sqrt{4 - x^2}$, $x \geq 0$
29. $y = -\sqrt{9 - x^2}$
30. $y = -\sqrt{25 - x^2}$

31. Georgia runs a sandwich shop. By studying past results, she has found that the cost of operating her shop is given by

$$C(x) = 2x^2 - 1200x + 180{,}100,$$

where x is the daily volume of sandwiches sold.

(a) Graph $C(x)$.
(b) Find the vertex of the parabola and determine the number of sandwiches Georgia must sell to produce minimum cost.

32. Harry sells tacos. He has found that the profits of his stand are approximated by

$$P(x) = -x^2 + 4000x - 3{,}999{,}900,$$

where x is the number of tacos sold daily. Repeat steps (a) and (b) from Exercise 31 to determine the number of tacos that Harry should sell daily to produce the maximum profit.

33. Suppose the demand for a certain item is given by

$$D(x) = \sqrt{4 - x^2},$$

where $D(x)$ is the demand in thousands and x is the price in dollars. Find the demand at a price of (a) 0, (b) 1, (c) 2. (d) Graph $D(x)$.

34. Suppose the supply in thousands of the item from Exercise 33 is given by

$$S(x) = \sqrt{3x}.$$

(a) Graph $S(x)$ on the graph of $D(x)$ from Exercise 33.
(b) Find the equilibrium price.

35. Suppose the supply and demand of a certain mathematics textbook are given in thousands by

$$D(x) = \sqrt{144 - x^2} \quad \text{and} \quad S(x) = \sqrt{10x},$$

where x represents the price of the book.

(a) Find the demand at a price of 5; at a price of 10; at a price of 12.
(b) Find the supply at a price of 5; at a price of 10; at a price of 12.
(c) Graph $D(x)$ and $S(x)$ on the same axes.
(d) Find the equilibrium price.

36. A charter flight charges a fare of $200 per person, plus $4 per person for each unsold seat on the plane. If the plane holds 100 passengers, and if x represents the number of unsold seats, find:

(a) An expression for the total revenue received for the flight. (Hint: Multiply the number of people flying, $100 - x$, and the price per ticket.)
(b) The graph for the expression from part (a).
(c) The number of unsold seats that will produce the maximum revenue.
(d) The maximum revenue.

37. The demand for a certain type of cosmetic is given by

$$D(x) = 500 - x,$$

where x is the price.

(a) Find the revenue, $R(x)$, that would be obtained at a price x. (Hint: Revenue = demand $\times$ price.)
(b) Graph the revenue function, $R(x)$.
(c) From the graph of the revenue function, estimate the price that will produce the maximum revenue.

2.4 Polynomial and Rational Functions

A function of the form

$$f(x) = a_n x^n + a_{n-1} x^{n-1} + \cdots + a_1 x + a_0, \quad a_n \neq 0,$$

is called a **polynomial function of degree n**. We have already discussed polynomial functions of degree $n = 1$ (which are linear functions) and $n = 2$ (quadratic functions), and in this section we shall discuss more general polynomial functions. To graph a function f of degree 3 such as $y = x^3 - 2x^2 - x + 2$, it is necessary to find several ordered pairs, as shown in the chart of Figure 2.23. If we plot the resulting ordered pairs, and draw a smooth curve through the points, we get the graph shown in Figure 2.23. This graph is typical of the graphs of polynomial functions of degree $n = 3$. As shown by the graph, both the domain and range of this function are the set of all real numbers.

Figure 2.23

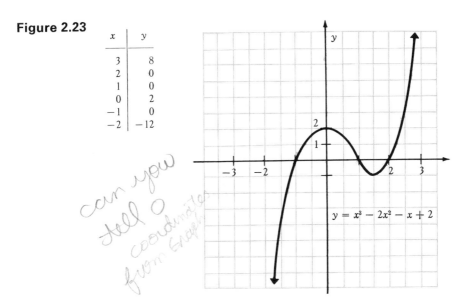

x	y
3	8
2	0
1	0
0	2
-1	0
-2	-12

It can be shown that a polynomial function P of degree n has at most n real **zeros**. That is, there are at most n real values of x for which $P(x) = 0$. Therefore, a polynomial function of degree 3 can have at most 3 zeros. This means that the graph will intersect the x-axis at most three times. From the graph in Figure 2.23, we can see that $y = x^3 - 2x^2 - x + 2$ has exactly three zeros, at $x = -1$, $x = 1$, and $x = 2$.

In general, it is difficult to graph polynomial functions of degree 3 or more. In Chapter 5 we shall discuss some techniques of calculus which will reduce the number of points that must be found, and will produce a more accurate graph. However, for now we must settle for plotting enough points to show the basic shape of the graph. In many cases, plotting the points obtained by letting $x = -3, -2, -1, 0, 1, 2$, and 3 is sufficient.

Example 21 Graph $y = 2x^4 + 3x^3 - 4x^2 - 3x + 2$.

If we again complete several ordered pairs and draw a smooth curve through the points determined by them, we get the graph shown in Figure 2.24. This graph is typical of polynomial functions of degree $n = 4$, where the coefficient of the fourth degree term is positive. The domain is the set of all real numbers, and it can be shown that the range is of the form $y \geq k$, where k is a real number. Note that the graph in Figure 2.24 indicates the maximum number of zeros for this function, four.

Figure 2.24

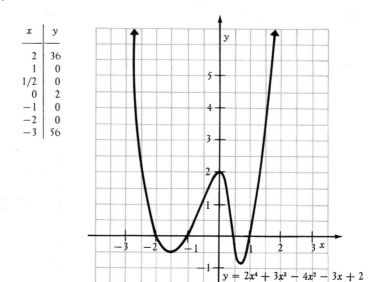

Example 22 Suppose the cost to produce x units of small transistor radios is given by

$$C(x) = 2x^3 - 21x^2 + 60x.$$

Find the cost to produce (a) 0 units, (b) 1 unit, (c) 2 units, (d) 3 units, (e) 4 units, (f) 5 units, (g) 6 units. (h) Graph $C(x)$. For $x \geq 1$, approximately what is the minimum cost?

The cost to produce 0 units of radios is given by $C(0)$:

$$C(0) = 2(0)^3 - 21(0)^2 + 60(0) = 0.$$

Verify that $C(1) = 41$, $C(2) = 52$, $C(3) = 45$, $C(4) = 22$, $C(5) = 25$, and $C(6) = 36$. Plotting these points and sketching a curve through them gives the graph of Figure 2.25. The graph shows that the minimum cost for $x \geq 1$ is about 20.

A function whose rule of correspondence can be expressed as the quotient of two polynomials is called a **rational function**. For example,

$$f(x) = \frac{1}{x}, \quad f(x) = \frac{x+1}{x-1}, \quad f(x) = \frac{3x^2 + 4x - 1}{x - 1}$$

are all rational functions.

To graph the rational function

$$f(x) = \frac{2}{1+x},$$

Figure 2.25

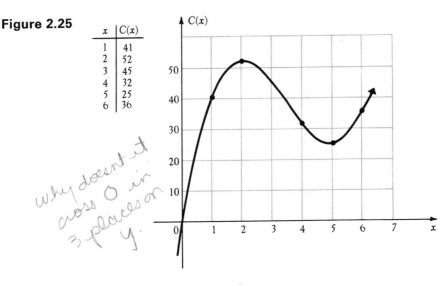

first note that the function is undefined for $x = -1$, since the denominator then becomes $1 + (-1) = 0$. Therefore, the graph of this function will not cross the vertical line $x = -1$. We can replace x with any number we wish, except -1, so that we can let x approach -1 as closely as we wish. From the following table, note that as x gets closer and closer to -1, the quantity $1 + x$ gets closer and closer to 0, and $|2/(1 + x)|$ gets larger and larger. When the resulting ordered pairs are graphed, as in Figure 2.26, we see that the value of x is getting closer and closer to the vertical line $x = -1$, while $|f(x)|$ gets larger and larger. For this reason, the line $x = -1$ is called a **vertical asymptote**.

x	$1 + x$	$\dfrac{2}{1+x}$	Ordered Pair
-0.5	0.5	4	$(-0.5, 4)$
-0.8	0.2	10	$(-0.8, 10)$
-0.9	0.1	20	$(-0.9, 20)$
-0.99	0.01	200	$(-0.99, 200)$
-1.01	-0.01	-200	$(-1.01, -200)$
-1.1	-0.1	-20	$(-1.1, -20)$
-1.5	-0.5	-4	$(-1.5, -4)$

Verify that as $|x|$ gets larger and larger, the quantity $2/(x + 1)$ gets closer and closer to 0. We say that the graph has a **horizontal asymptote** at the x-axis. By using the asymptotes and plotting points, we get the graph shown in Figure 2.26.

56 Functions

Figure 2.26

x	y
3	1/2
2	2/3
1	1
0	2
-1/2	4
-3/2	-4
-2	-2
-3	-1
-4	-2/3

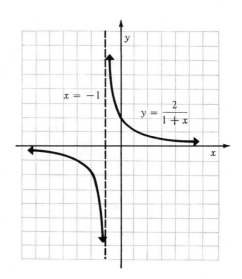

To graph the rational function

$$y = \frac{3x + 2}{2x + 4},$$

first note that the graph has a vertical asymptote at $x = -2$, since $x = -2$ makes the denominator 0. We can locate a horizontal asymptote by solving the equation for x. To do this, first multiply both sides of the equation by $2x + 4$. This gives

$$y(2x + 4) = 3x + 2$$
$$2xy + 4y = 3x + 2.$$

Collect all terms containing x on one side of the equation:

$$2xy - 3x = 2 - 4y.$$

Factor out x on the left, and solve for x:

$$x(2y - 3) = 2 - 4y$$

$$x = \frac{2 - 4y}{2y - 3}.$$

From this form of the equation, we see that y cannot take the value $y = 3/2$. In the same way that we identified $x = -2$ as a vertical asymptote, we have $y = 3/2$ as a horizontal asymptote. The graph is shown in Figure 2.27.

Figure 2.27

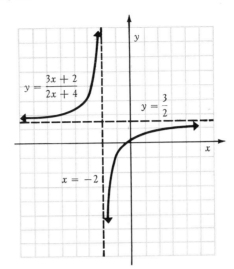

Example 23 In many situations involving environmental pollution, much of the pollutant can be removed from water or air at a fairly reasonable cost, but the last, small part of the pollutant can become increasingly expensive to remove. Cost as a function of percent of pollutant removed from the environment can be calculated for various percents of removal, with a curve then fitted through the resulting data points. Rational functions are often a good choice for these **cost-benefit curves**. For example, suppose a cost-benefit curve is given by

$$y = \frac{18x}{106 - x},$$

where y is the cost, in thousands of dollars, to remove x percent of a certain pollutant. The domain of x is the set $\{x \mid 0 \le x \le 100\}$; any amount of a

Figure 2.28

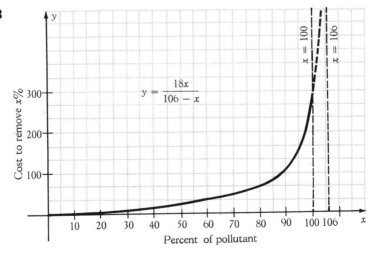

pollutant from 0% to 100% can be removed. To remove 100% of the pollutant here would cost

$$y = \frac{18(100)}{106 - 100} = 300,$$

or $300,000. Verify that 95% of the pollutant can be removed for $155,000, 90% for $101,000, and 80% for $55,000. Using these points, as well as others that we could obtain from the function above, we get the graph of Figure 2.28 (see p. 57).

2.4 Exercises

Graph each of the following functions.

1. $y = x^3$
2. $y = x^4$
3. $y = x^3 - 2x^2 - 5x + 6$
4. $y = x^3 + 3x^2 - 4x - 12$
5. $y = x^4 + 2x^3 - 7x^2 - 8x + 12$
6. $y = x^4 - x^3 - 7x^2 + x + 6$
7. $y = \dfrac{1}{x}$
8. $y = \dfrac{1}{x + 1}$
9. $y = \dfrac{3}{2x - 4}$
10. $y = \dfrac{-1}{3x + 6}$
11. $y = \dfrac{2x}{x + 1}$
12. $y = \dfrac{-3x}{x - 4}$
13. $y = \dfrac{x}{x + 3}$
14. $y = \dfrac{x}{2 - x}$
15. $y = \dfrac{2x - 1}{4x + 2}$
16. $y = \dfrac{3x - 6}{6x - 1}$
17. $y = \dfrac{-2x + 1}{5x + 2}$
18. $y = \dfrac{-3x - 6}{4x + 2}$

19. Suppose that the cost $C(x)$ of producing x units of margarine is given by

$$C(x) = \frac{500}{x + 30}.$$

Find the cost per unit to produce
(a) 10 units;
(b) 20 units;
(c) 50 units;
(d) 70 units.
(e) Graph $C(x)$.

20. Suppose a cost-benefit curve (see Example 23) is given by

$$y = \frac{6.5x}{102 - x},$$

where y is the cost, in thousands of dollars, of removing x percent of a certain pollutant. Find the cost of removing the following percents of pollution:
(a) 0
(b) 50
(c) 80
(d) 90
(e) 95
(f) 99
(g) 100
(h) Graph the function.

21. Suppose a cost-benefit curve is given by

$$y = \frac{6.7x}{100 - x},$$

where y is the cost, in thousands of dollars, of removing x percent of a given pollutant. Find the cost of removing each of the following percents of pollutants:
- (a) 50
- (b) 70
- (c) 80
- (d) 90
- (e) 92
- (f) 95
- (g) 98
- (h) 99
- (i) Is it possible, according to this function, to remove all the pollutant?
- (j) Graph the function.

Given the polynomial equation

$$x^n + a_{n-1}x^{n-1} + \cdots + a_1 + a_0 = 0,$$

in which the coefficient of the term of highest degree is 1, it can be shown that any integer solutions of the equation must divide evenly into a_0. Thus, any integer solutions of $x^3 - 4x^2 + x + 6 = 0$ must divide evenly into 6. The only possible integer solutions are therefore ± 1, ± 2, ± 3, or ± 6, the only integers that divide evenly into 6. By replacing x in the equation $x^3 - 4x^2 + x + 6 = 0$ by each of the possible solutions, we find that $x = 3$, $x = 2$, and $x = -1$ are the only integer solutions of $x^3 - 4x^2 + x + 6 = 0$.

Find all integer solutions of each of the following equations.

22. $x^3 - 6x^2 + 11x - 6 = 0$
23. $x^3 - x^2 - 10x - 8 = 0$
24. $x^3 + 3x^2 - 4x - 12 = 0$
25. $x^3 - 2x^2 - 9x + 18 = 0$
26. $x^3 + 3x^2 + 2x + 2 = 0$
27. $x^3 - 4x^2 + 2x + 3 = 0$
28. $x^4 - 4x^3 - x^2 + 16x - 12 = 0$
29. $x^4 - x^3 - 11x^2 + 9x + 18 = 0$
30. $x^4 + 4x^3 - 7x^2 - 22x + 24 = 0$

2.5 Operations on Functions

Many functions can be considered as combinations of simpler functions. For example, the function

$$f(x) = x^2 + 2x$$

can be thought of as the sum of the two functions $g(x) = x^2$ and $h(x) = 2x$. To express this idea, we write

$$f(x) = [g + h](x) = g(x) + h(x).$$

60 Functions

The domain of the new function is the set of numbers common to the domains of the simpler functions.

In the same way, by subtraction we can form a new function f from the functions g and h, where $g(x) = x^3 + 2$ and $h(x) = \sqrt{x}$, to get $f = g - h$. Thus,

$$f(x) = g(x) - h(x) = x^3 + 2 - \sqrt{x}.$$

Functions which are products and quotients of simpler functions can be formed similarly. For example, if $g(x) = |x|$ and $h(x) = x^3 - 1$, we have

$$f(x) = [gh](x) = |x| \cdot (x^3 - 1)$$

and

$$F(x) = \left[\frac{g}{h}\right](x) = \frac{|x|}{x^3 - 1}.$$

To form a quotient function like F, we must restrict the denominator function to nonzero values. In the example above, we would have

$$F(x) = \frac{g(x)}{h(x)}, \qquad h(x) \neq 0.$$

The domain of F is the set of all real numbers except 1, since both g and h have all real numbers in their domains, but $h(1) = 0$.

Example 24 Given $f(x) = \sqrt{x + 1}$ and $g(x) = x^2$, find (a) $f + g$; (b) $f \cdot g$; (c) f/g; and give their domains.

(a) $\qquad [f + g](x) = f(x) + g(x) = \sqrt{x + 1} + x^2.$

The domain of $\sqrt{x + 1}$ is $[-1, +\infty)$; the domain of x^2 is the real numbers. The numbers common to both domains, $[-1, +\infty)$, give the domain of $f + g$.

(b) $\qquad [fg](x) = f(x) \cdot g(x) = (\sqrt{x + 1})x^2.$

The domain is the same as in part (a).

(c) $\qquad \left[\frac{f}{g}\right](x) = \frac{f(x)}{g(x)} = \frac{\sqrt{x + 1}}{x^2}, \qquad x \neq 0.$

The domain is the same as in part (a) with the additional restriction that $x \neq 0$. Thus, the domain is $[-1, 0]$ or $[0, +\infty)$.

Another way to form a new function from two functions is to use the range of one function g as the domain of the other function f. This **composite function** is written as $f \circ g$ where

$$[f \circ g](x) = f[g(x)].$$

For example, if $f(x) = x^3$ and $g(x) = x^2 - 1$, then

$$[f \circ g](x) = f[g(x)] = f(x^2 - 1) = (x^2 - 1)^3.$$

Also,

$$[g \circ f](x) = g[f(x)] = g(x^3) = (x^3)^2 - 1 = x^6 - 1.$$

2.5 Operations on Functions

The domain of a composite function $f \circ g$ can be complicated to find. It is the set $\{x \mid g(x)$ belongs to the domain of $f\}$. In the example above, the domain of g, the set of real numbers, leads to a range $g(x) \geq -1$, which belongs to the domain of f (also the set of real numbers). Thus, the domain of $f \circ g$ is the set of real numbers.

Example 25 Suppose the demand for a product is given by
$$D(p) = 1000 - p^2,$$
where p is the price. Also suppose the price can be expressed as
$$p(c) = c + 25,$$
where c is the cost of the product to the merchant. Then we can express the demand for the product in terms of the cost by forming the composite function $D \circ p$ where
$$[D \circ p](c) = D[p(c)] = 1000 - (c + 25)^2$$
$$= 1000 - (c^2 + 50c + 625)$$
$$= 375 - 50c - c^2.$$

2.5 Exercises

Given $f(x) = \sqrt{x}$ and $g(x) = x^2 - 1$, find the following functions and give the domain of each.

1. $[f + g](x)$
2. $[g - f](x)$
3. $[fg](x)$
4. $\left[\dfrac{f}{g}\right](x)$
5. $\left[\dfrac{g}{f}\right](x)$
6. $[f \circ g](x)$
7. $[g \circ f](x)$
8. $[f \circ (g \circ f)](x)$

Each of the following functions is a sum, difference, product, or quotient of two functions, $g(x)$ and $f(x)$. Find $g(x)$ and $h(x)$. It is possible to do these problems in more than one way.

9. $\dfrac{1}{x}$
10. $x^3 - 3x$
11. $3|x|$
12. $\dfrac{x^2 + 1}{x - 3}$
13. $|x|^2$
14. $x + \sqrt{x}$
15. $\dfrac{x^3 + 1}{x}$
16. $2(x + 3)^2$

Each of the following functions is a composite function $[f \circ h]$. Find $f(x)$ and $h(x)$. It is possible to do these problems in more than one way.

17. $\sqrt{x^2 - 1}$
18. $|4x - 1|$
19. $\dfrac{1}{x^2 - 1}$
20. $(x^3 + 2)^2$
21. $\sqrt{2x^2 + 1} - 1$
22. $\dfrac{x^3 + 3x - 2}{2}$

62 Functions

23. Suppose the population P of a certain species of fish depends on the number x of a smaller kind of fish which serves as its food supply, so that
$$P(x) = 2x^2 + 1.$$
Suppose, also, that the number x of the smaller species of fish depends upon the amount a of its food supply, a kind of plankton. Suppose
$$x = f(a) = 3a + 2.$$
Find $[P \circ f](a)$, the relationship between the population P of the large fish and the amount a of plankton available.

24. Suppose the demand for a certain brand of vacuum cleaner is given by
$$D(p) = \frac{-p^2}{100} + 500,$$
where p is the price in dollars. If the price, in terms of the cost, c, is expressed as
$$p(c) = 2c - 10,$$
find the demand in terms of the cost.

● Case 2   Estimating Oil Tanker Construction Costs

The U.S. Maritime Administration estimated that in a recent year the cost of building an oil tanker of 50,000 deadweight tons in the United States was $409 per ton. The cost per ton for a 100,000-ton tanker was $310, while the cost per ton for a 400,000-ton tanker was $178.

Figure 1 shows these values plotted on a graph, where x represents tons (in thousands) and y represents the cost per ton. There is a gap in the information

Figure 1

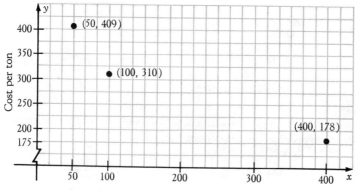

presented by the government agency: the data skips from $x = 100$ to $x = 400$, with no intermediate values. If we could fit a curve through the data points we do have, we could then approximate any desired intermediate values. Many different functions have graphs which provide a good fit for the given data. By studying the graph of Figure 1, we might decide that a rational function, like the ones graphed in Section 2.4, is one such function. Using methods that we shall not discuss here, we can obtain

$$y = \frac{110{,}000}{x + 225}$$

as a rational function which approximates the given data. We can substitute the known values $x = 50$, $x = 100$, and $x = 400$ to test the "goodness of fit" of the rational function. Doing this, we obtain the following results.

x	y (given)	y (from function)
50	409	400
100	310	338
400	178	176

As the chart shows, the rational function provides a reasonable fit to the data, so that we can use it to approximate y for intermediate values of x. If we let $x = 150$, we get

$$y = \frac{110{,}000}{150 + 225} = 293.$$

Verify that $y = \$259$ when $x = 200$, while if $x = 300$, $y = \$210$. We can use these points to graph the function, as shown in Figure 2.

Figure 2

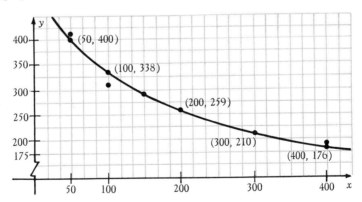

Case 2 Exercises

Estimate the cost per ton of building a tanker of:

1. 25,000 tons   **2.** 75,000 tons   **3.** 250,000 tons   **4.** 450,000 tons

Chapter 3 Pretest

1. Factor $x^2 - 5x + 6$. [1.3]
2. Simplify the rational expression $\dfrac{x^2 - 5x + 6}{x - 3}$. [1.4]
3. Rationalize the *numerator*. [1.4]
    (a) $\dfrac{\sqrt{x} - 4}{x - 16}$
    (b) $\dfrac{1 - \sqrt{x}}{1 - x}$
    (c) $\dfrac{\sqrt{x} - 5}{x - 25}$
4. Perform the indicated operations. [1.4]
    (a) $(x^2 - 2)\dfrac{1}{x}$
    (b) $\dfrac{1}{x^3}(2x^2 - 3x + 2)$
5. Write with exponents. [1.5]
    (a) $\sqrt{x}$
    (b) $\sqrt{x - 5}$
    (c) $\sqrt{(x+5)^3}$

Graph each of the following.

6. $f(x) = 3x - 1$ [2.2]
7. $f(x) = \dfrac{2}{3}x + 2$ [2.2]
8. $f(x) = 5$ [2.2]
9. $f(x) = \dfrac{1}{x}$ [2.4]
10. $f(x) = \dfrac{x + 2}{2x - 3}$ [2.4]
11. $f(x) = x^2 - 4$ [2.3]
12. $f(x) = x^3 + 1$ [2.4]
13. $f(x) = \begin{cases} 2x & \text{if } x < 0 \\ x^2 & \text{if } x \geq 0 \end{cases}$ [2.2, 2.3]
14. $f(x) = |2x - 1|$ [2.3]
15. $f(x) = \begin{cases} x^2 & \text{if } x < 4 \\ 1 & \text{if } x = 4 \\ \sqrt{x} & \text{if } x > 4 \end{cases}$ [2.3]

3
Limits

In Chapter 2 we introduced the concept of a function and discussed several useful types of functions. In this chapter we consider the value (if one exists) that $f(x)$ approaches as x gets closer and closer to some number a. We call any such value of $f(x)$ a *limit*. The limit concept is essential to the development of one of the central ideas of calculus, that of the derivative, which is introduced in the next chapter.

3.1 The Limit of a Function

Figure 3.1

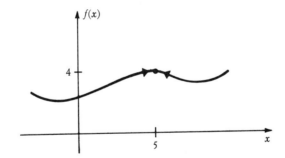

If we let the values of x in the domain of a function f get closer and closer to some number a, what happens to the values of $f(x)$? Do they, in some cases, get closer and closer to some one number? Figure 3.1 shows the graph of a function f. The arrowheads indicate that as x gets closer and closer to the number 5, $f(x)$ gets closer and closer to the number 4. This is true whether x approaches 5 from the left (below) or from the right (above). In such a situation, we say that the limit of $f(x)$ as x approaches 5 is the number 4, written as

$$\lim_{x \to 5} f(x) = 4.$$

The graph in Figure 3.2 shows another function g. The open circle at (3, 2) indicates the fact that the function is undefined at $x = 3$. (In other words, $g(3)$

does not exist.) What about $\lim_{x \to 3} g(x)$? The graph shows that as x gets closer and closer to 3, $g(x)$ gets closer and closer to the number 2, so that

$$\lim_{x \to 3} g(x) = 2,$$

even though $g(3)$ does not exist.

Figure 3.2

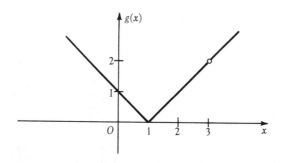

Now look at the graph of h in Figure 3.3. As x gets close to 4 from the left, $h(x)$ approaches 1. However, as x gets close to 4 from the right, $h(x)$ approaches the number 3. Since $h(x)$ approaches two different numbers depending upon whether x approaches 4 from the left or the right, we say

$$\lim_{x \to 4} h(x) \text{ does not exist.}$$

(In a situation like this, we sometimes refer to a right-hand limit and a left-hand limit.) From the graph of Figure 3.3 we can also see that $\lim_{x \to 6} h(x) = 5$ and $\lim_{x \to 8} h(x)$ does not exist.

Figure 3.3

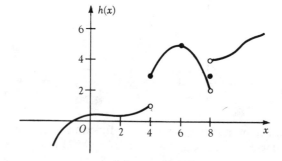

Example 1 Graph $f(x) = \begin{cases} 2x + 1 & \text{if } x \leq 0 \\ 2x & \text{if } x > 0. \end{cases}$

Use the graph to find (a) $\lim_{x \to -2} f(x)$; (b) $\lim_{x \to 1} f(x)$; (c) $\lim_{x \to 0} f(x)$.

From the graph of Figure 3.4, we see that

$$\lim_{x \to -2} f(x) = -3 \quad \text{and} \quad \lim_{x \to 1} f(x) = 2.$$

As x approaches 0 from the left, $f(x)$ gets closer and closer to 1. However, as x approaches 0 from the right, $f(x)$ approaches 0. Thus, $\lim_{x \to 0} f(x)$ does not exist.

Figure 3.4

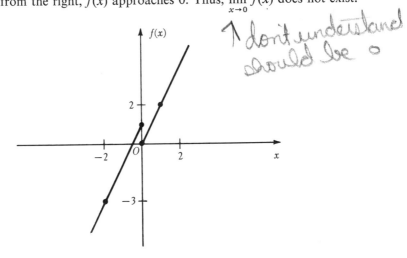

Example 2 Graph $f(x) = \dfrac{1}{x}$ and use the graph to find $\lim_{x \to 0} f(x)$.

The graph in Figure 3.5 shows that as x gets close to 0 from the left, the values of $f(x)$ get smaller and smaller. On the other hand, as x approaches 0 from the right, the values of $f(x)$ get larger and larger. Thus, the values of $f(x)$ approach no one number as x gets closer and closer to 0. Therefore,

$$\lim_{x \to 0} \frac{1}{x} \text{ does not exist.}$$

Figure 3.5

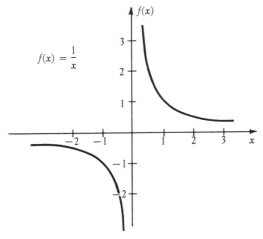

In general, we can say that

$$\lim_{x \to a} f(x)$$

exists and equals the number L, only if we can guarantee that $f(x)$ will be as close to L as we wish whenever x is sufficiently close to a. In other words, we must be able to make $|f(x) - L|$ less than any preassigned small, positive number by choosing x so that $|x - a|$ is suitably small. If these conditions are met, then we can write

$$\lim_{x \to a} f(x) = L.$$

The definition is not useful for finding limits—only for verifying that a given number L is the desired limit.

Without graphing, we can find the limit of a function f as x approaches the number a. We begin by finding several values of $f(x)$ as x gets closer and closer to a. For example, given the function f where

$$f(x) = \frac{x^2 - 4}{x - 2},$$

we can find $\lim_{x \to 1} f(x)$ by choosing several values of x that are close to 1 and calculating $f(x)$ for each of these values. We need to look at values of x close to 1 on either side of 1—that is, values which are smaller than 1 and also values which are greater than 1. The results for several selected values of x are shown in the following chart.

					Closer & closer to 1				
x	0.8	0.9	0.99	0.9999	↓	1.0000001	1.0001	1.001	1.05	1.1
$f(x)$	2.8	2.9	2.99	2.9999	↑	3.0000001	3.0001	3.001	3.05	3.1
					Closer & closer to 3				

From the results given in the chart, we see that the values of $f(x)$ get closer and closer to 3 as the values of x get closer and closer to 1. In fact, we can make $f(x)$ take on a value as close to 3 as we might like, by selecting a value of x sufficiently close to 1. By the definition of limit which we gave above, this is exactly what we mean when we say

$$\lim_{x \to 1} \frac{x^2 - 4}{x - 2} = 3.$$

The value $x = 2$ is not an element of the domain of $f(x) = (x^2 - 4)/(x - 2)$, so that $f(2)$ does not exist. Even though $f(2)$ does not exist,

$$\lim_{x \to 2} \frac{x^2 - 4}{x - 2}$$

may exist. Recall that in Figure 3.2 we saw that $\lim_{x \to 3} g(x)$ existed (and was equal to 2) even though $g(3)$ did not exist. Does the limit of $f(x)$ as $x \to 2$ exist, and if it does, what is its value? To answer this question, we can make a chart, as we did above, showing values of $f(x)$ for selected values of x close to 2.

3.1 The Limit of a Function

		Closer & closer to 2				
x	1.8 1.9 1.99	1.9999 ↓ 2.0000001	2.0001	2.001	2.05	2.1
$f(x)$	3.8 3.9 3.99	3.9999 ↑ 4.0000001	4.0001	4.001	4.05	4.1
		Closer & closer to 4				

From the chart, we see that the values of $f(x)$ get closer and closer to 4 as x gets closer and closer to 2. In fact, we can find values of $f(x)$ as close to 4 as we wish, by selecting x close enough to 2. Thus,

$$\lim_{x \to 2} \frac{x^2 - 4}{x - 2} = 4.$$

By plotting a number of points, we get the graph of f shown in Figure 3.6. You may have noticed that

$$f(x) = \frac{x^2 - 4}{x - 2} = \frac{(x + 2)(x - 2)}{x - 2} = x + 2, \quad \text{if } x \neq 2,$$

which indicates a simple way to calculate the values of $f(x)$ in the chart above.

Figure 3.6

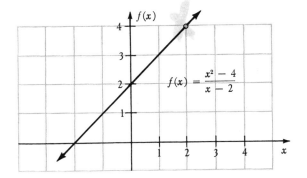

Example 3 Given $f(x) = \dfrac{x^2 - x - 2}{x + 1}$, find the following: (a) $\lim_{x \to 0} f(x)$; (b) $\lim_{x \to -1} f(x)$.

(a) Using the fact that

$$\frac{x^2 - x - 2}{x + 1} = \frac{(x + 1)(x - 2)}{x + 1} = x - 2, \quad \text{for } x \neq -1,$$

we get the following values of $f(x)$ for values of x close to 0:

x	−0.1	−0.01	−0.001	0.001	0.01	0.1
$f(x)$	−2.1	−2.01	−2.001	−1.999	−1.99	−1.9

From the chart, we see that

$$\lim_{x \to 0} \frac{x^2 - x - 2}{x + 1} = -2.$$

Note that $f(0) = -2$ for this function.

(b) Again, we use the fact that

$$\frac{x^2 - x - 2}{x + 1} = x - 2, \quad \text{for } x \neq -1,$$

to complete a table of values for x close to -1.

x	-1.1	-1.01	-1.001	-0.999	-0.99	-0.9
$f(x)$	-3.1	-3.01	-3.001	-2.999	-2.99	-2.9

The table indicates that

$$\lim_{x \to -1} \frac{x^2 - x - 2}{x + 1} = -3.$$

Example 4 Find $\lim_{x \to 4} \frac{1}{x - 4}$.

We can make a chart, as we did above, showing values of $1/(x - 4)$ for values of x close to 4.

x	3.9	3.99	3.999	4.001	4.01	4.1
$f(x)$	-10	-100	-1000	1000	100	10

From the results shown in this chart, $f(x)$ approaches no one number as a limit as x approaches 4 from either side. Hence,

$$\lim_{x \to 4} \frac{1}{x - 4} \text{ does not exist.}$$

Example 5 The *entropy* of a biological or chemical system is a measure of the disorder in the system. In general, a substance has a lower entropy at a lower temperature, and conversely. If a solid is heated, the molecules of the substance vibrate more and more rapidly, causing more and more disorder in

Figure 3.7

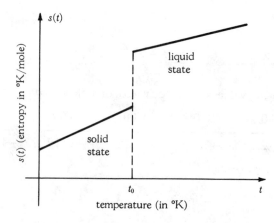

the system, with a corresponding rise in entropy. When the temperature of the substance reaches the correct point (the melting point), the bonds holding the molecules together are broken, permitting the substance to change from solid to liquid. This change from solid to liquid causes a large increase in entropy. The entropy of a typical substance is shown in Figure 3.7. As shown by the graph, $\lim_{t \to t_0} s(t)$ does not exist.

3.1 Exercises

Use the following graphs to determine the indicated limits if they exist.

1. $\lim_{x \to 3} f(x)$

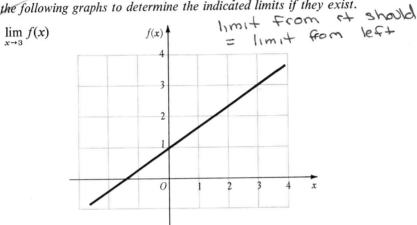

2. $\lim_{x \to 2} F(x)$

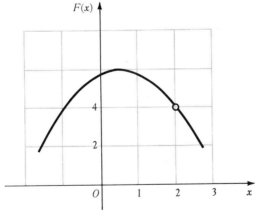

3. $\lim_{x \to -2} f(x)$

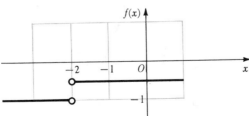

72 Limits

4. $\lim\limits_{x \to 3} g(x)$

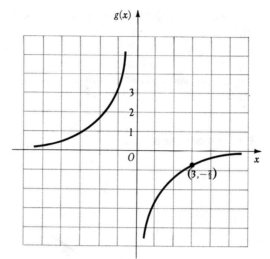

5. $\lim\limits_{x \to 0} f(x)$

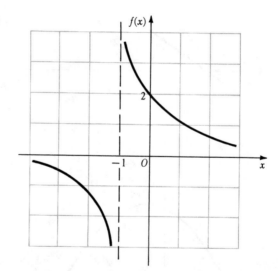

6. $\lim\limits_{x \to 1} h(x)$

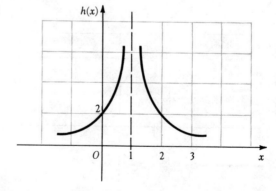

3.1 The Limit of a Function 73

7. $\lim_{x \to 0} f(x)$

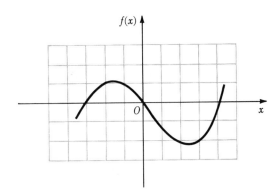

8. $\lim_{x \to 1} g(x)$

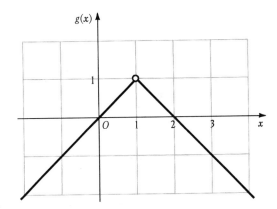

9. $\lim_{x \to 3} F(x)$

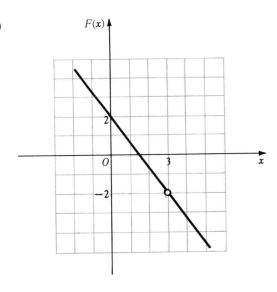

10. $\lim\limits_{x \to 0} f(x)$

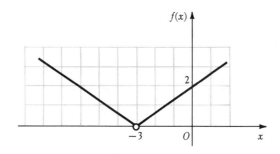

Sketch the graph of each of the following functions and use it to find the indicated limit, if it exists.

11. $\lim\limits_{x \to -1} (3x - 1)$
12. $\lim\limits_{x \to 2} (5x - 3)$
13. $\lim\limits_{x \to 1} (x^2 + 2)$
14. $\lim\limits_{x \to 0} (-x^2 + 3)$
15. $\lim\limits_{x \to 0} x^3$
16. $\lim\limits_{x \to 2} (x - 2)^3$

17. $\lim\limits_{x \to 1} f(x);\quad f(x) = \begin{cases} x + 2 & \text{if } x \le 1 \\ x & \text{if } x > 1 \end{cases}$

18. $\lim\limits_{x \to 0} f(x);\quad f(x) = \begin{cases} x^2 & \text{if } x \le 0 \\ x & \text{if } x > 0 \end{cases}$

19. $\lim\limits_{x \to 2} f(x);\quad f(x) = \begin{cases} |x - 1| & \text{if } x \le 2 \\ \frac{1}{2}x & \text{if } x > 2 \end{cases}$

20. $\lim\limits_{x \to -1} f(x);\quad f(x) = \begin{cases} x^2 - 1 & \text{if } x \le 0 \\ -x^2 & \text{if } x > 0 \end{cases}$

21. $\lim\limits_{x \to -1} \dfrac{x^2 - 1}{x + 1}$
22. $\lim\limits_{x \to 2} \dfrac{x^2 - 4}{x - 2}$

Use a chart of values of $f(x)$ for x close to the given values of a to find $\lim\limits_{x \to a} f(x)$. See Examples 3 and 4.

23. $f(x) = (4 + 3x);\quad a = -1, 0, 2$
24. $f(x) = (2x - 5);\quad a = 1, 2, 3$
25. $f(x) = (x^2 + 2x + 1);\quad a = -1, 0, 1$
26. $f(x) = (x^3 - x^2);\quad a = 0, 2$
27. $f(x) = \dfrac{2x - 3}{x};\quad a = 2, 3, -1, 0$
28. $f(x) = \dfrac{x^2 - 16}{x + 4};\quad a = 2, 4, -2, -4$
29. $f(x) = \dfrac{2}{x - 1};\quad a = 2, 1, -1, 5$
30. $f(x) = -\dfrac{1}{x};\quad a = -1, 0, 1, 2$

Use a calculator or computer to complete the table and determine the indicated limit in each of the following.

31. $\lim\limits_{x \to 2} \dfrac{x^3 - 2x - 4}{x - 2}$

x	1.9	1.99	1.999	2.001	2.01	2.1
$f(x)$						

32. $\lim\limits_{x \to -1} \dfrac{2x^3 + 3x^2 - 4x - 5}{x + 1}$

x	-1.1	-1.01	-1.001	-0.999	-0.99	-0.9
$f(x)$						

33. $\lim\limits_{x \to 1} \dfrac{x^4 - x^3 - x + 1}{x - 1}$

x	1.1	1.01	1.001	0.999	0.99	0.9
$f(x)$						

34. $\lim\limits_{x \to 2} \dfrac{x^3 - 2x^2 - 2x + 4}{x - 2}$

x	1.9	1.99	1.999	2.001	2.01	2.1
$f(x)$						

3.2 Properties of Limits

The methods we used in Section 3.1 to find $\lim\limits_{x \to a} f(x)$ (graphing and finding values of $f(x)$ for values of x close to a) were useful for illustration but awkward to use. In this section, we present without proof several theorems which simplify the process of finding limits for those functions which are most commonly used.*

Theorem 3.1 For any constant k, $\lim\limits_{x \to a} k = k$.

To see that Theorem 3.1 is true, we note that for the function $f(x) = k$, every value of x gives $f(x) = k$. Thus, for any value of x close to a, $f(x) = k$, and $\lim\limits_{x \to a} k = k$ as shown in Figure 3.8.

Figure 3.8

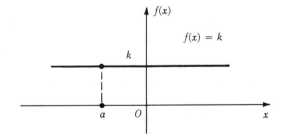

---

* The proofs of these limit theorems can be found in *The Calculus with Analytic Geometry*, second edition, Part 1, Louis Leithold, Harper & Row, New York, 1972, p. 85 ff.

Example 6 Find $\lim_{x \to 4} -3$.

The number -3 is a constant real number. Hence, by Theorem 3.1,
$$\lim_{x \to 4} -3 = -3.$$

Theorem 3.2 $\lim_{x \to a} x = a$.

Here we have $f(x) = x$, so as the value of x gets closer and closer to a, the values of $f(x)$ also get closer and closer to a. Thus, $\lim_{x \to a} x = a$. See Figure 3.9.

Figure 3.9

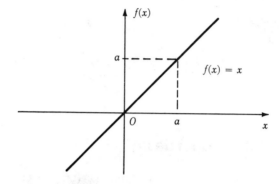

Example 7 Find $\lim_{x \to -2} x$.

By Theorem 3.2, $\lim_{x \to -2} x = -2$.

Theorem 3.3 If all indicated limits exist, then
$$\lim_{x \to a} [f(x) \pm g(x)] = \lim_{x \to a} f(x) \pm \lim_{x \to a} g(x).$$

Example 8 Find $\lim_{x \to -1} (x + 2)$.

Using Theorem 3.3, we have
$$\lim_{x \to -1} (x + 2) = \lim_{x \to -1} x + \lim_{x \to -1} 2.$$

Now we can use Theorems 3.1 and 3.2 to write
$$\lim_{x \to -1} x + \lim_{x \to -1} 2 = -1 + 2 = 1.$$

Theorem 3.4 If all indicated limits exist, then
$$\lim_{x \to a} [f(x) \cdot g(x)] = \lim_{x \to a} f(x) \cdot \lim_{x \to a} g(x).$$

Example 9 Find $\lim_{x \to -1} 4x$.

By Theorem 3.4,
$$\lim_{x \to -1} 4x = \left(\lim_{x \to -1} 4\right)\left(\lim_{x \to -1} x\right),$$

and by Theorems 3.1 and 3.2,

$$\left(\lim_{x \to -1} 4\right)\left(\lim_{x \to -1} x\right) = 4(-1) = -4.$$

Theorem 3.5 If all indicated limits exist and if $\lim_{x \to a} g(x) \neq 0$, then

$$\lim_{x \to a} \frac{f(x)}{g(x)} = \frac{\lim_{x \to a} f(x)}{\lim_{x \to a} g(x)}.$$

Example 10 Find $\lim_{x \to -1} \frac{x+2}{4x}$.

By Theorem 3.5,

$$\lim_{x \to -1} \frac{x+2}{4x} = \frac{\lim_{x \to -1} (x+2)}{\lim_{x \to -1} 4x}.$$

Using the results of Examples 8 and 9, we have

$$\frac{\lim_{x \to -1} (x+2)}{\lim_{x \to -1} 4x} = \frac{1}{-4}$$

Theorem 3.6 If all indicated limits exist, then for any real number n,

$$\lim_{x \to a} [f(x)]^n = \left[\lim_{x \to a} f(x)\right]^n.$$

Example 11 Find $\lim_{x \to 8} \sqrt{x-4}$.

We can write $\sqrt{x-4}$ as $(x-4)^{1/2}$. Then by Theorem 3.6, we have

$$\lim_{x \to 8} (x-4)^{1/2} = \left[\lim_{x \to 8} (x-4)\right]^{1/2}.$$

Now we use Theorems 3.3, 3.1, and 3.2:

$$\left[\lim_{x \to 8} (x-4)\right]^{1/2} = \left[\lim_{x \to 8} x - \lim_{x \to 8} 4\right]^{1/2}$$
$$= (8-4)^{1/2}$$
$$= 4^{1/2}$$
$$= 2.$$

Example 12 Find $\lim_{x \to 2} (4x^2 - 2x + 3)$.

By Theorem 3.3, we have

$$\lim_{x \to 2} (4x^2 - 2x + 3) = \lim_{x \to 2} 4x^2 + \lim_{x \to 2} (-2x) + \lim_{x \to 2} 3.$$

If we now use Theorem 3.4, the right side becomes

$$= \left(\lim_{x\to 2} 4\right)\left(\lim_{x\to 2} x^2\right) + \left(\lim_{x\to 2} -2\right)\left(\lim_{x\to 2} x\right) + \lim_{x\to 2} 3$$
$$= (4)(2^2) + (-2)(2) + 3$$
$$= 15.$$

We used Theorems 3.1, 3.2, and 3.6 in the next-to-last step.

Example 13 Find $\lim_{x\to -3} (2x^2)(3 - x)$.

Using the appropriate theorems, we have

$$\lim_{x\to -3} (2x^2)(3 - x) = \left[\lim_{x\to -3} 2x^2\right]\left[\lim_{x\to -3} (3 - x)\right]$$
$$= \left(\lim_{x\to -3} 2\right)\left(\lim_{x\to -3} x^2\right)\left(\lim_{x\to -3} 3 - \lim_{x\to -3} x\right)$$
$$= 2(-3)^2[3 - (-3)]$$
$$= 108.$$

Example 14 Find $\lim_{x\to -1} \frac{3x - 5}{4 + 2x}$.

By Theorem 3.5,

$$\lim_{x\to -1} \frac{3x - 5}{4 + 2x} = \frac{\lim_{x\to -1} (3x - 5)}{\lim_{x\to -1} (4 + 2x)} = \frac{-8}{2} = -4.$$

Example 15 Find $\lim_{x\to -1} \frac{x^2 - 1}{x + 1}$.

Here, $g(x)$, the denominator, is 0 for $x = -1$, so Theorem 3.5 does not apply. Since we are interested in the value of the function for x near -1 but not equal to -1, we use the fact that

$$\frac{x^2 - 1}{x + 1} = \frac{(x + 1)(x - 1)}{x + 1} = x - 1, \quad \text{for } x \neq -1.$$

Then,

$$\lim_{x\to -1} \frac{x^2 - 1}{x + 1} = \lim_{x\to -1} x - 1 = -2.$$

Example 16 Find $\lim_{x\to 4} \frac{\sqrt{x} - 2}{x - 4}$.

Again, Theorem 3.5 does not apply since the denominator equals 0 when $x = 4$. We need to rewrite the expression in some other equivalent form. Multiplying an expression like $\sqrt{x} - 2$ by its conjugate, $\sqrt{x} + 2$, is often helpful, so we try

$$\frac{\sqrt{x} - 2}{x - 4} \cdot \frac{\sqrt{x} + 2}{\sqrt{x} + 2} = \frac{x - 4}{(x - 4)(\sqrt{x} + 2)} = \frac{1}{\sqrt{x} + 2}, \quad \text{for } x \neq 4$$

Now Theorem 3.5 applies and we can find the limit:

$$\lim_{x \to 4} \frac{\sqrt{x} - 2}{x - 4} = \lim_{x \to 4} \frac{1}{\sqrt{x} + 2} = \frac{1}{4}.$$

3.2 Exercises

Find each of the following limits that exist.

1. $\lim_{x \to 4} 3x$
2. $\lim_{x \to -3} -4x$
3. $\lim_{x \to -3} (4x^2 + 2x - 1)$
4. $\lim_{x \to 2} (2x^2 - 3x + 5)$
5. $\lim_{x \to 0} (6x^3 - 4x^2 + 6x - 4)$
6. $\lim_{x \to 0} (3x^4 - 5x^2 + 6)$
7. $\lim_{x \to 2} (2x^2 - 1)(3x + 2)$
8. $\lim_{x \to -1} (4x^2 + 3x - 2)(x + 2)$
9. $\lim_{x \to 3} \frac{5x - 6}{2x + 1}$
10. $\lim_{x \to -2} \frac{2x + 1}{3x - 4}$
11. $\lim_{x \to 0} \frac{2x^2 - 6x + 3}{3x^2 - 4x + 2}$
12. $\lim_{x \to 0} \frac{-4x^2 + 6x - 8}{3x^2 + 7x - 2}$
13. $\lim_{x \to 0} \frac{x^2 - x}{x - 1}$
14. $\lim_{x \to 0} \frac{x^3 + 2x}{x + 2}$
15. $\lim_{x \to 3} \frac{x^2 - 9}{x - 3}$
16. $\lim_{x \to -2} \frac{x^2 - 4}{x + 2}$
17. $\lim_{x \to -2} \frac{x^2 - x - 6}{x + 2}$
18. $\lim_{x \to 5} \frac{x^2 - 3x - 10}{x - 5}$
19. $\lim_{x \to 6} \frac{x^2 - 5x - 6}{x - 6}$
20. $\lim_{x \to -4} \frac{x^2 - 2x - 24}{x + 4}$
21. $\lim_{x \to 0} \frac{x^2 - x}{x}$
22. $\lim_{x \to 0} \frac{3x^2 - 4x}{x}$
23. $\lim_{x \to 12} \sqrt{x + 4}$
24. $\lim_{x \to 3} \sqrt{x^2 - 5}$
25. $\lim_{x \to 2} \sqrt{x^2 - 4}$
26. $\lim_{x \to -1} \sqrt{x^4}$
27. $\lim_{x \to 25} \frac{\sqrt{x} - 5}{x - 25}$
28. $\lim_{x \to 9} \frac{x - 9}{\sqrt{x} - 3}$
29. $\lim_{x \to 16} \frac{4 - \sqrt{x}}{16 - x}$
30. $\lim_{x \to 4} \frac{\sqrt{x} - 8}{x - 64}$

The limit given by

$$\lim_{\Delta x \to 0} \frac{f(x + \Delta x) - f(x)}{\Delta x}$$

is of special importance in mathematics. For further practice in finding limits, we can calculate this limit which we will use in the next chapter. This limit is found by going through the following sequence of steps (the necessary steps are on the left, and an example is given on the right).

80   Limits

Steps	Example
1. A function f must be given.	1. We shall use $f(x) = 3x^2$.
2. Calculate $f(x + \Delta x)$, where Δx (read "delta x") is thought of as a variable.	2. $f(x + \Delta x)$ $= 3(x + \Delta x)^2$ $= 3[x^2 + 2x(\Delta x) + (\Delta x)^2]$ $= 3x^2 + 6x(\Delta x) + 3(\Delta x)^2$.
3. Calculate the difference: $f(x + \Delta x) - f(x)$.	3. Since $f(x) = 3x^2$, we have $f(x + \Delta x) - f(x)$ $= 3x^2 + 6x(\Delta x) + 3(\Delta x)^2 - 3x^2$ $= 6x(\Delta x) + 3(\Delta x)^2$.
4. Form the "difference quotient": $\dfrac{f(x + \Delta x) - f(x)}{\Delta x}.$	4. Here we get $\dfrac{6x(\Delta x) + 3(\Delta x)^2}{\Delta x} = 6x + 3(\Delta x).$
5. Find the limit of the difference quotient as $\Delta x \to 0$: $\lim_{\Delta x \to 0} \dfrac{f(x + \Delta x) - f(x)}{\Delta x}.$	5. We have $\lim_{\Delta x \to 0} \dfrac{f(x + \Delta x) - f(x)}{\Delta x}$ $= \lim_{\Delta x \to 0} [6x + 3(\Delta x)] = 6x.$

Find $\lim_{\Delta x \to 0} \dfrac{f(x + \Delta x) - f(x)}{\Delta x}$ for each of the following functions.

31. $f(x) = 4x + 3$ 32. $f(x) = 6 - 2x$ 33. $f(x) = -3x^2 - 4$
34. $f(x) = 3x^2 + 5$ 35. $f(x) = 4x^2 - 3x + 2$ 36. $f(x) = 2x^2 + 5x - 6$
37. $f(x) = x^3 + 2x$ 38. $f(x) = -2x^3 + 3x + 4$
39. $f(x) = 8$ (Hint: $8 = 0x + 8$)   40. $f(x) = -5$

Up to for quiz

3.3 Continuity

In Chapter 2 we studied linear, quadratic, and polynomial functions. The graphs in Figure 3.10 show one example of each of these kinds of functions. Note that the graphs have no breaks or sudden jumps and each can be drawn without lifting the pencil from the paper.

Figure 3.10

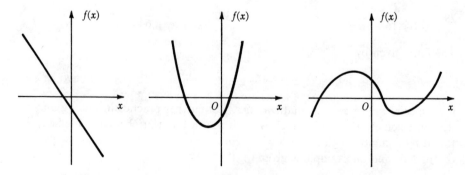

On the other hand, some rational functions have graphs with at least one gap or break. Figure 3.11 shows the graph of the rational function f where

$$f(x) = \frac{1}{x-6}.$$

From the graph we can see that there is a break at $x = 6$ because $f(6)$ does not exist. Also, from the graph we see that

$$\lim_{x \to 6} \frac{1}{x-6}$$

does not exist.

Figure 3.11

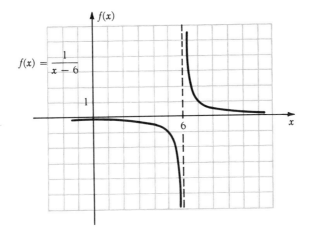

Figure 3.12 shows the graph of the function F defined by

$$F(x) = \begin{cases} x + 1 & \text{for } x < 3 \\ 5 & \text{for } x = 3 \\ -2x + 12 & \text{for } x > 3. \end{cases}$$

The graph of this function has no breaks or sudden jumps and can be drawn without lifting the pencil from the paper except at the point where $x = 3$. At the point $x = 3$ the function makes a sudden jump. Verify that $\lim_{x \to 3} F(x)$ does not exist. On the other hand, note that the graph in Figure 3.12 is smooth and has no gaps or jumps around the point where $x = 5$, and the limit as $x \to 5$ exists; in fact, $\lim_{x \to 5} F(x) = 2$. By substitution, we find that $F(5) = 2$. Thus,

$$F(5) = \lim_{x \to 5} F(x).$$

Figure 3.12

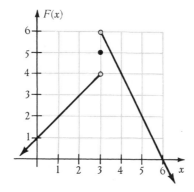

Similar results hold for the function g given by

$$g(x) = \begin{cases} -1 & \text{for } x < 3 \\ 1 & \text{for } x \geq 3. \end{cases}$$

The graph of g is shown in Figure 3.13. Note that $\lim_{x \to 3} g(x)$ does not exist, and that the graph of g has a break at $x = 3$. On the other hand,

$$\lim_{x \to 4} g(x) = 1 = g(4)$$

and the graph is smooth and has no gaps at $x = 4$.

Figure 3.13

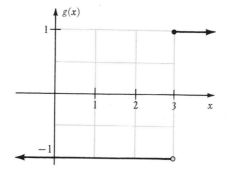

In general, if the graph of a function $y = f(x)$ has no gaps or jumps around the point where $x = a$, then $\lim_{x \to a} f(x)$ will exist and will equal $f(a)$. In such a case, the function is said to be *continuous* at $x = a$. Stated more formally, a function $y = f(x)$ is **continuous** at $x = a$ if

(a) $f(a)$ exists;
(b) $\lim_{x \to a} f(x)$ exists;
(c) $\lim_{x \to a} f(x) = f(a)$.

If a function is not continuous at $x = a$, it is said to be **discontinuous** at $x = a$.

Example 17 Graph the following function and find any values of x at which it is discontinuous:

$$h(x) = \begin{cases} x^2 & \text{if } x \neq 4 \\ 6 & \text{if } x = 4. \end{cases}$$

We see that $\lim_{x \to 4} h(x) = 16$. However, $h(4) = 6$, so that $\lim_{x \to 4} h(x) \neq h(4)$. Thus $h(x)$ is discontinuous at $x = 4$, as shown in the graph of Figure 3.14.

Figure 3.14

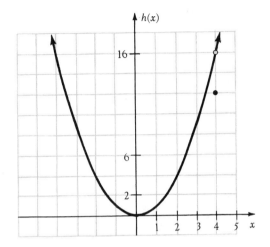

Example 18 Graph $F(x)$ and find the x values of any points of discontinuity, given

$$F(x) = \frac{1}{(x-4)(x+3)}.$$

The graph is given in Figure 3.15. Since $F(4)$ and $F(-3)$ do not exist, the function is discontinuous at $x = 4$ and $x = -3$.

Figure 3.15

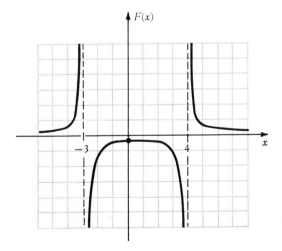

Example 19 The post office currently charges 10¢ per ounce or fraction of an ounce to mail a letter. If $f(x)$ represents the cost in cents of mailing a letter weighing x ounces, then $f(\frac{1}{2}) = 10$, $f(4) = 40$, $f(4.5) = 50$, and so on.
(a) Graph $f(x)$.
(b) Find $\lim_{x \to 1.5} f(x)$. Is the function continuous at 1.5?
(c) Find $\lim_{x \to 2} f(x)$. Is the function continuous at 2?

(a) The function is constant between integers, but is discontinuous at integer points, with the graph as shown in Figure 3.16. The solid circles are used to show that those points are included in the graph, while the open circles indicate points that are not part of the graph.

Figure 3.16

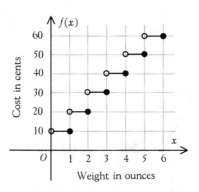

(b) The function is continuous at the point $x = 1.5$, since
$$\lim_{x \to 1.5} f(x) = f(1.5) = 20.$$
This is verified by the graph.

(c) The graph makes a jump at $x = 2$, and so is not continuous at $x = 2$. From the graph, we see that as $x \to 2$ from the left, $f(x)$ approaches 20, while as $x \to 2$ from the right, $f(x)$ approaches 30. Since $20 \ne 30$,
$$\lim_{x \to 2} f(x) \text{ does not exist.}$$

An interval on the number line where the endpoints are excluded, such as $\{x \mid -2 < x < 3\}$, is called an **open interval**. Open intervals are often written in an abbreviated notation using parentheses. For example, the open interval $\{x \mid -2 < x < 3\}$ can be written as $(-2, 3)$. Although this is the same notation used to indicate an ordered pair, the context in which it is given makes clear which is intended. In the same way, we can write $(-1, 5)$ to indicate the open interval $\{x \mid -1 < x < 5\}$ and $(5, +\infty)$ to indicate the interval $\{x \mid x > 5\}$. If both endpoints are included in an interval, it is called a **closed interval**. Brackets are used as an abbreviated notation for closed intervals. Some examples of closed intervals are $\{x \mid -5 \le x \le -1\}$, written as $[-5, -1]$, and $\{x \mid 4 \le x \le 10\}$, which could be written as [4, 10]. An interval which includes only one endpoint, such as [2, 4) or (−1, 3], is called a **half-open interval**. If a function is continuous at every point in an open interval, it is said to be continuous on the open interval.

Example 20 Graph $g(x)$ and find all the open intervals on which $g(x)$ is continuous if

$$g(x) = \begin{cases} -1 & \text{if } x < -2 \\ \dfrac{1}{x} & \text{if } -2 \le x \le 2 \\ 1 & \text{if } x > 2. \end{cases}$$

The graph of $g(x)$ is shown in Figure 3.17. At $x = -2$, $\lim_{x \to -2} g(x)$ does not exist even though $g(-2)$ exists. The same thing holds true at $x = 2$. At $x = 0$, $g(0)$ does not exist. Thus, there are discontinuities at -2, 0, and 2, so that the intervals on which $g(x)$ is continuous are $(-\infty, -2)$, $(-2, 0)$, $(0, 2)$, and $(2, +\infty)$.

Figure 3.17

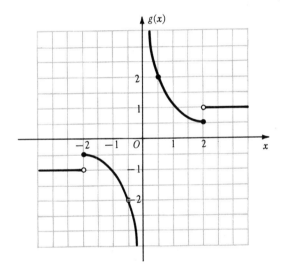

3.3 Exercises

Find any values of x at which the following functions are discontinuous.

1. $f(x) = \dfrac{2}{x - 3}$
2. $f(x) = \dfrac{5}{3 + x}$
3. $f(x) = \dfrac{6}{x}$
4. $f(x) = \dfrac{-2}{x}$
5. $f(x) = \dfrac{1}{(x - 3)(x + 5)}$
6. $f(x) = \dfrac{1}{(x + 6)(x - 2)}$
7. $f(x) = \dfrac{-2x}{(2x + 1)(3x - 2)}$
8. $f(x) = \dfrac{6x + 2}{(3x + 4)(5x - 2)}$

9. $f(x) = \dfrac{x^2 - 4}{x + 2}$

10. $g(x) = \dfrac{x^2 - 25}{x + 5}$

11. $g(x) = x^2 - 4x + 4$

12. $g(x) = -8x^3 + 2x^2 + 5$

13. $g(x) = \dfrac{4}{x^2 + 4}$

14. $g(x) = \dfrac{-8}{x^2 + 1}$

15. $h(x) = \begin{cases} 2x + 1 & \text{if } x \leq 2 \\ 7 - x & \text{if } x > 2 \end{cases}$

16. $h(x) = \begin{cases} -3x + 2 & \text{if } x \leq 1 \\ 2x - 1 & \text{if } x > 1 \end{cases}$

17. $h(x) = \begin{cases} 4x & \text{if } x < 2 \\ 3x + 1 & \text{if } x \geq 2 \end{cases}$

18. $h(x) = \begin{cases} -2x + 1 & \text{if } x < 5 \\ 3x + 1 & \text{if } x \geq 5 \end{cases}$

19. $h(x) = \begin{cases} -3 & \text{if } x < 2 \\ 2 & \text{if } x = 2 \\ 3 & \text{if } x > 2 \end{cases}$

20. $f(x) = \begin{cases} 4 & \text{if } x < 0 \\ 5 & \text{if } x = 0 \\ 6 & \text{if } x > 0 \end{cases}$

For each of the following functions, determine whether the function is continuous on the interval $(-2, 2)$. If not, give the intervals between -2 and 2 where each function is continuous.

21. $g(x) = \dfrac{1}{x - 1}$

22. $h(x) = x^2 - 2$

23. $f(x) = x^3$

24. $G(x) = \dfrac{4}{x + 4}$

25. $f(x) = \dfrac{x}{x - 2}$

26. $h(x) = \dfrac{x + 1}{x + 2}$

27. $F(x) = \begin{cases} -1 & \text{if } x < 1 \\ 0 & \text{if } x = 1 \\ 1 & \text{if } x > 1 \end{cases}$

28. $g(x) = \begin{cases} 0 & \text{if } x \leq 2 \\ 2 & \text{if } x > 2 \end{cases}$

29. $F(x) = \begin{cases} 3x + 2 & \text{if } x \leq 0 \\ \sqrt{4 - x} & \text{if } x > 0 \end{cases}$

30. $F(x) = \begin{cases} x^2 & \text{if } x \leq 1 \\ 2x & \text{if } x > 1 \end{cases}$

31. Give the intervals over which the post office function of Example 19 is continuous.

32. A taxi company charges a flat fee of 50¢ plus 10¢ per mile or fraction of a mile.
    (a) Write an expression for the function which gives the charges C in terms of the number of miles traveled, x.
    (b) Graph the function C.
    (c) Give the intervals on which C is continuous.

3.4 Limits to Infinity

So far we have studied $\lim\limits_{x \to a} f(x)$, where a is a fixed real number. We are often interested in finding a limit as x gets large without bound, written

$$\lim_{x \to \infty} f(x)$$

(read "the limit of $f(x)$ as x increases without bound"). On the other hand, we may want to find a limit as x gets small without bound, written

$$\lim_{x \to -\infty} f(x).$$

Let us investigate these two limits for

$$f(x) = \frac{1}{x+2}.$$

We can make a chart showing values of $f(x)$ as x increases and also x decreases:

x	10	1000	10,000
$f(x)$	0.0833	0.0010	0.0001

x	-10	-1000	$-10,000$
$f(x)$	-0.1250	-0.0010	-0.0001

From the chart, it seems reasonable to conclude that

$$\lim_{x \to +\infty} \frac{1}{x+2} = 0 \quad \text{and} \quad \lim_{x \to -\infty} \frac{1}{x+2} = 0.$$

In a similar way, we could verify the result of the following theorem, for which the proof is omitted.*

Theorem 3.7 If n is any positive integer and k is any constant, then

$$\lim_{x \to +\infty} \frac{k}{x^n} = 0 \quad \text{and} \quad \lim_{x \to -\infty} \frac{k}{x^n} = 0.$$

The following examples illustrate the use of Theorem 3.7 to find limits to infinity.

Example 21 Find $\lim_{x \to +\infty} \frac{2x^2 + 2}{x^2 - x}$.

To use Theorem 3.7, we need to write the function in an alternate form. We multiply both the numerator and denominator by $1/x^n$ where n is the highest power of x in the fraction. In our example $n = 2$, so we multiply by $1/x^2$ as follows:

$$\frac{(2x^2 + 2)\frac{1}{x^2}}{(x^2 - x)\frac{1}{x^2}} = \frac{2 + \frac{2}{x^2}}{1 - \frac{1}{x}}.$$

---

* For a proof of this theorem, see *The Calculus with Analytic Geometry*, second edition, Part 1, Louis Leithold, Harper & Row, New York, p. 97.

We can now use limit Theorem 3.1, together with Theorem 3.7, to find the limit:

$$\lim_{x \to +\infty} \frac{2 + \frac{2}{x^2}}{1 - \frac{1}{x}} = \frac{2 + 0}{1 - 0} = 2.$$

Example 22 Find $\lim_{x \to -\infty} \frac{2x^3 - 3x - 1}{3x^4 + 2x^2 + x}$.

We begin by multiplying both the numerator and denominator of the fraction by $1/x^4$ to get

$$\frac{\frac{2}{x} - \frac{3}{x^3} - \frac{1}{x^4}}{3 + \frac{2}{x^2} + \frac{1}{x^3}}.$$

Then, using the limit theorems, we have

$$\lim_{x \to -\infty} \frac{\frac{2}{x} - \frac{3}{x^3} - \frac{1}{x^4}}{3 + \frac{2}{x^2} + \frac{1}{x^3}} = \frac{0 - 0 - 0}{3 + 0 + 0} = 0.$$

Example 23 Find $\lim_{x \to +\infty} \frac{2x^5}{x^3 + 3}$.

Multiplying both numerator and denominator by $1/x^5$, we get

$$\lim_{x \to +\infty} \frac{2x^5}{x^3 + 3} = \lim_{x \to +\infty} \frac{2}{\frac{1}{x^2} + \frac{3}{x^5}} = \frac{2}{0 + 0}.$$

Since the denominator has a limit of zero, the assumptions of Theorem 3.5 are violated, and the limit does not exist.

Limits as x increases or decreases without bound are useful in finding horizontal asymptotes in curve sketching. For example, to graph the function f given by

$$f(x) = \frac{3x^2 + 2x}{x^2 - 1},$$

we can determine any horizontal asymptotes by finding

$$\lim_{x \to +\infty} \frac{3x^2 + 2x}{x^2 - 1} \quad \text{and} \quad \lim_{x \to -\infty} \frac{3x^2 + 2x}{x^2 - 1}.$$

Since $f(x)$ can be written as

$$\frac{3 + \frac{2}{x}}{1 - \frac{1}{x^2}}$$

both limits are 3. That is, as x either increases or decreases without bound, $f(x)$ approaches 3. Thus, the line $y = 3$ is a horizontal asymptote.

3.4 Exercises

Find the indicated limits.

1. $\lim\limits_{x \to +\infty} \dfrac{3x}{5x - 1}$
2. $\lim\limits_{x \to +\infty} \dfrac{2x + 3}{4x - 7}$
3. $\lim\limits_{x \to +\infty} \dfrac{2x}{x + 1}$
4. $\lim\limits_{x \to -\infty} \dfrac{5x}{3x - 1}$
5. $\lim\limits_{x \to +\infty} \dfrac{8x + 2}{2x - 5}$
6. $\lim\limits_{x \to +\infty} \dfrac{x + 4}{x - 1}$
7. $\lim\limits_{x \to +\infty} \dfrac{x - 2}{3x + 1}$
8. $\lim\limits_{x \to -\infty} \dfrac{x^2 + 2x}{2x^2 - 2x + 1}$
9. $\lim\limits_{x \to +\infty} \dfrac{x^2 + 2x - 5}{3x^2 + 2}$
10. $\lim\limits_{x \to -\infty} \dfrac{3x^2 + 2x}{5x^2 + x}$
11. $\lim\limits_{x \to -\infty} \dfrac{3x^3 + 2x - 1}{2x^4 - 3x^3 + 2}$
12. $\lim\limits_{x \to -\infty} \dfrac{2x^2 + 2x}{5x^3 + 3x^2 + 2x}$
13. $\lim\limits_{x \to -\infty} \dfrac{2x^4 + 3x + 4}{x^2 - 4x - 1}$
14. $\lim\limits_{x \to -\infty} \dfrac{x^2 - 1}{2x^4 + 2}$
15. $\lim\limits_{x \to +\infty} \dfrac{x^3 + 3x}{x^2 - 1}$
16. $\lim\limits_{x \to +\infty} \dfrac{2x^2 - 3}{x^3 + 2}$
17. $\lim\limits_{x \to +\infty} \dfrac{x^3 + 1}{x^4 - 2x + 1}$
18. $\lim\limits_{x \to -\infty} \dfrac{x^5 - 3}{x^2 + 1}$
19. $\lim\limits_{x \to -\infty} \dfrac{x^6 - 1}{x^2}$
20. $\lim\limits_{x \to +\infty} \dfrac{x^2 + 4}{x - 3}$

For each of the following functions, find any horizontal asymptotes.

21. $f(x) = \dfrac{x - 5}{2x - 2}$
22. $f(x) = \dfrac{x^2 + 2}{(x - 1)^2}$
23. $f(x) = \dfrac{3x^2 + 2x - 5}{(x + 3)(x - 1)}$
24. $f(x) = \dfrac{2x^2 + 3x}{(x + 2)(2x - 1)}$
25. $f(x) = \dfrac{1}{x^2 + 5}$
26. $f(x) = \dfrac{3x - 5}{x^2 + 2}$
27. $f(x) = \dfrac{x^2 - x}{x + 3}$
28. $f(x) = \dfrac{x^2 + 2x + 7}{x + 1}$

● **Case 3 Limit of a Sequence**

This case is a little different from the others in this book. Here we introduce a new idea which is closely related to the work of this chapter. Then we discuss several applications of this new idea.

A **sequence** is a function whose domain is a set of positive integers. For example, if $a(n) = 4n - 3$ is a sequence having as domain the set of all positive integers, then

$$a(1) = 4(1) - 3 = 1,$$
$$a(2) = 4(2) - 3 = 5,$$
$$a(3) = 4(3) - 3 = 9,$$
$$a(4) = 4(4) - 3 = 13,$$

and so on. The range elements, 1, 5, 9, 13, and so on, are called the **terms** of the sequence. To express the fact that 9 is the third term of this sequence, or that $a(3) = 9$, we use the symbol $a_3 = 9$ (a_3 is read "a-sub three"). In the same way, $a_1 = 1$, $a_2 = 5$, $a_4 = 13$, $a_5 = 17$, $a_6 = 21$, and so on. In this example, $a(n) = 4n - 3$, which can be written as $a_n = 4n - 3$. The term a_n is called the *general term* or *nth term* of the sequence.

The following example shows how a sequence might occur in a practical situation. The limit of the terms of the sequence can be found to get an estimate of the long-range trend of the sequence.

Example 1 There are two kinds of sex chromosomes in humans, the X and Y chromosome. Females inherit X chromosomes from both parents (and are called XX individuals), while males are XY individuals. Suppose A and a are contrasting characteristics inherited alternatively—that is, an individual inherits either the A or the a from each parent to form one of the three possible genotypes (hereditary factors): AA, Aa, or aa. Now suppose A and a are associated only with the X chromosome so that there is no corresponding characteristic for the Y chromosome. Since males are XY, males can only be of the genotypes A· or a·, while females may be AA, Aa, or aa. Under the assumption of random mating, the frequencies of A and a in the population oscillate from generation to generation.

For example, it can be shown that under certain conditions, if q_1 is the initial frequency of the male genotype a·, and $q_2, q_3, \ldots$ are the corresponding frequencies for subsequent generations, then the nth term, q_n, is given by

$$q_n = 0.40 + 0.20(-\tfrac{1}{2})^{n-1}$$

for $n = 1, 2, 3, \ldots$. In this example, we have

$$q_1 = 0.40 + 0.20 = 0.60,$$
$$q_2 = 0.40 + 0.20(-\tfrac{1}{2}) = 0.30,$$
$$q_3 = 0.40 + 0.20(\tfrac{1}{4}) = 0.45,$$
$$q_4 = 0.40 + 0.20(-\tfrac{1}{8}) = 0.375,$$

and so on. As n increases without bound, verify that $(-\tfrac{1}{2})^n$ approaches 0, so that

$$\lim_{n \to \infty} q_n = 0.40.$$

Thus, over the years, the frequency of the male genotype will stabilize at about 0.40, or 40%.

A sequence in which each term after the first is a constant multiple of the preceding term is called a **geometric sequence**. For example,

$$6, 12, 24, 48, 96, 192, \ldots$$

is a geometric sequence. Here, each term after the first is obtained by multiplying the preceding term by 2. The number 2 is called the **common ratio** of the sequence. The sequence $8, -4, 2, -1, 1/2, -1/4, 1/8, \ldots$ is a geometric sequence with common ratio $-1/2$.

The sum of the first n terms of a geometric sequence, denoted S_n, is given by the formula

$$S_n = \frac{a_1(1 - r^n)}{1 - r}.$$

For example, the sum of the first five terms of the sequence $4, 8, 16, 32, 64, \ldots$ is found by letting $n = 5$, $a_1 = 4$, and $r = 2$ in the formula above:

$$S_5 = \frac{4(1 - 2^5)}{1 - 2} = \frac{4(1 - 32)}{-1} = 124.$$

Using the formula above, the sum of the first 6 terms of the geometric sequence

$$1/2, 1/4, 1/8, 1/16, \ldots$$

is given by

$$S_6 = \frac{\frac{1}{2}[1 - (\frac{1}{2})^6]}{1 - \frac{1}{2}} = \frac{\frac{1}{2}(1 - \frac{1}{64})}{\frac{1}{2}} = \frac{63}{64}.$$

In general, the sum of the first n terms of this sequence is given by

$$S_n = \frac{\frac{1}{2}[1 - (\frac{1}{2})^n]}{1 - \frac{1}{2}}.$$

We know that $S_6 = 63/64$. Verify that $S_8 = 255/256$ and $S_{10} = 1023/1024$. As n increases, the values of S_n get closer and closer to 1. To see why this is so, note that as n gets larger and larger, $(\frac{1}{2})^n$ gets closer and closer to 0. Thus, as $n \to \infty$, we have

$$\lim_{n \to \infty} S_n = \lim_{n \to \infty} \frac{\frac{1}{2}[1 - (\frac{1}{2})^n]}{1 - \frac{1}{2}} = \frac{\frac{1}{2}(1)}{\frac{1}{2}} = 1.$$

We shall define the *sum* of the infinite geometric sequence $1/2, 1/4, 1/8, 1/16, 1/32, \ldots$ by writing

$$\frac{1}{2} + \frac{1}{4} + \frac{1}{8} + \frac{1}{16} + \frac{1}{32} + \cdots = 1.$$

In general, consider a geometric sequence having first term a_1 and common ratio r, where $|r| < 1$. Then we have $\lim_{n \to \infty} r^n = 0$ (which is true only if $|r| < 1$). Hence

$$\lim_{n \to \infty} S_n = \lim_{n \to \infty} \frac{a_1(1 - r^n)}{1 - r} = \frac{a_1(1)}{1 - r} = \frac{a_1}{1 - r} \quad \text{if } |r| < 1.$$

Thus, the **sum of an infinite geometric sequence** with first term a_1 and common ratio r, where $|r| < 1$, is defined as $a_1/(1 - r)$.

Example 2 Evaluate the sum $3 + \frac{3}{2} + \frac{3}{4} + \frac{3}{8} + \cdots$.
Here $a_1 = 3$ and $r = 1/2$. Thus, the sum is given by

$$\frac{3}{1 - \frac{1}{2}} = \frac{3}{\frac{1}{2}} = 6.$$

Therefore,

$$3 + \frac{3}{2} + \frac{3}{4} + \frac{3}{8} + \cdots = 6.$$

Example 3 Evaluate the sum $\frac{2}{3} - \frac{4}{9} + \frac{8}{27} - \frac{16}{81} + \cdots$.
Here $a_1 = 2/3$ and $r = -2/3$. Thus, the sum is

$$\frac{\frac{2}{3}}{1 - (-\frac{2}{3})} = \frac{\frac{2}{3}}{1 + \frac{2}{3}} = \frac{\frac{2}{3}}{\frac{5}{3}} = \frac{2}{5}.$$

Example 4 Evaluate the sum $\frac{3}{4} + 1 + \frac{4}{3} + \frac{16}{9} + \frac{64}{27} + \cdots$.
In this example, $a_1 = 3/4$ and $r = 4/3$. Since $r \geq 1$, the formula above does not apply, and we must say that the sum $\frac{3}{4} + 1 + \frac{4}{3} + \frac{16}{9} + \frac{64}{27} + \cdots$ does not exist.

Example 5 Suppose a company spends $100,000 for a payroll in a certain city. Suppose also that the employees of the company reside in the city. Assume that on the average the inhabitants of this city spend 80% of their income in the same city, with the remaining 20% either saved or spent elsewhere. Then 80% of the original $100,000, or (0.80)($100,000) = $80,000, will also be spent in that city. An additional 80% of this $80,000, or $64,000, will in turn be spent in the city, as will 80% of the $64,000, and so on. These amounts, $100,000, $80,000, $64,000, $51,200, and so on, are the terms of an infinite geometric sequence with $a_1 = \$100,000$ and $r = 0.80$. The sum of these amounts is thus

$$\frac{a_1}{1 - r} = \frac{\$100,000}{1 - 0.80} = \$500,000.$$

The original $100,000 payroll leads to a total expenditure of $500,000 in the city. In economics, the quotient of these numbers, $500,000/$100,000 = 5, is called the **multiplier**.

Example 6 Retinoblastoma is a kind of cancer of the eye in children. Medical researchers believe that the disease depends on a single dominant gene, say A. Let a be the normal gene. It is believed that a fraction of the population $m = 2 \times 10^{-5}$ per generation will experience *mutation*, a sudden unaccountable change, of a into A. (We exclude the possibility of back mutations of A into a.) With medical care, approximately 70% of those affected with the disease survive. According to past data, the survivors reproduce at half the normal rate. The net fraction of affected persons who produce offspring is thus $r = 35\% = 0.35$. Since gene A is extremely rare, practically all the affected persons are of genotype Aa, so that we may neglect the few individuals of genotype AA. Starting with 0 inherited cases in an early generation, we have

- m fraction of population with disease due to mutation in the nth generation
- mr fraction of population with disease due to mutation in the $(n-1)$st generation
- mr^2 fraction of population with disease due to mutation in the $(n-2)$nd generation

    (and so on)

- mr^n fraction of population with disease due to mutation in the original generation (number 0)

The total fraction of the population having the disease in the nth generation, p_n, is thus

$$p_n = m + mr + mr^2 + \cdots + mr^n.$$

As $n \to \infty$, we can use the formula for the sum of the terms of an infinite geometric sequence to write

$$\lim_{n \to \infty} p_n = \frac{m}{1-r} = \frac{2 \times 10^{-5}}{1 - (0.35)} = 3 \times 10^{-5}.$$

Thus, the fraction of the population having retinoblastoma will finally stabilize at about 3×10^{-5}, which is about 50% more than the rate of mutation.

Case 3 Exercises

Write the first five terms of each of the following geometric sequences.

1. $a_1 = 5, r = 2$
2. $a_1 = 4, r = 3$
3. $a_1 = 16, r = -1/2$
4. $a_1 = 243, r = -1/3$
5. $a_4 = 1, r = 1/4$
6. $a_3 = 2, r = 1/2$

Find the sum of the first five terms for each of the following geometric sequences.

7. $1, 2, 4, 8, \ldots$
8. $3, 6, 12, 24, \ldots$
9. $12, 6, 3, \ldots$
10. $400, 200, 100, \ldots$

Find the sum of each of the following infinite sequences for which the formula of this section is valid.

11. $\dfrac{3}{4} + \dfrac{3}{8} + \dfrac{3}{16} + \cdots$
12. $\dfrac{4}{5} + \dfrac{2}{5} + \dfrac{1}{5} + \cdots$
13. $3 - \dfrac{3}{2} + \dfrac{3}{4} - \cdots$
14. $9 - 3 + 1 - \cdots$

15. $\dfrac{1}{3} - \dfrac{2}{9} + \dfrac{4}{27} - \dfrac{8}{81} + \cdots$

16. $1 + \dfrac{1}{1.01} + \dfrac{1}{1.01^2} + \cdots$

17. $\dfrac{1}{36} + \dfrac{1}{30} + \dfrac{1}{25} + \cdots$

18. $1 + \dfrac{1}{2^2} + \dfrac{1}{2^4} + \cdots$

19. $\left(\dfrac{1}{4}\right) + \left(\dfrac{1}{4}\right)^2 + \left(\dfrac{1}{4}\right)^3 + \cdots$

20. $\left(\dfrac{9}{10}\right) + \left(\dfrac{9}{10}\right)^2 + \left(\dfrac{9}{10}\right)^3 + \cdots$

21. $(1.2) + (1.2)^2 + (1.2)^3 + \cdots$
22. $(1.001) + (1.001)^2 + (1.001)^3 + \cdots$
23. $(0.99) + (0.99)^2 + (0.99)^3 + \cdots$
24. $(0.999) + (0.999)^2 + (0.999)^3 + \cdots$

25. Joann drops a ball from a height of 12 feet and notices that on each bounce the ball returns to about 3/4 of its previous height. About how far will the ball travel before it comes to rest? (Hint: Consider the sum of *two* sequences.)

26. A sugar factory receives an order for 1000 units of sugar. The production manager thus orders production of 1000 units of sugar. He forgets, however, that the production of sugar requires some sugar (to prime the machines, for example), and so he ends up with only 900 units of sugar. He then orders an additional 100 units, and receives only 90 units. A further order for 10 units produces 9 units. Finally realizing his error, the manager decides to try mathematics. He views the production process as an infinite geometric sequence with $a_1 = 1000$ and $r = 0.1$. Using this, find the number of units of sugar that he should have ordered originally.

Find the multiplier for each of the following sets of conditions. (*See Example 5.*)

27. Original expenditure $50,000; 75% spent locally.
28. Original expenditure $300,000; 60% spent locally.

Chapter 4 Pretest

1. Given $f(x) = 3x + 1$, find $f(2); f(10); f(m)$. [2.1]
2. Simplify the following rational expressions. [1.4]

    (a) $\dfrac{5x^3 + 3x^2 - 2x}{x}$

    (b) $\dfrac{(x + h)^2 - x^2}{h}$

    (c) $\dfrac{\dfrac{1}{x + h} - \dfrac{1}{x}}{h}$

3. Find the following limits. [3.2]

    (a) $\lim\limits_{x \to -2} 2x + 3$

    (b) $\lim\limits_{x \to 1} \dfrac{1}{x}$

    (c) $\lim\limits_{x \to 0} \dfrac{x^2 - 2}{x + 1}$

4. Given $f(x) = 2x^2 + 3x - 5$, find $f(1); f(2); f(2) - f(1); \dfrac{f(2) - f(1)}{2 - 1}$.

    [1.4, 2.1]

5. Given $f(x) = 2 - x^2$, find $f(2); f(a); f(a) - f(2)$; $\dfrac{f(a) - f(2)}{2 - a}; \lim\limits_{a \to 0} \dfrac{f(a) - f(2)}{2 - a}$. [1.4, 2.1, 3.2]

6. Find the slope of the line through (2, 3) and (−1, 4). [2.2]
7. Find the equation of the line through the point (−2, 3) with slope of −3/4. [2.2]
8. Find any points at which $f(x)$ is not continuous. [3.3]

    (a) $f(x) = 1/x$

    (b) $f(x) = \dfrac{1}{x - 3}$

    (c) $f(x) = |2x + 1|$

9. Given $f(x) = (x + 1)^2$ and $g(x) = 1/x$, find $f[g(x)]$ and $g[f(x)]$. [2.5]

4
The Derivative

The management of a firm is willing to increase its advertising budget for one of its products as long as the total profit earned by the product also increases. The expenditure on advertising that will provide the maximum possible profit is at the point where the rate of growth of total profit slows to zero. However, too much would be spent on advertising if an increase in expenditure on advertising caused the rate of growth of the profit to become negative. The task of finding an optimum level of expenditure is, in general, not at all easy. Calculus is a tool that is often used to optimize such variables. We shall see calculus used to determine maximum profit, minimum cost, and maximum enclosed area for a given amount of fence, to mention some of the examples of this text. In social science and biology, calculus is most often used to describe rates of change, and we shall see examples of this also.

Historically, calculus has been broken down into two parts: differential calculus, discussed in this chapter; and integral calculus, discussed in Chapter 7. **Differential calculus** is used to find rates of change, and slopes of lines tangent to curves. **Integral calculus**, closely connected to differential calculus by the *fundamental theorem of calculus*, is used to find areas which are bounded by curves. Both these branches of calculus provide many examples of practical applications, as we shall see.

The basic idea of differential calculus is the *derivative*, which we discuss in the first section of this chapter. Methods of finding derivatives of functions are often called *techniques of differentiation*, which we discuss in Section 4.2.

4.1 Definition of the Derivative

Let us suppose that a company finds that its profit is related to the volume of production by the function

$$f(x) = 16x - x^2,$$

where $f(x)$ represents the total profit from the production of x units, and where we assume that the company can produce any nonnegative number of units. A graph of this function is shown in Figure 4.1. An inspection of the graph shows that an increase in production from 1 unit to 2 units will increase profit more than an increase in production from 6 units to 7 units. An increase from 7 units to 8 units produces very little increase in profit, while an increase in production from 8 units to 9 units actually produces a decline in total profit. In other words, profit increases at a decreasing rate until production reaches 8 units.

Figure 4.1

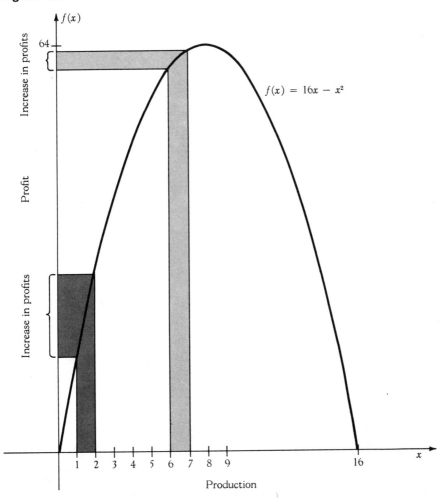

Using the given profit function, $f(x) = 16x - x^2$, we see that the profit from 1 unit of production is given by

$$f(1) = 16(1) - 1^2 = 15.$$

An increase to 2 units of production increases profit to a total of

$$f(2) = 16(2) - 2^2 = 28.$$

This 1 unit increase of production (from 1 unit to 2 units) increased profit by $28 - 15 = 13$. If 4 units are produced, the total profit is $f(4) = 48$. An increase in production from 1 unit to 4 units is a $4 - 1 = 3$ unit increase in production. The corresponding increase in profit is $f(4) - f(1) = 48 - 15 = 33$. The average rate of increase in profit for this 3 unit increase in production is given by the **difference quotient** found by dividing the difference of the two profits by the difference of the corresponding productions:

$$\frac{f(4) - f(1)}{4 - 1} = \frac{33}{3} = 11.$$

We have seen that an increase in production from 1 unit to 4 units leads to an average rate of increase in profit of 11, while an increase from 1 unit to 2 units leads to an average rate of increase in profit of 13. In general, if we let production increase from 1 unit to $1 + \Delta x$ units, where Δx is a small positive number, what happens to the rate of increase in profit? The chart below shows the average rate of increase in profit for some selected values of Δx close to 0.

Δx	0.1	0.01	0.001	0.0001
$1 + \Delta x$	1.1	1.01	1.001	1.0001
$f(1 + \Delta x)$	16.39	15.1399	15.013999	15.00139999
$f(1)$	15	15	15	15
Increase in profit	1.39	0.1399	0.013999	0.00139999
Average rate of increase in profit	**13.9**	**13.99**	**13.999**	**13.9999**

Again, we found the average rate of increase in profit for the various values of Δx by dividing the difference in profit by the difference in production. That is, if $1 + \Delta x$ represents the increased production, then the numbers in the bottom line of the chart (the average rate) can be found by evaluating the difference quotient

$$\frac{f(1 + \Delta x) - f(1)}{(1 + \Delta x) - 1} = \frac{f(1 + \Delta x) - f(1)}{\Delta x}.$$

Here $f(1 + \Delta x)$ is the profit from producing $1 + \Delta x$ units, while $f(1 + \Delta x) - f(1)$ is the increase in profit when production increases from 1 unit to $1 + \Delta x$ units.

From the chart above, it appears that as production gets closer and closer to 1 (or, equivalently, as $\Delta x \to 0$), the average rate of increase in profit approaches the limit 14. This limit,

$$\lim_{\Delta x \to 0} \frac{f(1 + \Delta x) - f(1)}{\Delta x} = 14,$$

is called the *instantaneous rate of change* of the profit at a production level of $x = 1$.

Instantaneous rate of change is significant in many other situations besides profit. For example, if $f(x) = 16x - x^2$ represents the population of a colony of microorganisms at time x, then, as shown above, 14 is the instantaneous rate of change of the population at time $x = 1$.

To continue with the profit function $f(x) = 16x - x^2$ given above, we could find the instantaneous rate of change of profit at some other level of production, say $x = 4$, by going through the above steps, using $x = 4$ instead of $x = 1$. However, it is easier in the long run to find a formula for the instantaneous rate of change for any value x, and not just for some specific value of x. To do this, note that if production increases from x units to $x + \Delta x$ units, profit increases a total of $f(x + \Delta x) - f(x)$ units. The average rate of change in profit is thus given by the difference quotient

$$\frac{f(x + \Delta x) - f(x)}{\Delta x}.$$

To get the instantaneous rate of change in profit, take the limit as Δx approaches 0:

$$\lim_{\Delta x \to 0} \frac{f(x + \Delta x) - f(x)}{\Delta x}.$$

In our example, where $f(x) = 16x - x^2$, the necessary limit can be found as follows. (Here we use the fact that $(a + b)^2 = a^2 + 2ab + b^2$.)

$$\lim_{\Delta x \to 0} \frac{f(x + \Delta x) - f(x)}{\Delta x}$$

$$= \lim_{\Delta x \to 0} \frac{[16(x + \Delta x) - (x + \Delta x)^2] - (16x - x^2)}{\Delta x}$$

$$= \lim_{\Delta x \to 0} \frac{16x + 16(\Delta x) - x^2 - 2x(\Delta x) - (\Delta x)^2 - 16x + x^2}{\Delta x}$$

$$= \lim_{\Delta x \to 0} \frac{16(\Delta x) - 2x(\Delta x) - (\Delta x)^2}{\Delta x}$$

$$= \lim_{\Delta x \to 0} (16 - 2x - \Delta x)$$

$$= 16 - 2x.$$

The new function $16 - 2x$ can be used to find the instantaneous rate of change in profit at any level of production x. For example, at $x = 4$ the instantaneous rate of change in profit is

$$16 - 2 \cdot 4 = 16 - 8 = 8,$$

while at $x = 10$ the instantaneous rate of change is

$$16 - 2 \cdot 10 = 16 - 20 = -4.$$

We might call the function obtained above, namely $16 - 2x$, the *instantaneous rate of change of profit function*. However, the procedure used to obtain

$16 - 2x$ occurs in so many diverse practical situations that a more general name is given to a function derived in this way. The function $16 - 2x$ is called the derivative of the function $f(x) = 16x - x^2$, and denoted

$$f'(x) = 16 - 2x,$$

read "f-prime of x." Using this new notation, verify that, for example,

$$f'(3) = 16 - 2 \cdot 3 = 10,$$
$$f'(5) = 6,$$
$$f'(8) = 0,$$

and

$$f'(12) = -8.$$

By these results, we see that the company can increase total profit by increasing production up to 8 units. However, any increase in production over 8 units results in a decrease in total profit. (To see this, find $f'(9) = -2$.)

We used the notation $f'(x) = 16 - 2x$ to represent the derivative of the function $f(x) = 16x - x^2$. Other common notations include

$$y' = 16 - 2x,$$

$$\frac{dy}{dx} = 16 - 2x,$$

$$\frac{d(16x - x^2)}{dx} = 16 - 2x,$$

and

$$D_x[16x - x^2] = 16 - 2x.$$

The derivative of the function f is sometimes written as f'. We shall often use these other notations, especially dy/dx, which shows that the derivative is taken with respect to the variable x.

In summary,

if the function f' represents the **derivative** of a function f, then

$$f'(x) = \lim_{\Delta x \to 0} \frac{f(x + \Delta x) - f(x)}{\Delta x},$$

for all values of x for which this limit exists. If this limit does not exist for a value of x, the function has no derivative at x.

Virtually all the functions used in this text have derivatives. Note, however, that the absolute value function $f(x) = |x|$ does not have a derivative at $x = 0$ since

$$f'(0) = \lim_{\Delta x \to 0} \frac{|0 + \Delta x| - |0|}{\Delta x}$$

does not exist.

We have seen one interpretation of this limit, as the instantaneous rate of change of profit, and we shall see two other interpretations in the third section of this chapter.

4.1 Definition of the Derivative

Example 1 Let $f(x) = -3x^2 + 5x + 6$. Find $f'(x)$.

To use the definition of derivative, it is helpful to organize the work as shown in the following four steps.

Step 1. Find $f(x + \Delta x)$. In this example we have

$$\begin{aligned} f(x + \Delta x) &= -3(x + \Delta x)^2 + 5(x + \Delta x) + 6 \\ &= -3[x^2 + 2x(\Delta x) + (\Delta x)^2] + 5(x + \Delta x) + 6 \\ &= -3x^2 - 6x(\Delta x) - 3(\Delta x)^2 + 5x + 5(\Delta x) + 6. \end{aligned}$$

Step 2. Find the difference $f(x + \Delta x) - f(x)$.

$$\begin{aligned} f(x + \Delta x) - f(x) &= -3x^2 - 6x(\Delta x) - 3(\Delta x)^2 + 5x + 5(\Delta x) + 6 - (-3x^2 + 5x + 6) \\ &= -6x(\Delta x) - 3(\Delta x)^2 + 5(\Delta x). \end{aligned}$$

Step 3. Form the difference quotient $\dfrac{f(x + \Delta x) - f(x)}{\Delta x}$.

$$\begin{aligned} \frac{f(x + \Delta x) - f(x)}{\Delta x} &= \frac{-6x(\Delta x) - 3(\Delta x)^2 + 5(\Delta x)}{\Delta x} \\ &= -6x - 3(\Delta x) + 5. \end{aligned}$$

Step 4. Find $\lim\limits_{\Delta x \to 0} \dfrac{f(x + \Delta x) - f(x)}{\Delta x}$. If this limit exists, it is the desired derivative. Here we have

$$\begin{aligned} \lim_{\Delta x \to 0} \frac{f(x + \Delta x) - f(x)}{\Delta x} &= \lim_{\Delta x \to 0} [-6x - 3(\Delta x) + 5] \\ &= -6x + 5. \end{aligned}$$

Therefore, if $f(x) = -3x^2 + 5x + 6$, then the derivative of f is $f'(x) = -6x + 5$.

Example 2 Let $f(x) = 4/x$ and find $f'(x)$.

Step 1. Find $f(x + \Delta x)$. Here

$$f(x + \Delta x) = \frac{4}{x + \Delta x}.$$

Step 2. Find the difference $f(x + \Delta x) - f(x)$.

$$\begin{aligned} f(x + \Delta x) - f(x) &= \frac{4}{x + \Delta x} - \frac{4}{x} \\ &= \frac{4x - 4(x + \Delta x)}{x(x + \Delta x)} \\ &= \frac{4x - 4x - 4(\Delta x)}{x(x + \Delta x)} \\ &= \frac{-4(\Delta x)}{x(x + \Delta x)}. \end{aligned}$$

Step 3. Form the difference quotient

$$\frac{f(x + \Delta x) - f(x)}{\Delta x} = \frac{\frac{-4(\Delta x)}{x(x + \Delta x)}}{\Delta x}$$

$$= \frac{-4}{x(x + \Delta x)}.$$

Step 4. See if $\lim_{\Delta x \to 0} \frac{f(x + \Delta x) - f(x)}{\Delta x}$ exists; if it does, this limit is the derivative. Here we have

$$\lim_{\Delta x \to 0} \frac{f(x + \Delta x) - f(x)}{\Delta x} = \lim_{\Delta x \to 0} \frac{-4}{x(x + \Delta x)}$$

$$= \frac{-4}{x \cdot x}$$

$$= \frac{-4}{x^2}.$$

Therefore, if $f(x) = 4/x$, then $f'(x) = -4/x^2$. Note that neither this function nor its derivative exist at $x = 0$.

In summary, the derivative of a function $f(x)$ can be found by performing the following four steps.

Step 1. Find $f(x + \Delta x)$.

Step 2. Find $f(x + \Delta x) - f(x)$.

Step 3. Find the difference quotient $\frac{f(x + \Delta x) - f(x)}{\Delta x}$.

Step 4. $f'(x) = \lim_{\Delta x \to 0} \frac{f(x + \Delta x) - f(x)}{\Delta x}$ if this limit exists.

4.1 Exercises

Find the derivatives of each of the following functions.

1. $f(x) = 3x + 5$
2. $f(x) = 9 - 5x$
3. $f(x) = 2x^2$
4. $f(x) = -5x^2$
5. $f(x) = -6$ (Hint: $-6 = -6 + 0x$.)
6. $f(x) = 2$
7. $f(x) = 3x^2 - 4x + 2$
8. $f(x) = 8 - 5x - 3x^2$
9. $f(x) = -3/x$
10. $f(x) = 5/x$
11. $f(x) = x^3 + x^2$
12. $f(x) = 5x^3 + 1$
13. $f(x) = \sqrt{x}$ (Hint: In Step 3 multiply numerator and denominator by $\sqrt{x + \Delta x} + \sqrt{x}$.)
14. $f(x) = 1/\sqrt{x}$

4.2 Techniques of Differentiation

For each of the following functions, find

(a) $f'(0)$ (b) $f'(2)$ (c) $f'(-3)$

The derivatives of all these functions were found in Exercises 1–14 above.

15. $f(x) = 2x^2$
16. $f(x) = -5x^2$
17. $f(x) = 3x^2 - 4x + 2$
18. $f(x) = 8 - 5x - 3x^2$
19. $f(x) = 3x + 5$
20. $f(x) = 9 - 5x$
21. $f(x) = -3/x$
22. $f(x) = 5/x$
23. $f(x) = x^3 + x^2$
24. $f(x) = 5x^3 + 1$
25. $f(x) = \sqrt{x}$
26. $f(x) = 1/\sqrt{x}$

For each of the following functions, find all values of x for which $f'(x) = 0$.

27. $f(x) = 5x^2 - 10x$
28. $f(x) = x^2 - 12x$
29. $f(x) = 5x^2 - 25$
30. $f(x) = 2x + 8x^2$

31. Suppose that the demand for a certain item, $D(x)$, is given by the function

$$D(x) = -2x^2 + 4x + 6,$$

where x represents the price of the item. Find the instantaneous rate of change of the demand with respect to price at the point when the price is $x = 3$. (In other words, find $D'(3)$.)

32. Suppose profit is related to advertising expenditure according to the function

$$P(x) = 1000 + 32x - 2x^2,$$

where $P(x)$ represents profit when x dollars are spent on advertising. Given that the company is spending the following amounts on advertising, should they increase these expenditures?

(a) $x = 8$ (b) $x = 12$
(c) $x = 6$ (d) $x = 20$

33. A biologist estimates that when a bactericide is introduced into a culture of bacteria, the number of bacteria present is given by

$$B(t) = 1000 + 50t - 5t^2,$$

where $B(t)$ is the number of bacteria present at time t. Find the rate of change of the number of bacteria at each of the following times:

(a) $t = 2$ (b) $t = 3$ (c) $t = 4$
(d) $t = 5$ (e) $t = 6$
(f) When does the population of bacteria start to decline?

4.2 Techniques of Differentiation

In the previous section we defined the derivative of the function f as the limit

$$\lim_{\Delta x \to 0} \frac{f(x + \Delta x) - f(x)}{\Delta x}$$

for all values of x for which this limit exists. If derivatives of functions always had to be found by using this definition, calculus would probably not be nearly as useful today as it is. However, the definition above can be used to obtain a number of shortcut methods for finding derivatives. By using these shortcuts, derivatives of the most common functions can be found in a mechanical, straightforward way, as we shall see. We shall develop several of these shortcuts in this section, and then look at two applications of the derivative in the next section. Further shortcuts will then be developed in the remainder of the chapter. (A summary of all derivative formulas of this chapter is given in Section 4.5.)

A function of the form $f(x) = k$, where k is a real number, has perhaps the simplest derivative of any function, as shown by the next theorem.

Theorem 4.1 If $f(x) = k$, where k is any real number, then $f'(x) = 0$. (The derivative of a constant is 0.)

We can use the definition of the derivative to show that this result is true. To do this, notice that the function $f(x) = k$ always takes on the value k for any value of x. Hence, $f(x) = k$ and $f(x + \Delta x) = k$. Using these results, and the definition of derivative, we have

$$f'(x) = \lim_{\Delta x \to 0} \frac{f(x + \Delta x) - f(x)}{\Delta x} = \lim_{\Delta x \to 0} \frac{k - k}{\Delta x} = 0.$$

For example, if $f(x) = 4$, then $f'(x) = 0$. Also, if $f(x) = 12\pi$, then $f'(x) = 0$.

Functions of the form $f(x) = x^n$, where n is a real number, occur often in practical applications. To see how to find a shortcut formula for the derivative of $f(x) = x^n$, let us first consider two special cases: $f(x) = x^2$ and $f(x) = x^3$.

First, we can use the definition to find the derivative of $f(x) = x^2$. If we use the fact that $(a + b)^2 = a^2 + 2ab + b^2$, we have

$$f'(x) = \lim_{\Delta x \to 0} \frac{f(x + \Delta x) - f(x)}{\Delta x}$$

$$= \lim_{\Delta x \to 0} \frac{(x + \Delta x)^2 - x^2}{\Delta x}$$

$$= \lim_{\Delta x \to 0} \frac{x^2 + 2x(\Delta x) + (\Delta x)^2 - x^2}{\Delta x}$$

$$= \lim_{\Delta x \to 0} \frac{2x(\Delta x) + (\Delta x)^2}{\Delta x}$$

$$= \lim_{\Delta x \to 0} (2x + \Delta x)$$

$$= 2x.$$

Thus, if $f(x) = x^2$, we have $f'(x) = 2x$. We can find the derivative of the function $f(x) = x^3$ in much the same way. Using the fact that $(a + b)^3 = a^3 + 3a^2b + 3ab^2 + b^3$, we have

$$\lim_{\Delta x \to 0} \frac{(x + \Delta x)^3 - x^3}{\Delta x} = \lim_{\Delta x \to 0} \frac{x^3 + 3x^2(\Delta x) + 3x(\Delta x)^2 + (\Delta x)^3 - x^3}{\Delta x}$$

$$= \lim_{\Delta x \to 0} [3x^2 + 3x(\Delta x) + (\Delta x)^2]$$

$$= 3x^2.$$

By this result, the derivative of $f(x) = x^3$ is $f'(x) = 3x^2$. Both these derivatives are special cases of the result given in the next theorem.

Theorem 4.2 If $f(x) = x^n$, for any real number n, then $f'(x) = n \cdot x^{n-1}$. (Multiply by the exponent, and reduce the exponent by 1.)

This theorem was proven above for the cases $n = 2$ and $n = 3$. More general proofs can be given for all positive integers n or for all real numbers n. As an example of the use of Theorem 4.2, let $f(x) = x^5$. Then $f'(x) = 5 \cdot x^{5-1} = 5x^4$. Also, if $f(x) = 1/x^4 = x^{-4}$, then $f'(x) = -4 \cdot x^{-4-1} = -4x^{-5}$, and if $f(x) = \sqrt{x} = x^{1/2}$, then $f'(x) = \frac{1}{2}x^{-1/2}$.

Theorem 4.2, together with the following theorem, allows us to find the derivative of any product of a constant and a variable to a given power.

Theorem 4.3 If $f(x) = k \cdot g(x)$, for a real number k and a function g having derivative g', then

$$f'(x) = k \cdot g'(x).$$

Thus, if $f(x) = 12x^3$, then $f'(x) = 12(3x^2) = 36x^2$. Also, if

$$f(x) = \frac{-8}{x^{3/2}} = -8x^{-3/2},$$

then $f'(x) = -8(-\frac{3}{2}x^{-5/2}) = 12x^{-5/2}$. The proof of Theorem 4.3 depends on the definition of derivative and the properties of limits from Chapter 3.

The two theorems above, together with the next theorem, permit us to take the derivative of any polynomial, or other sum of functions with known derivatives.

Theorem 4.4 If $f(x) = g(x) + h(x)$, for any functions g and h, having derivatives g' and h', then

$$f'(x) = g'(x) + h'(x).$$

(The derivative of a sum is the sum of the derivatives.)

Thus, if $f(x) = x^3 - 6x^2 + 1$, then $f'(x) = 3x^2 + (-6)2x + 0 = 3x^2 - 12x$.

This result can be proved using properties of limits and the definition of derivative. To do this, let $f(x) = g(x) + h(x)$. Then

$$f'(x) = \lim_{\Delta x \to 0} \frac{f(x + \Delta x) - f(x)}{\Delta x}$$

$$f'(x) = \lim_{\Delta x \to 0} \frac{g(x + \Delta x) + h(x + \Delta x) - g(x) - h(x)}{\Delta x}$$

$$= \lim_{\Delta x \to 0} \frac{g(x + \Delta x) - g(x)}{\Delta x} + \lim_{\Delta x \to 0} \frac{h(x + \Delta x) - h(x)}{\Delta x}$$

$$= g'(x) + h'(x),$$

provided these two derivatives exist.

Example 3 A company believes that its profit, $P(x)$, in thousands of dollars, is given by

$$P(x) = -x^2 + 10x + 50,$$

where x represents the number of thousands of dollars spent on advertising. Suppose the company is now spending $x = 3$ thousand dollars on advertising. Should this amount be increased?

As mentioned before, the derivative of the profit function gives the rate of change of the profit with respect to the amount spent on advertising. Here

$$P'(x) = -2x + 10.$$

At $x = 3$, we have $P'(3) = -2(3) + 10 = 4$. Thus, the rate of change of profit with respect to the amount spent on advertising is positive (so that profit is increasing), and it would be wise to spend more on advertising. At an expenditure of $x = 6$ thousand dollars, the rate of change of profit with respect to the amount spent on advertising is $P'(6) = -2$, which is negative. Thus, profit is declining, and the company is probably spending too much on advertising.

To find the ideal amount to spend on advertising, find the value of x that makes $P'(x) = 0$:

$$P'(x) = -2x + 10$$
$$0 = -2x + 10$$
$$x = 5.$$

Thus, an expenditure of $5000 on advertising will increase profit to the maximum.

4.2 Exercises

Use the theorems of this section to find the derivatives of each of the following functions.

1. $f(x) = 5x^2$
2. $f(x) = -8x^4$
3. $f(x) = 8x + 12$
4. $f(x) = -4x + 6$
5. $f(x) = 3x^2 - 4x + 6$
6. $f(x) = 2x^2 + 3x - 8$
7. $f(x) = -4x^3 - x^2 + 2x - 6$
8. $f(x) = 3x^3 - 4x^2 + 6x - 8$
9. $f(x) = x^4 - x^3 + x^2 - 4x + 1$
10. $f(x) = -3x^4 + 2x^3 - 5x^2 + 6x - 2$
11. $f(x) = 4x^{1/2}$
12. $f(x) = 3x^{3/2}$
13. $f(x) = 3x^{1/2} + 2x^{3/2}$

4.2 Techniques of Differentiation

14. $f(x) = -11\sqrt{x}$ (Hint: $\sqrt{x} = x^{1/2}$)
15. $f(x) = 4\sqrt{x} - 3x$
16. $f(x) = -4x^3 + \sqrt{x}$
17. $f(x) = 2\sqrt{x} - x^{3/2}$
18. $f(x) = -6\sqrt{x} + 2x^{5/2}$
19. $f(x) = -6x^{-2}$
20. $f(x) = 2x^{-4}$
21. $f(x) = -3x^{-3}$
22. $f(x) = -5x^{-3}$
23. $f(x) = 2x^{-1} + 3x^{-2}$
24. $f(x) = 3x^{-2} - 4x^{-1}$
25. $f(x) = 8x^{-3/2}$
26. $f(x) = 4x^{-1/2} + 2x^{-3} + 5^{-2}$
27. $f(x) = 8x^{-3/2} - 4x^{1/2} + x^{-1/2}$
28. $f(x) = 4/x^6$
29. $f(x) = -3/x^5$
30. $f(x) = 1/x + x$
31. $f(x) = 3 - (3/x^3)$ *have to convert $3x^{-3}$*
32. $f(x) = 5x - (2/x^2)$

33. The sales, $S(x)$, of a certain company are related to the total amount spent, x, in thousands of dollars, on salaries of sales people by the function

$$S(x) = -\tfrac{2}{5}x^{5/2} + 8x^{3/2} + 24\sqrt{x} + 200.$$

Find the rate of change of sales with respect to salaries at each of the following levels of total salaries:
(a) $x = 4$
(b) $x = 9$
(c) $x = 16$
(d) $x = 25$

34. The number of foxes, $F(b)$, in hundreds, in a certain field is related to the number of bunnies, b, in thousands, by the function

$$F(b) = -b^3 + 15b^2 + 600b + 500.$$

Find the rate of change of the number of foxes with respect to the number of bunnies at each of the following levels of bunny population:
(a) $b = 5$
(b) $b = 10$
(c) $b = 20$
(d) $b = 30$

35. An oil well off the Gulf Coast is leaking, with the leak spreading oil over the surface as a circle. At any time t, in minutes, after the beginning of the leak, the radius of the circular oil slick on the surface is $4t$ feet. The formula for the area of a circle is $A = \pi r^2$, where r is the radius of the circle.
    (a) Use the formula for the area of a circle to find a formula for the area of the oil slick in terms of t.
    (b) Find the rate of change of the area with respect to time at time $t = 4$, $t = 20$, $t = 60$, and $t = 100$.

36. When a thermal inversion layer is over a city (such as happens often in Los Angeles) pollutants cannot rise vertically, but are trapped below the layer and must disperse horizontally. Assume that a factory smokestack begins emitting a pollutant at 8 A.M. Assume that the pollutant disperses horizontally, forming a circle. If t represents time, in hours, since the factory began emitting pollutants ($t = 0$ represents 8 A.M.), assume that the radius of the circle of pollution is $2t$ miles.
    (a) Find a formula for the area of the circle of pollution in terms of t.
    (b) Find the rate of change of the area of the circle with respect to time at 10 A.M., noon, 2 P.M., and 5 P.M.

4.3 Applications of the Derivative

In this section we discuss two different applications of the derivative—the derivative as the slope of a line tangent to a curve, and the derivative as a marginal cost function.

Slope of a Tangent Line A *tangent line to a circle* at a point on the circle is a line which touches the circle in only one point, as shown in Figure 4.2(a). The idea of a derivative can be used to define the tangent line to a more general curve, as we shall do in this section. Figure 4.2(b) shows a typical curve and tangent lines to the curve at three different points.

We shall define the tangent line to a curve by first considering an example. Figure 4.3 shows the portion of the graph of the function $f(x) = x^3$ that lies in

Figure 4.2a

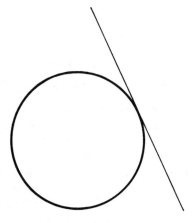

(a)

Figure 4.2b

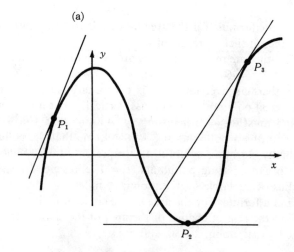

(b)

the first quadrant. We can use the derivative to obtain a definition for a tangent line to the curve at some point on the curve. As an example, let us define the tangent line at the point $P = (1, 1)$ of the curve. If we let $Q = (3/2, 27/8)$ and $R = (2, 8)$ be two other points on the curve, as shown in Figure 4.3, we can draw the *secant lines* (lines cutting a graph in at least two points) PQ and PR.

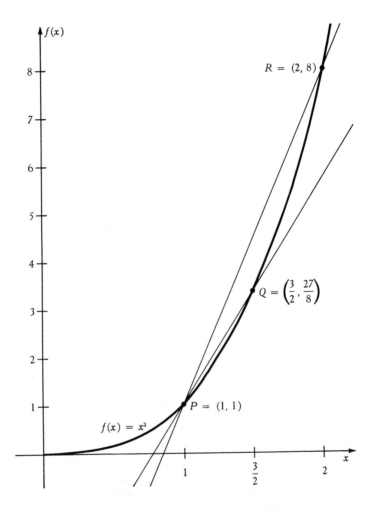

Figure 4.3

Recall from Section 2.2 that the slope of a line is given by the quotient of the difference of the y-values and the difference of the x-values for any two distinct points that lie on the line. Using this definition, we find that the

$$\text{slope of } PR = \frac{8-1}{2-1} = 7,$$

$$\text{slope of } PQ = \frac{\frac{27}{8}-1}{\frac{3}{2}-1} = \frac{19}{4} = 4.75.$$

In the same way, the slope of the line through (1, 1) and (5/4, 125/64) is given by 61/16 = 3.8125. By selecting more such points, with x and y each closer and closer to 1, we get the results shown in the following chart.

Point on graph	(1.1, 1.331)	(1.01, 1.030301)	(1.001, 1.003003001)
Slope of line through the given point and (1, 1)	3.31	3.0301	3.003001

These slopes seem to approach a numerical limit of 3, while a series of carefully drawn graphs would show that the secant lines tend to approach a "limiting line." To verify that 3 is the limit of the slopes, let us calculate the slope of the line through the point (1, 1) and the point $(x, f(x)) = (1 + \Delta x, f(1 + \Delta x))$. Using the definition of slope, we have the difference quotient

$$\frac{f(1 + \Delta x) - f(1)}{(1 + \Delta x) - 1} = \frac{f(1 + \Delta x) - f(1)}{\Delta x}.$$

Since $f(x) = x^3$ in this example, we have $f(1 + \Delta x) = (1 + \Delta x)^3$, which equals

$$(1 + \Delta x)^3 = 1 + 3(\Delta x) + 3(\Delta x)^2 + (\Delta x)^3.$$

Moreover, $f(1) = 1$. Thus, the difference quotient becomes

$$\frac{f(1 + \Delta x) - f(1)}{\Delta x} = \frac{[1 + 3(\Delta x) + 3(\Delta x)^2 + (\Delta x)^3] - 1}{\Delta x}$$

$$= \frac{3(\Delta x) + 3(\Delta x)^2 + (\Delta x)^3}{\Delta x}$$

$$= 3 + 3(\Delta x) + (\Delta x)^2.$$

If we now let Δx approach 0, the difference quotient approaches 3, or

$$\lim_{\Delta x \to 0} \frac{f(1 + \Delta x) - f(1)}{\Delta x} = 3.$$

The line through the point $(1, f(1))$ with slope $m = 3$ is called the *tangent line* to the graph of $f(x) = x^3$ at the point where $x = 1$. The tangent line is shown in Figure 4.4. (Think of the tangent line as a "limiting line" for the secants through (1, 1) and $(1 + \Delta x, f(1 + \Delta x))$ as Δx approaches 0.)

To find the slope of the tangent line to $f(x) = x^3$ for any arbitrary value of x, we follow the example above and calculate

$$\lim_{\Delta x \to 0} \frac{f(x + \Delta x) - f(x)}{\Delta x}$$

for the function $f(x) = x^3$. But this limit is exactly the same as the definition of the derivative of the given function. Since $f(x) = x^3$, we can use Theorem 4.2 of the last section to write $f'(x) = 3x^2$. Thus, the slope of the tangent line to $f(x) = x^3$ at any point $(x, f(x))$ on the graph is given by $3x^2$. For example, the slope of the tangent line at the point on the graph where $x = 4$ is given by $3 \cdot 4^2 = 48$, and so on.

Figure 4.4

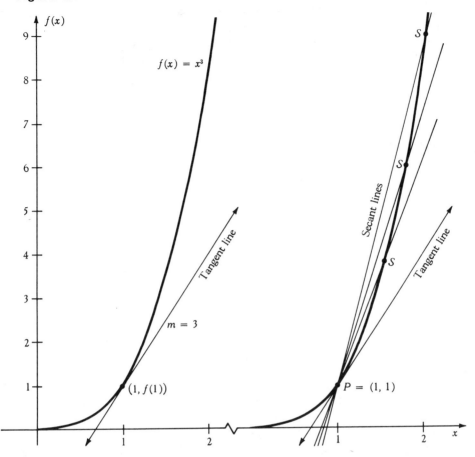

This example has now led to another application of the derivative—the derivative of a function can be used to find the slope of the tangent line to the graph of the function at any point on the graph. Thus, the **slope of the tangent line** to the graph of a function f at the point $(x_0, f(x_0))$ on the graph is given by $f'(x_0)$, provided this derivative exists.

Example 4 Find the equation of the tangent line to the curve $f(x) = x^3$ at the point where $x = 2$.

As seen above, the slope of the tangent line to the curve $f(x) = x^3$ is given by $f'(x) = 3x^2$. For $x = 2$, the slope is $f'(2) = 3 \cdot 2^2 = 12$. When $x = 2$, we have $f(2) = 2^3 = 8$. Hence, the tangent line goes through the point $(2, 8)$ and has slope 12. Using the point-slope form of the equation of a line (see Section 2.2), we have

$$y - y_1 = m(x - x_1)$$
$$y - 8 = 12(x - 2)$$
$$y = 12x - 16$$

as the equation of the line tangent to $f(x) = x^3$ at the point where $x = 2$.

The derivative of a function gives the slope of the tangent line at any point on the graph of a function, provided the derivative exists. However, not all functions have derivatives for all points in the domain of the function. For example, if $f(x) = \sqrt{x}$, then $f'(x) = 1/(2\sqrt{x})$, which does not exist at $x = 0$, even though 0 is in the domain of the function. Figure 4.5 shows a graph of $f(x) = \sqrt{x}$; note that the y-axis is the tangent line at the point on the graph where $x = 0$. The y-axis is a vertical line, and a vertical line has no slope.

Figure 4.5

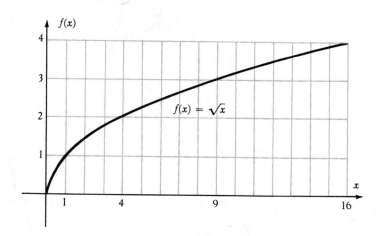

It is shown in more advanced courses that only a function which is continuous at a point can have a derivative at that point. Just because a function is continuous at a point, however, we have no assurance that the function will have a derivative at that point. For example, the function $f(x) = |x - 1|$ is continuous at $x = 1$, and yet has no derivative at $x = 1$. The graph of the function is shown in Figure 4.6. As shown by the graph, it is not possible to define a unique tangent line to the graph at the point where $x = 1$.

Figure 4.6

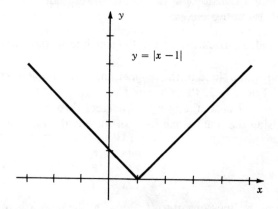

Marginal Cost The derivative also has a useful interpretation in economics. For example, suppose a factory is producing a certain number of items at a certain total production cost. The cost of producing one more item is called the *marginal cost* of that item. For example, if $C(x)$, the cost of producing x units of a certain item, is given by

$$C(x) = x^2 - 18x + 100,$$

then to produce 50 units costs a total of

$$C(50) = 50^2 - 18 \cdot 50 + 100 = 1700,$$

while to produce 51 units costs a total of

$$C(51) = 51^2 - 18 \cdot 51 + 100 = 1783.$$

Hence, by definition, the marginal cost of the fifty-first unit is given by

$$C(51) - C(50) = 1783 - 1700 = 83.$$

In the same way, verify that the marginal cost of the twelfth unit is

$$C(12) - C(11) = 28 - 23 = 5.$$

Thus, the twelfth unit is much less expensive to produce than the fifty-first unit.

We can give examples of such a situation in actual practice. For example, a ton of iron processed at the beginning of production in an iron mine is likely to cost less than a ton of iron produced from the lower grade of ore processed at the end of the mine's useful life. A factory might be old, and have machines that can be operated efficiently only at low volume. Any attempt to increase production beyond a certain low level will result in more breakdowns and will increase the cost of maintenance at a greater rate than the production rate increases.

On the other hand, marginal costs often decline with additional volume, as, for example, in the publishing industry. A publisher has a substantial investment involved in advance payments to authors, editorial costs, and the costs involved in the actual production and printing process. (For example, this text required an investment of around $8000 just for typesetting.) Thus, the marginal costs of producing more and more copies of a given book get smaller and smaller.

The definition of marginal cost as a difference of cost functions corresponds closely to the definition of derivative. In fact, we shall define the *marginal cost at a level of production x* (sometimes called the instantaneous marginal cost) by saying that C' is the marginal cost function for a cost function C. Hence, the marginal cost at some specific level of production, x_0, is simply $C'(x_0)$.

For example, suppose that the cost, in dollars, of producing beer is approximated by

$$C(x) = 20 + 5\sqrt{x},$$

where x is the number of units produced. Using the theorems of Section 4.2, we can show that

$$C'(x) = \frac{5}{2\sqrt{x}}.$$

114 The Derivative

Thus, the marginal cost at a level of production of $x = 16$ units is

$$C'(16) = \frac{5}{2\sqrt{16}} = \frac{5}{8} = 0.625.$$

Usually (but not always) it is safe to say that 5/8 (or $0.625) is the approximate cost of producing the seventeenth unit of beer. The actual marginal cost of the seventeenth unit in this case is given by

$$C(17) - C(16) = (20 + 5\sqrt{17}) - (20 + 5\sqrt{16})$$
$$\approx 0.616,$$

which is close to $C'(16)$, as found above.

Example 5 Suppose that the cost, $C(x)$, of producing x units of a chemical is approximated by

$$C(x) = 5000 + 2x^2 - 6x^{3/2}.$$

Find the marginal cost at $x = 64$; at $x = 90$.

To find the marginal cost, we use the derivative $C'(x)$. Using the theorems of Section 4.2, we have

$$C'(x) = 4x - 9x^{1/2}.$$

If $x = 64$, then

$$C'(64) = 4(64) - 9(64)^{1/2}$$
$$= 256 - 72$$
$$= 184.$$

Thus, the approximate cost of producing a sixty-fifth unit of the chemical is 184. If $x = 90$, we have

$$C'(90) = 4(90) - 9(90)^{1/2}$$
$$= 360 - 9(9.487)$$
$$= 274.6.$$

The idea of a marginal cost also applies to revenue, or profit. For example, if $R(x) = 80x + 10x^2 - 60$ gives the total revenue obtained from the sale of x units of a low energy consuming light bulb, then $R'(x) = 80 + 20x$ gives the instantaneous marginal revenue, or simply the *marginal revenue* at a level of sales x. For example, $R'(10) = 80 + 20(10) = 80 + 200 = 280$. Thus, the approximate additional revenue received from selling 11 units instead of 10 is $280.

4.3 Exercises

Find the slope of the tangent lines to the graphs of the following functions at the points with x values as given.

1. $f(x) = 3x + 5$ at $x = 2$
2. $f(x) = 9 - 5x$ at $x = 2$
3. $f(x) = 2x^2$ at $x = 1$
4. $f(x) = -5x^2$ at $x = 1$

5. $f(x) = 2x^2$ at $x = 6$
6. $f(x) = -5x^2$ at $x = 6$
7. $f(x) = -6x^2 + 3$ at $x = -3$
8. $f(x) = 2x^2 + 1$ at $x = 0$
9. $f(x) = -3/x$ at $x = 2$
10. $f(x) = 5/x$ at $x = -3$
11. $f(x) = \sqrt{x}$ at $x = 4$
12. $f(x) = 5x^3 + 1$ at $x = 2$
13. $f(x) = \sqrt{x}$ at $x = 16$
14. $f(x) = 5x^3 + 1$ at $x = 10$

Find an equation of the tangent line to each of the following curves at the point whose x value is given.

15. $f(x) = 2x^2$ at $x = -3$
16. $f(x) = -5x^2$ at $x = -1$
17. $f(x) = -4x^2$ at $x = 2$
18. $f(x) = 3x^2 - 2x$ at $x = 3$
19. $f(x) = 4x^2 + 3x - 5$ at $x = -1$
20. $f(x) = -6x^2 + 14x + 3$ at $x = 2$
21. $f(x) = 2x^3 + 3x^2$ at $x = -1$
22. $f(x) = -4x^3 + 8x^2 + 3$ at $x = 1$
23. $f(x) = 3x^3 - 4x + 2$ at $x = -2$
24. $f(x) = 2x + \sqrt{x}$ at $x = 4$

25. In the text we considered the <u>cost function</u>
$$C(x) = x^2 - 18x + 100.$$
   (a) Find the marginal cost at $x = 50$. Compare this with the marginal cost of the fifty-first item given on page 113.
   (b) Find the marginal cost at $x = 11$. Compare this with the marginal cost of the twelfth unit given on page 113.

26. Suppose the cost of producing a total of x copies of a certain textbook, $C(x)$, is given by
$$C(x) = 25{,}000 + 100\sqrt{x}.$$
   (a) Find the marginal cost of the seventeenth book produced by calculating $C(17) - C(16)$.
   (b) Find the marginal cost at $x = 16$ by calculating $C'(16)$.
   (c) Find the marginal cost of the 901st copy of the text by calculating $C(901) - C(900)$. (Hint: $\sqrt{901} = 30.0167$.) Calculate $C'(900)$.

27. Let $T(x)$ represent the cost to manufacture x units of tacos, where
$$T(x) = 500 + 2x^{1/2} + 4x^{3/2}.$$
   (a) Find the marginal cost at $x = 9$.
   (b) Find the marginal cost at $x = 25$.

28. Let $W(x)$ represent the cost of manufacturing x units of low-priced cheese, where
$$W(x) = 6000 + 8x^{1/2} + 6x^{3/2}.$$
   (a) Find the marginal cost at $x = 16$.
   (b) Find the marginal cost at $x = 36$.

29. The owner of Harry's Hamburgers had a management consultant analyze his operations. The consultant told Harry that his profit, $P(x)$, is related to sales by
$$P(x) = -x^2 + 20x - 50,$$
where x is the sales volume, in thousands, of hamburgers.

(a) Find the marginal profit of the ninth unit of hamburgers by calculating $P(9) - P(8)$.
(b) Find the marginal profit at $x = 8$ by calculating $P'(8)$.
(c) Find the marginal profit of the twelfth unit by finding $P(12) - P(11)$.
(d) Find $P'(11)$.

4.4 Derivatives of Products and Quotients

The derivative of a sum of two functions is the sum of the derivatives. What about products? Is the derivative of a product equal to the product of the derivatives? That is, if $f(x) = g(x) \cdot h(x)$, does $f'(x) = g'(x) \cdot h'(x)$? To find out, let us consider an example. Let $g(x) = 2x + 3$, while $h(x) = 3x^2$. Then we have $g'(x) = 2$ and $h'(x) = 6x$. Note that

$$f(x) = g(x) \cdot h(x) = (2x + 3)(3x^2) = 6x^3 + 9x^2,$$

so that $f'(x) = 18x^2 + 18x$. On the other hand, $g'(x) \cdot h'(x) = 2(6x) = 12x$, which is not the same as $f'(x)$. Hence, we must conclude that, in general, the derivative of a product is *not* the product of the derivatives. The formula which must be used to find derivatives of products is given in the next theorem.

Theorem 4.5 (*Product Rule*) If $f(x) = g(x) \cdot h(x)$, and if both g' and h' exist, then

$$f'(x) = g(x) \cdot h'(x) + h(x) \cdot g'(x).$$

(The derivative of a product of two functions is the first function times the derivative of the second, plus the second function times the derivative of the first.)

In the example above, we had $g(x) = 2x + 3$ and $h(x) = 3x^2$. If $f(x) = g(x) \cdot h(x)$, we can use this theorem to write

$$\begin{aligned} f'(x) &= g(x) \cdot h'(x) + h(x) \cdot g'(x) \\ &= (2x + 3)(6x) + (3x^2)(2) \\ &= 12x^2 + 18x + 6x^2 \\ &= 18x^2 + 18x, \end{aligned}$$

which is the same as the result obtained above.

To prove the product rule, let

$$f(x) = g(x) \cdot h(x).$$

Then $f(x + \Delta x) = g(x + \Delta x) \cdot h(x + \Delta x)$, and the derivative of the function f is given by

$$\begin{aligned} f'(x) &= \lim_{\Delta x \to 0} \frac{f(x + \Delta x) - f(x)}{\Delta x} \\ &= \lim_{\Delta x \to 0} \frac{g(x + \Delta x) \cdot h(x + \Delta x) - g(x) \cdot h(x)}{\Delta x}. \end{aligned}$$

4.4 Derivatives of Products and Quotients

Now we apply a step which is not obvious at first glance, but which works. We add and subtract $g(x + \Delta x) \cdot h(x)$ in the numerator. This gives

$$f'(x) = \lim_{\Delta x \to 0} \frac{g(x + \Delta x) \cdot h(x + \Delta x) - g(x + \Delta x) \cdot h(x) + g(x + \Delta x) \cdot h(x) - g(x)h(x)}{\Delta x}$$

$$= \lim_{\Delta x \to 0} g(x + \Delta x) \left[\frac{h(x + \Delta x) - h(x)}{\Delta x} \right] + \lim_{\Delta x \to 0} h(x) \left[\frac{g(x + \Delta x) - g(x)}{\Delta x} \right]$$

$$= \lim_{\Delta x \to 0} g(x + \Delta x) \cdot \lim_{\Delta x \to 0} \frac{h(x + \Delta x) - h(x)}{\Delta x} + \lim_{\Delta x \to 0} h(x) \cdot \lim_{\Delta x \to 0} \frac{g(x + \Delta x) - g(x)}{\Delta x}. \quad (1)$$

Here we used the fact that the limit of a sum is the sum of the limits and the limit of a product is the product of the limits. If h' and g' both exist, then

$$\lim_{\Delta x \to 0} \frac{h(x + \Delta x) - h(x)}{\Delta x} = h'(x) \quad \text{and} \quad \lim_{\Delta x \to 0} \frac{g(x + \Delta x) - g(x)}{\Delta x} = g'(x).$$

Also, $\lim_{\Delta x \to 0} g(x + \Delta x) = g(x)$. Substituting these results into equation (1) gives

$$f'(x) = g(x) \cdot h'(x) + h(x) \cdot g'(x),$$

which is the desired result.

The next two examples illustrate the use of the product rule in finding derivatives.

Example 6 Let $f(x) = (x^2 - 4x)(3x + 5)$. Find $f'(x)$.
Define $g(x) = x^2 - 4x$ and $h(x) = 3x + 5$. Then $g'(x) = 2x - 4$ and $h'(x) = 3$. By the product rule, f' is

$$f'(x) = (x^2 - 4x)(3) + (3x + 5)(2x - 4)$$
$$= 3x^2 - 12x + 6x^2 - 2x - 20$$
$$= 9x^2 - 14x - 20.$$

Example 7 Let $f(x) = (3x - 5)(4x^{1/2} + 1)$, and find $f'(x)$.
Here we have

$$f'(x) = (3x - 5)(2x^{-1/2}) + (4x^{1/2} + 1)(3)$$
$$= \frac{(3x - 5)(2)}{\sqrt{x}} + 12\sqrt{x} + 3$$
$$= \frac{(3x - 5)(2)}{\sqrt{x}} + \frac{12x}{\sqrt{x}} + \frac{3\sqrt{x}}{\sqrt{x}}$$
$$= \frac{6x - 10 + 12x + 3\sqrt{x}}{\sqrt{x}}$$
$$= \frac{18x - 10 + 3\sqrt{x}}{\sqrt{x}}.$$

Just as the derivative of a product is not the product of the derivatives, the derivative of a quotient is not the quotient of the derivatives. In fact, the derivative of a quotient is the somewhat complicated result given by the next theorem.

Theorem 4.6 (*Quotient Rule*) If $f(x) = \dfrac{g(x)}{h(x)}$, where g and h are functions whose derivatives exist, and where $h(x) \neq 0$, then

$$f'(x) = \frac{h(x) \cdot g'(x) - g(x) \cdot h'(x)}{[h(x)]^2}.$$

(The derivative of a quotient is the denominator times the derivative of the numerator minus the numerator times the derivative of the denominator, all over the denominator squared.)

The proof of the quotient rule is similar to the proof of the product rule given above.

Example 8 Let $f(x) = \dfrac{2x - 1}{4x + 3}$. Find $f'(x)$.

Let $g(x) = 2x - 1$ and $h(x) = 4x + 3$. Then $g'(x) = 2$ and $h'(x) = 4$. By the quotient rule, we have

$$f'(x) = \frac{(4x + 3)(2) - (2x - 1)(4)}{(4x + 3)^2}$$

$$= \frac{8x + 6 - 8x + 4}{(4x + 3)^2}$$

$$= \frac{10}{(4x + 3)^2}.$$

Example 9 Find $f'(x)$ if $f(x) = \dfrac{3x^2 - 4}{2x^2 + x}$.

Using the quotient rule, we have

$$f'(x) = \frac{(2x^2 + x)(6x) - (3x^2 - 4)(4x + 1)}{(2x^2 + x)^2}$$

$$= \frac{(12x^3 + 6x^2) - (12x^3 + 3x^2 - 16x - 4)}{(2x^2 + x)^2}$$

$$= \frac{12x^3 + 6x^2 - 12x^3 - 3x^2 + 16x + 4}{(2x^2 + x)^2}$$

$$= \frac{3x^2 + 16x + 4}{(2x^2 + x)^2}.$$

4.4 Derivatives of Products and Quotients

Example 10 Let $f(x) = \dfrac{3x^3}{2\sqrt{x}}$. Find $f'(x)$.

With the quotient rule, f' becomes

$$f'(x) = \frac{(2\sqrt{x})(9x^2) - (3x^3)(x^{-1/2})}{[2\sqrt{x}]^2}$$

$$= \frac{(2x^{1/2})(9x^2) - (3x^3)(x^{-1/2})}{2^2(\sqrt{x})^2}$$

$$= \frac{18x^{5/2} - 3x^{5/2}}{4x}$$

$$= \frac{15x^{5/2}}{4x}$$

$$= \frac{15x^{3/2}}{4}.$$

Since $(3x^3)/(2\sqrt{x}) = (3x^{5/2})/2$, f' could have been found more readily by the formulas of Section 4.2. We used the quotient rule here to illustrate its use.

4.4 Exercises

Find the derivative of each of the following functions.

1. $f(x) = x(x - 1)$
2. $f(x) = (3x - 1)(2x + 3)$
3. $f(x) = (5 - 2x)(x + 4)$
4. $f(x) = (8 - 5x)(6 + 3x)$
5. $f(x) = (x^2 - x)(3x - 2)$
6. $f(x) = (2x^2 + 1)(4x^2 - 3)$
7. $f(x) = (-5x + 6x^2)(2x^3 - 1)$
8. $f(x) = (3x^3 + 2x)(-4x^2 + 2x)$
9. $f(x) = (x^2 - 4)^2$ (Hint: $(x^2 - 4)^2 = (x^2 - 4)(x^2 - 4)$.)
10. $f(x) = (x^3 - 3x^2)^2$
11. $f(x) = (2x - 1)(x^{1/2})$
12. $f(x) = (3x - 2)(2x^{1/2})$
13. $f(x) = (4x^{1/2} + 1)(2x^{1/2})$
14. $f(x) = (3x^{1/2} - 4x)(3x^{-1/2})$
15. $f(x) = (8x^3 - 4x^{1/2})(2x^{-3/2})$
16. $f(x) = (-8x^{-3/2} + x^{2/3})(3x^{-1/2})$
17. $f(x) = \dfrac{x + 1}{x - 1}$
18. $f(x) = \dfrac{1}{x + 1}$
19. $f(x) = \dfrac{2x}{3 + x}$
20. $f(x) = \dfrac{4 + 3x}{3 + 2x}$
21. $f(x) = \dfrac{5 - 3x}{4 + 3x}$
22. $f(x) = \dfrac{3 + 2x}{5 - x}$
23. $f(x) = \dfrac{6x^2 + 2x}{x - 1}$
24. $f(x) = \dfrac{4x^2 + 3x}{x + 3}$
25. $f(x) = \dfrac{5x^2 - x}{2x^2 + 1}$
26. $f(x) = \dfrac{6x^2 - 3x}{x^2 - 4}$
27. $f(x) = \dfrac{x^2 - 3x - 4}{x + 1}$
28. $f(x) = \dfrac{x^2 + 2x - 48}{x - 6}$

29. $f(x) = \dfrac{\sqrt{x}}{2x+1}$

30. $f(x) = \dfrac{\sqrt{x}}{3x-5}$

31. $f(x) = \dfrac{\sqrt{x}\,(2x-3)}{x-1}$

32. $f(x) = \dfrac{\sqrt{x}\,(x+1)}{2x-1}$

33. Suppose you are the manager of a trucking firm, and one of your drivers reports that, according to her calculations, her truck burns fuel at the rate of

$$G(x) = \frac{1}{200}\left(\frac{800}{x} + x\right)$$

gallons per mile when traveling at x miles per hour on a smooth, dry road. (If a driver did report this to you, it would probably be a good idea to make the driver your mathematics consultant.)

(a) If the driver tells you that she wants to travel 20 miles per hour, what should you tell her? (Hint: Take the derivative of G, and evaluate it for $x = 20$. Then interpret your results.)

(b) If the driver wants to go 40 miles per hour, what should you say? (Hint: Find $G'(40)$. In Example 9 of Section 5.3 we will find the speed that will produce the lowest possible cost.)

4.5 The Chain Rule

So far we have no method other than the definition of the derivative for finding the derivative of functions such as $f(x) = \sqrt{3x-5}$. In this section we discuss the *chain rule* which provides a useful method for finding the derivatives of such functions. The chain rule can be used to find derivatives of composite functions which are, loosely speaking, functions of functions. For example, suppose $y = 3u^2$, but $u = 4x - 5$. Then by substitution, $y = 3(4x-5)^2$. In the same way, if $f(u) = \sqrt{u}$, while $u = g(x) = 3x - 5$, then $[f \circ g](x) = f[g(x)] = \sqrt{3x-5}$. This notation, called the composition of two functions, was discussed in Section 2.5. To find the derivative of these functions, we use the chain rule, as given in the next theorem.

Theorem 4.7 (*The Chain Rule*) Let $y = f(u)$ and $u = g(x)$. Then $y = f[g(x)]$ and

$$y' = f'[g(x)] \cdot g'(x),$$

provided the derivatives of f and g both exist.

Perhaps an example will clarify the notation. If $y = 3(4x-5)^2$, with $u = g(x) = 4x - 5$, then $y = f(u) = 3u^2$. Here $f'(u) = 6u$ and $g'(x) = 4$.

Using the chain rule, we have

$$y' = f'[g(x)] \cdot g'(x)$$
$$= f'(u) \cdot g'(x)$$
$$= 6u(4)$$
$$= 6(4x - 5)(4)$$
$$= 24(4x - 5).$$

To verify this result, expand the original function. Write $y = 3(4x - 5)^2$ as $y = 3(16x^2 - 40x + 25) = 48x^2 - 120x + 75$, so that we have $y' = 96x - 120 = 24(4x - 5)$.

Example 11 Given that $y = \sqrt{4x^2 - 3}$, find y'.

Let $y = f(u) = \sqrt{u}$, where $u = g(x) = 4x^2 - 3$. Verify that $f'(u) = (u^{-1/2})/2$, while $g'(x) = 8x$. Using the chain rule, we have

$$y' = \tfrac{1}{2}(4x^2 - 3)^{-1/2}(8x)$$
$$= \frac{4x}{\sqrt{4x^2 - 3}}.$$

Example 12 Let $y = (1/x + x)^3$, and find y'.

Here $y = f(u) = u^3$, where $u = g(x) = 1/x + x$. Also, $f'(u) = 3u^2$, and $g'(x) = -1/x^2 + 1$. Thus,

$$y' = 3(1/x + x)^2(-1/x^2 + 1).$$

Example 13 Suppose the revenue, $R(x)$, produced by selling x units (where x is at least 2) of agar-agar to biologists is given by

$$R(x) = \sqrt{3x^2 - 11} + 4x^{-2}.$$

Find the marginal revenue at $x = 5$.

As we saw in Section 4.3, the marginal revenue is given by the derivative of the revenue function. Using the chain rule, we have

$$R'(x) = \frac{6x}{2\sqrt{3x^2 - 11}} + \frac{-8}{x^3}.$$

If $x = 5$, we have

$$R'(5) = \frac{30}{2\sqrt{64}} + \frac{-8}{125}$$
$$= \frac{30}{16} + \frac{-8}{125}$$
$$= \frac{1811}{1000}$$
$$= 1.811.$$

Hence, selling a sixth unit will increase revenues by about 1.8. To find the actual increase in revenue after selling the sixth unit, we would have to evaluate $R(6) - R(5)$. (Verify that $R(6) - R(5) \approx 1.8$.)

> We can now summarize the formulas for derivatives that we have developed so far. If all indicated derivatives exist, then
>
> 1. If $f(x) = k$, where k is a real number, then
> $$f'(x) = 0.$$
> 2. If $f(x) = x^n$, for any real number n, then
> $$f'(x) = n \cdot x^{n-1}.$$
> 3. If $f(x) = k \cdot g(x)$, for any real number k, then
> $$f'(x) = k \cdot g'(x).$$
> 4. If $f(x) = g(x) + h(x)$, then
> $$f'(x) = g'(x) + h'(x).$$
> 5. (product rule) If $f(x) = g(x) \cdot h(x)$, then
> $$f'(x) = g(x) \cdot h'(x) + h(x) \cdot g'(x).$$
> 6. (quotient rule) If $f(x) = \dfrac{g(x)}{h(x)}$, where $h(x) \neq 0$, then
> $$f'(x) = \frac{h(x) \cdot g'(x) - g(x) \cdot h'(x)}{[h(x)]^2}.$$
> 7. (chain rule) If $y = f(u)$, and $u = g(x)$, then
> $$y' = f'[g(x)] \cdot g'(x).$$

4.5 Exercises

Find the derivatives of each of the following functions.

1. $f(x) = (x - 5)^2$
2. $f(x) = (3x - 2)^4$
3. $f(x) = (5x^2 + 6)^3$
4. $f(x) = (2x^2 - 5x + 1)^3$
5. $f(x) = (x^2 - 2x)^{1/2}$
6. $f(x) = (3x^2 - 5)^{1/2}$
7. $f(x) = (4x - 1)^{3/2}$
8. $f(x) = (2x + 5)^{3/2}$
9. $f(x) = 4x(3x - 2)^2$
10. $f(x) = -8x(3x - 1)^2$
11. $f(x) = \sqrt{2x - 1}$
12. $f(x) = \sqrt{4x + 3}$
13. $f(x) = 5x\sqrt{2x - 1}$
14. $f(x) = -3x\sqrt{5x + 2}$
15. $f(x) = -2x^2\sqrt{3x + 1}$
16. $f(x) = x^2\sqrt{x - 1}$
17. $f(x) = \dfrac{2x}{(x - 5)^3}$
18. $f(x) = \dfrac{-4x}{(3x - 5)^3}$
19. $f(x) = (3x^2 - 3x)^{3/2}$
20. $f(x) = (4x^2 - 2)^{3/2}$
21. $f(x) = x(x + 1/x)^{1/2}$
22. $f(x) = (1/x)(x - 1/x)^{1/2}$

23. $f(x) = \left(\dfrac{x+1}{x-1}\right)^{1/2}$ 24. $f(x) = \left(\dfrac{3x-1}{2x+1}\right)^{1/2}$

25. Assume that the total receipts, $R(x)$, from the sale of x television sets is given by
$$R(x) = 100x\left(1 - \dfrac{x}{5000}\right).$$
Find the marginal receipts when $x = 200$. What does $R'(200)$ approximate?

26. Assume that the total number of bacteria, $N(t)$, in thousands, present at a certain time is given by
$$N(t) = (t-10)^2\sqrt{5t+9} + 12,$$
where t represents time in hours since the beginning of the experiment. At what rate is the population of bacteria changing when $t = 8$? When $t = 11$? The second answer is positive, while the first answer is negative. What does this mean in terms of the population of bacteria?

27. Assume that the relationship between the demand, $D(x)$, for a certain product and the price, x, of the product is given by
$$D(x) = \sqrt{16 + 4x - x^2}.$$
Find the rate of change of demand with respect to price when the price is $x = 6$.

28. If the demand, $D(x)$, for a product is related to the price, x, by the function
$$D(x) = \dfrac{1000}{\sqrt{2x+5}},$$
find the rate at which demand is changing with respect to the price when the price is $x = 10$. Is the demand increasing or decreasing at that point?

4.6* Implicit Differentiation

In almost every example so far in this text, the functions that we have used have been written in the form $y = f(x)$, in which y is expressed as a function of x. Where $y = f(x)$, the variable y is said to be expressed as an **explicit function** of x. For example, $y = 3x - 2$, $y = x^2 + x + 6$, and $y = -x^3 + 2$ are all explicit functions of x. We can express $4xy - 3x = 6$ as an explicit function of x by solving for y. This gives
$$4xy - 3x = 6$$
$$4xy = 3x + 6$$
$$y = \dfrac{3x+6}{4x}.$$

* None of the material in this section will be used in later chapters.

On the other hand, some functions have equations which cannot be solved readily for y, and some cannot be solved for y at all. For example, it would be possible (but tedious) to use the quadratic formula to solve for y in the equation $y^2 + 2yx + 4x^2 = 0$. On the other hand, it would not be possible to solve for y in the equation $y^5 + 8y^3x + 6y^2x^2 + 2yx^3 + 6 = 0$. Functions such as these last two are said to be expressed as **implicit functions**.

Even though a function is written implicitly, it is still often possible to calculate dy/dx, the derivative of the function. (For convenience here, we use dy/dx rather than $f'(x)$.) For example, to find dy/dx for the function $3xy + 4y^2 = 10$, first take the derivative of both sides of the equation. In doing so, we are assuming that there exists some function f (which we may or may not be able to find) such that $y = f(x)$, and that f' exists. To take the derivative of both sides, we must treat $3xy$ as a product, and write the derivative of y with respect to x as dy/dx. Doing this, we have

$$\frac{d[3xy + 4y^2]}{dx} = \frac{d[10]}{dx},$$

where we use the notation $d[g(x)]/dx$ to represent the derivative of the function $g(x) = 3xy + 4y^2$. Taking the indicated derivatives term by term, we have

$$3x\frac{dy}{dx} + 3y + 8y\frac{dy}{dx} = 0.$$

To complete the process, solve this result for dy/dx:

$$(3x + 8y)\frac{dy}{dx} = -3y$$

$$\frac{dy}{dx} = \frac{-3y}{3x + 8y}.$$

Finding the derivative by this procedure is called **implicit differentiation**.

Example 14 Given the function $xy = 2$, find dy/dx by two methods:

(a) Solve for y as an explicit function of x, and then find dy/dx.

(b) Find dy/dx by implicit differentiation.

Compare the two results.

(a) If $xy = 2$, then $y = 2/x = 2x^{-1}$. Thus,

$$\frac{dy}{dx} = \frac{-2}{x^2}. \tag{1}$$

(b) If $xy = 2$, then dy/dx is found as follows:

$$\frac{d[xy]}{dx} = \frac{d[2]}{dx}$$

$$x\frac{dy}{dx} + y = 0$$

$$\frac{dy}{dx} = \frac{-y}{x}. \tag{2}$$

At first glance, these results for dy/dx seem different. However, we know that $xy = 2$, so that $y = 2/x$. Substituting this result into equation (2), we have

$$\frac{dy}{dx} = \frac{-y}{x} = \frac{\frac{-2}{x}}{x} = \frac{-2}{x^2},$$

the same result as given by equation (1).

Example 15 Find dy/dx, given that $x^2 + 2xy^2 + 3x^2y = 0$.
We have

$$\frac{d[x^2 + 2xy^2 + 3x^2y]}{dx} = \frac{d[0]}{dx}.$$

Now we treat $2xy^2$ and $3x^2y$ as products and differentiate term by term to get

$$2x + 2x\left(2y\frac{dy}{dx}\right) + 2y^2 + 3x^2\left(\frac{dy}{dx}\right) + 6xy = 0$$

or

$$2x + 4xy\left(\frac{dy}{dx}\right) + 2y^2 + 3x^2\left(\frac{dy}{dx}\right) + 6xy = 0,$$

from which

$$(4xy + 3x^2)\frac{dy}{dx} = -2x - 2y^2 - 6xy,$$

or, finally,

$$\frac{dy}{dx} = \frac{-2x - 2y^2 - 6xy}{4xy + 3x^2}.$$

Example 16 Find dy/dx if $x + \sqrt{xy} = y^2$.
We have

$$\frac{d[x + \sqrt{xy}]}{dx} = \frac{d[y^2]}{dx}.$$

Since $\sqrt{xy} = \sqrt{x} \cdot \sqrt{y} = x^{1/2} \cdot y^{1/2}$, we have, upon taking derivatives,

$$1 + x^{1/2}\left(\frac{1}{2}y^{-1/2} \cdot \frac{dy}{dx}\right) + y^{1/2}\left(\frac{1}{2}x^{-1/2}\right) = 2y\frac{dy}{dx}$$

$$1 + \frac{x^{1/2}}{2y^{1/2}} \cdot \frac{dy}{dx} + \frac{y^{1/2}}{2x^{1/2}} = 2y\frac{dy}{dx}.$$

Multiply both sides by $2x^{1/2} \cdot y^{1/2}$:

$$2x^{1/2} \cdot y^{1/2} + x\frac{dy}{dx} + y = 4x^{1/2} \cdot y^{3/2} \cdot \frac{dy}{dx}.$$

Upon combining terms we have

$$2x^{1/2} \cdot y^{1/2} + y = (4x^{1/2} \cdot y^{3/2} - x)\frac{dy}{dx},$$

or

$$\frac{dy}{dx} = \frac{2x^{1/2} \cdot y^{1/2} + y}{4x^{1/2} \cdot y^{3/2} - x}.$$

Example 17 Find an equation of the tangent line to the curve $x^2 + 4y^2 = 17$ at the point on the curve (1, 2).

To find the equation of the tangent line, we must first find the slope of the tangent to the curve at the point (1, 2). This can be done by finding dy/dx. Using implicit differentiation, we have

$$2x + 8y\frac{dy}{dx} = 0$$

$$\frac{dy}{dx} = \frac{-2x}{8y}$$

$$= \frac{-x}{4y}.$$

The slope of the tangent line to the curve at the point (1, 2) is thus

$$m = \frac{-x}{4y} = \frac{-1}{4(2)} = \frac{-1}{8}.$$

The equation of the tangent line is then found by using the point slope form of the equation of a line.

$$y - y_1 = m(x - x_1)$$

$$y - 2 = \frac{-1}{8}(x - 1)$$

$$8y - 16 = -x + 1$$

$$8y + x = 17.$$

Example 18 A biologist has placed a 50-foot ladder against a large building to enable him to reach certain rare wasp nests on the side of the building. In his haste to reach the nests, he neglects to notice that the base of the ladder is resting on an oil spill. Thus, as he climbs the ladder, the base slips (to the right in Figure 4.7) at the rate of 3 feet per minute. Find the rate of change of the height of the top of the ladder above the ground at the instant when the base of the ladder is 30 feet from the base of the building.

Figure 4.7

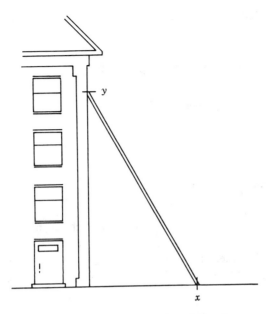

Let y be the height of the top of the ladder above the ground, and let x be the distance of the base of the ladder from the base of the building. By the Pythagorean theorem we have

$$50^2 = x^2 + y^2. \tag{3}$$

Both x and y are functions of time, t, measured from the moment that the ladder starts slipping. If we now take the derivative of both sides of equation (3) with respect to time, we get

$$\frac{d[50^2]}{dt} = \frac{d[x^2 + y^2]}{dt},$$

or

$$0 = 2x\frac{dx}{dt} + 2y\frac{dy}{dt}. \tag{4}$$

In the statement of the problem, we are told that the base is sliding at the rate of 3 feet per minute, so that

$$\frac{dx}{dt} = 3.$$

We are told that the base of the ladder is 30 feet from the base of the building. This fact can be used to find y. We know that $50^2 = x^2 + y^2$. Thus,

$$50^2 = 30^2 + y^2$$
$$2500 = 900 + y^2$$
$$1600 = y^2$$
$$y = 40 \text{ feet.}$$

128 The Derivative

We now know that $y = 40$, $x = 30$, and $dx/dt = 3$. Substituting these values into equation (4), we have

$$0 = 2x\frac{dx}{dt} + 2y\frac{dy}{dt}$$

$$0 = 2(30)(3) + 2(40)\frac{dy}{dt}$$

$$0 = 180 + 80\frac{dy}{dt}$$

$$80\frac{dy}{dt} = -180$$

$$\frac{dy}{dt} = \frac{-180}{80} = \frac{-9}{4} = -2.25 \text{ feet per minute.}$$

Thus, at the instant when the base of the ladder is 30 feet from the base of the building, the top of the ladder is sliding down the building at the rate of 2.25 feet per minute.

4.6 Exercises

Find dy/dx by implicit differentiation for each of the following.

1. $4x^2 + 3y^2 = 6$
2. $2x^2 - 5y^2 = 4$
3. $2xy + y^2 = 8$
4. $-3xy - 4y^2 = 2$
5. $6xy^2 - 8y + 1 = 0$
6. $-4y^2x^2 - 3x + 2 = 0$
7. $y^2 = 4x + 1$
8. $y^2 - 2x = 6$
9. $x^2 + 2xy + y^2 = 6$
10. $2x^2 - 3xy + 2y^2 = 10$
11. $6x^2 + 8xy + y^2 = 6$
12. $8x^2 = 6y^2 + 2xy$
13. $x^3 - y^2 + 4$
14. $x^3 - 6y^2 = 10$
15. $x^2y = 4$
16. $-2x^2y = 3$
17. $x^2y^2 = 6$
18. $5 - x^2y^2 = 0$
19. $x^2y + 2xy^2 + y^3 = 0$
20. $2xy^2 - 4x^2y + 2y^3 = 0$
21. $\sqrt{x} + \sqrt{y} = 4$
22. $2\sqrt{x} - \sqrt{y} = 1$
23. $\sqrt{xy} + y = 1$
24. $\sqrt{2xy} - 1 = 3y^2$

Find the equation of the tangent line at the given point on each of the following curves.

25. $x^2 + y^2 = 25$; $(-3, 4)$
26. $x^2 + y^2 = 25$; $(-3, -4)$
27. $x^2y^2 = 1$; $(-1, 1)$
28. $x^2y^3 = 8$; $(-1, 2)$

29. A 25-foot ladder is placed against a building. The base of the ladder is slipping away from the building at the rate of 4 feet per minute. Find the rate at which the top of the ladder is sliding down the building at the instant when the bottom of the ladder is 7 feet from the base of the building.

30. One car leaves a given point and travels north at 30 miles per hour. Another car leaves the same point at the same time and travels west at 40 miles per hour. At what rate is the distance between the two cars changing at the instant when the cars have traveled 2 hours?

● Case 4 Marginal Cost—Booz, Allen and Hamilton*

Booz, Allen and Hamilton is a large management consulting firm. One of the services they provide to client companies is profitability studies, in which they show ways in which the client can increase profit levels. The client company requesting the analysis presented in this case is a large producer of a staple food. The company buys from farmers, and then processes the food in its mills, resulting in a finished product. The company sells both at retail under its own brands, and in bulk to other companies who use the product in the manufacture of convenience foods.

The client company has been reasonably profitable in recent years, but the management retained Booz, Allen and Hamilton to see whether its consultants could suggest ways of increasing company profits. The management of the company had long operated with the philosophy of trying to process and sell as much of its product as possible, since, they felt, this would lower the average processing cost per unit sold. However, the consultants found that the client's fixed mill costs were quite low, and that, in fact, processing extra units made the cost per unit start to increase. (There are several reasons for this: the company must run three shifts, machines break down more often, and so on.)

To find the marginal cost for the various possible levels of production, the Booz, Allen and Hamilton analysts found several data points through study of past data. From these data points, the function

$$C(x) = 0.067x^2 + 10.09x + F$$

can be obtained, where $C(x)$ is the total cost of producing x million units of product A. The letter F represents the fixed cost for product A, a number we shall not need.

Using the total cost function, the marginal cost function $C'(x)$ can be found by taking the derivative of $C(x)$. We get

$$C'(x) = 0.134x + 10.09.$$

For example, at a level of production of 3.1 million units, an additional unit of product A would cost about

$$C'(3.1) = 0.134(3.1) + 10.09$$
$$= \$10.50.$$

*This case is based on material supplied by John R. Dowdle of the Booz, Allen and Hamilton firm of management consultants.

At a level of production of 5.7 million units, an extra unit costs $10.85. Figure 1 shows a graph of the marginal cost function from $x = 3.1$ to $x = 5.7$, the domain over which the function above was found to apply.

The selling price for product A is $10.73 per unit, so that, as shown on the graph of Figure 1, the company was losing money on many units of the product that it sold. Since the selling price could not be raised if the company was to remain competitive, the consultants recommended that production of product A be cut.

Figure 1

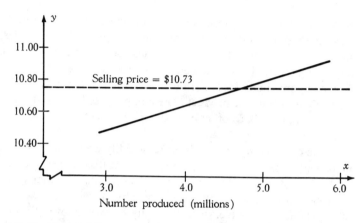

Number produced (millions)

For product B, the Booz, Allen and Hamilton consultants found a total cost function of

$$D(x) = 0.0334x^2 + 10.29x + G,$$

where $D(x)$ is the total cost of producing x million units of product B. Again, G is the fixed cost of producing product B. To find the marginal cost of a unit of product B, take the derivative of $D(x)$:

$$D'(x) = 0.0668x + 10.29.$$

Verify that at a production level of 3.1 million units the marginal cost is $10.50; while at a production level of 5.7 million units, the marginal cost is $10.67. Since the selling price of this product is $9.65, the consultants again recommended a cutback in production.

The consultants ran similar cost analyses of other products made by the company, and then issued their recommendation to the company: The company should reduce total production by 2.1 million units. The analysts predicted that this would raise profits for the products under discussion from $8.3 million annually to $9.6 million—which is very close to what actually happened when the client took the advice.

Case 4 Exercises

1. At what level of production, x, was the marginal cost of a unit of product A equal to the selling price?

2. Graph the marginal cost function for product B from $x = 3.1$ million units to $x = 5.7$ million units.
3. Find the number of units for which marginal cost equals the selling price for product B.
4. For product C, the total cost of production of x million units is

$$E(x) = 0.067x^2 + 9.46x + H.$$

   (a) Find the marginal cost at a level of production of 3.1 million units; of 5.7 million units.
   (b) Graph the marginal cost function.
   (c) For a selling price of $9.57, find the level of production for which the cost equals the selling price.

Chapter 5 Pretest

Solve.

1. $5 - 3x = 8$ [1.2]
2. $4x^2 - 7x + 2 = 0$ [1.3]
3. $\dfrac{4}{x^2} - 9 = 0$ [1.3]

Find the derivative.

4. $f(x) = -4$ [4.2]
5. $f(x) = 6x - 4$ [4.2]
6. $f(x) = 3x^2 - 2x + 7$ [4.2]
7. $f(x) = 3x^3 - 4x^2 + 2x + 5$ [4.2]
8. $f(x) = 2x^{1/2} + x^{3/2}$ [4.2]
9. $f(x) = \dfrac{5}{x + 1} + 3x$ [4.4]

Find all values of x for which the derivative does not exist.

10. $f(x) = \dfrac{2x + 1}{x^2 - 1}$ [4.4]
11. $f(x) = x^2 - 2x + 3$ [4.2]
12. $f(x) = \sqrt{x - 1}$ [4.5]

Find the limit if it exists. [3.2]

13. $\lim\limits_{x \to 1} 2x - 3$
14. $\lim\limits_{x \to -2} \dfrac{3x^2 + 5x - 4}{2}$
15. $\lim\limits_{x \to 0} \dfrac{5}{6}$
16. $\lim\limits_{x \to 3} \dfrac{2x}{3}$
17. $\lim\limits_{x \to 2} \dfrac{x^2 + x - 6}{x - 2}$
18. $\lim\limits_{x \to -5} \dfrac{x^2 + 10x + 25}{x + 5}$

5
Further Applications of the Derivative

Throughout history, most results in mathematics have been developed to aid in the solution of practical problems in various fields. The beginnings of modern calculus can be traced back to a series of practical problems from physics. These problems were recognized as being important and timely, and attracted the attention of the best scientific minds of the time—Isaac Newton of England (1642–1727) and Gottfried Wilhelm von Leibniz of Germany (1646–1716). In attempting to solve these problems, these two individuals independently discovered the basic ideas of calculus. While calculus has been used very successfully to solve problems in physical science for the past three hundred years, it has only been in the last thirty years or so that calculus has been successfully applied to significant problems in management and life science. One important application of calculus in these fields is in the area of *optimization*—finding the maximum or minimum values for a function. In this chapter we shall first discuss the theory behind optimization, and then discuss some practical examples. The chapter ends with a discussion of the derivative as an aid to sketching graphs, solving equations, and finding limits.

5.1 Optimization Theory—The First Derivative Test

Recall from Section 3.3 that an open interval, denoted (a, b), is the set of points $\{x \mid a < x < b\}$, while a closed interval, denoted $[a, b]$, is the set $\{x \mid a \leq x \leq b\}$. The open interval $\{x \mid a < x\}$ is denoted $(a, +\infty)$. An open interval with center at the point x_0 is called a **neighborhood** of x_0. Normally, the term neighborhood is used to imply a *small* open interval, centered at x_0. Intervals and neighborhoods will be used throughout our discussion of optimization.

A function f has a **relative maximum** at a value $x = x_0$ if there exists a neighborhood of x_0 such that $f(x_0) \geq f(x)$ for all values of x in that neighborhood. We can define a **relative minimum** similarly. The function whose graph is shown in Figure 5.1 has a relative minimum at $x = x_1$, with relative maxima at $x = x_0$ and $x = x_2$. As shown on the graph, there are values of x such that $f(x) < f(x_1)$, but the definition of relative minimum requires only that we be able to find at least one neighborhood of the point x_1 such that $f(x_1) \leq f(x)$ for all x in that neighborhood. The term **relative extrema** applies to either relative maxima or relative minima.

Figure 5.1

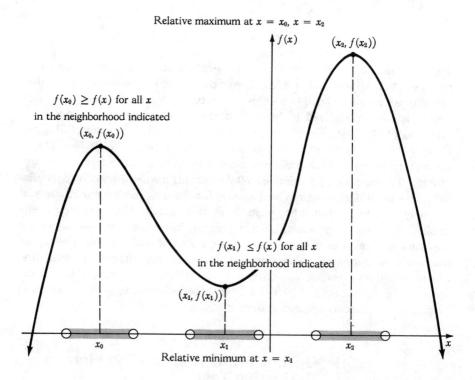

To find a point at which a function has a relative extremum, we use the fact that the derivative can be used to find the slope of a nonvertical line tangent to the graph of a function. At a point where a function has a relative maximum or a relative minimum value, any line tangent to the graph of the function is horizontal. (See Figure 5.2; recall that the letter m is used to denote slope.)

Figure 5.2

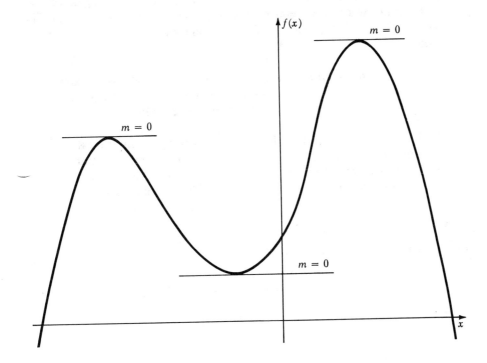

A horizontal line has slope 0. Thus, if a graph has a tangent line at a relative maximum or minimum, then the slope of the tangent line is 0, and $f'(x) = 0$ at that point. It is also possible that at a given point x_0, the derivative $f'(x_0)$ may not exist, even though the function has a relative maximum or minimum at x_0. For example, the function $f(x) = |x + 3|$ has a relative minimum value at $x = -3$, even though $f'(-3)$ does not exist. (See Figure 5.3.)

Figure 5.3

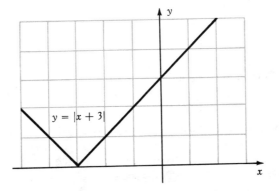

Hence, in looking for relative maxima or relative minima, it is necessary to check any value x_0 at which $f'(x_0) = 0$ and any value x_0 at which $f'(x_0)$ does not exist. Points where $f'(x) = 0$ or where $f'(x)$ does not exist are called **critical points** of the function f. The following theorem summarizes this discussion.

> **Theorem 5.1** If a function has any relative maxima or relative minima, then these relative maxima or relative minima occur only at critical points of the function.

Note first what the theorem does *not* say. It does not say that if x_0 is a critical point, then the function has a relative maximum or relative minimum at $x = x_0$. Figure 5.4 shows the graph of $f(x) = (x - 1)^3 + 2$. Verify that $f'(1) = 0$, even though the function has neither a relative maximum nor a relative minimum at $x = 1$.

Figure 5.4

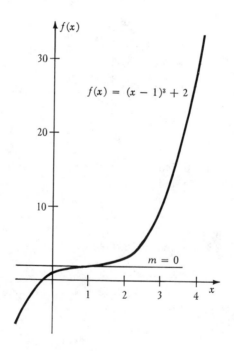

Example 1 Find all relative maxima and minima values of the function
$$f(x) = 3x^2 - 8x + 2.$$

Here $f'(x) = 6x - 8$. If we set this derivative equal to 0, we have
$$6x - 8 = 0$$
$$x = 4/3.$$

5.1 Optimization Theory—The First Derivative Test

There are no points at which the derivative does not exist, and so $x = 4/3$ is the only critical point. A sketch of the graph of $f(x) = 3x^2 - 8x + 2$, shown in Figure 5.5, shows that the function has a relative minimum at $x = 4/3$; this relative minimum value for the function is given by $f(4/3) = -10/3$.

Figure 5.5

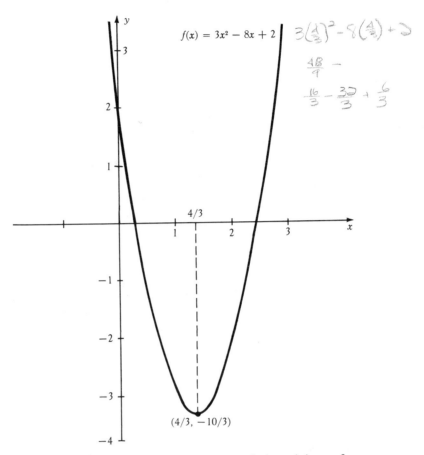

In the example above, we found that $x = 4/3$ led to a relative minimum for the given function by sketching a graph. It would be useful to have a more precise method to determine whether or not a given critical point leads to a relative maximum, or a relative minimum, or to neither. As shown in Figure 5.6, the slope of a tangent line just to the left of a relative minimum is negative, while a tangent line just to the right has a positive slope. Since the slope of a tangent line can be obtained from the derivative of a function, we have the following *first derivative test* for determining extrema.

Theorem 5.2 (First Derivative Test) If x_0 is a critical point of a function f, then
(a) the function has a relative maximum at $x = x_0$ if there exists a neighborhood of x_0 such that $f'(x) > 0$ for all values

Figure 5.6

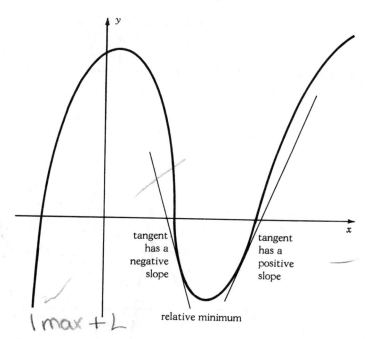

of x in the neighborhood to the left of x_0, and $f'(x) < 0$ for all values of x in the neighborhood to the right of x_0.

(b) The function has a relative minimum at $x = x_0$ if there exists a neighborhood of x_0 such that $f'(x) < 0$ for all values of x in the neighborhood to the left of x_0, and $f'(x) > 0$ for all values of x in the neighborhood to the right of x_0.

(c) If there exists a neighborhood of x_0 such that for all values of x (except $x = x_0$) in the neighborhood, both to the right and to the left of x_0, $f'(x)$ has the same sign, then the function has neither a relative maximum nor a relative minimum at $x = x_0$.

We shall discuss another such test in the next section.

Example 2 Find all relative extrema of the function

$$f(x) = 2x^3 - 3x^2 - 72x + 15.$$

Here we have

$$f'(x) = 6x^2 - 6x - 72.$$

If we set this derivative equal to 0, we have

$$6x^2 - 6x - 72 = 0$$
$$x^2 - x - 12 = 0$$
$$(x + 3)(x - 4) = 0$$
$$x = -3 \text{ or } x = 4.$$

Since there are no values of x for which $f'(x)$ does not exist, the only critical points here are $x = -3$ and $x = 4$. Let us use the first derivative test on $x = -3$. If we pick $x = -4$, a number to the left of -3 (it is usually safe to select any number that does not go past any other critical point) we have $f'(-4) = 6(-4)^2 - 6(-4) - 72 > 0$, while $x = -2$, which is to the right of -3, gives $f'(-2) = 6(-2)^2 - 6(-2) - 72 < 0$. Hence, by the first derivative test, the function has a relative maximum at $x = -3$. The relative maximum is $f(-3) = 150$. In the same way, we can show that the function has a relative minimum at $x = 4$. This relative minimum is given by $f(4) = -193$.

A function f has an **absolute maximum** at the point $x = x_0$ in case $f(x_0) \geq f(x)$ for all x in the domain of the function. The function has an **absolute minimum** at $x = x_1$ in case $f(x_1) \leq f(x)$ for all x in the domain of the function. The function whose graph is shown in Figure 5.7 has an absolute maximum at $x = x_0$ and an absolute minimum at $x = x_1$.

Figure 5.7

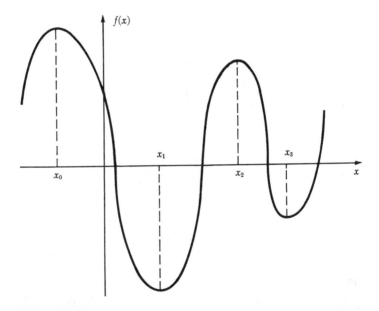

For a function f whose domain includes all real number values of x, and which has an absolute maximum or absolute minimum, the absolute maximum is the largest of any relative maxima, while an absolute minimum is the smallest of any relative minima (assuming only a finite number of such relative extrema). For a function whose domain is a closed interval, $[x_0, x_1]$, it is possible that an absolute extremum could occur at the endpoints of the domain of the function. For example, the function whose graph is shown in Figure 5.8 has a relative maximum at $x = 2$, but an absolute maximum at $x = 4$, the right endpoint of the domain of the function. The function has relative minima at $x = 0$ and $x = 3$, with an absolute minimum at $x = 0$. The left endpoint of the domain of the function, $x = -1$, leads to neither a relative extremum nor an absolute extremum. Intuitively, it seems plausible that $x = -1$ should lead to some sort of maximum for the function. However, the definition of relative extremum requires a neighborhood of the point in question, and we cannot find a neighborhood of -1 which is also completely inside the domain.

Figure 5.8

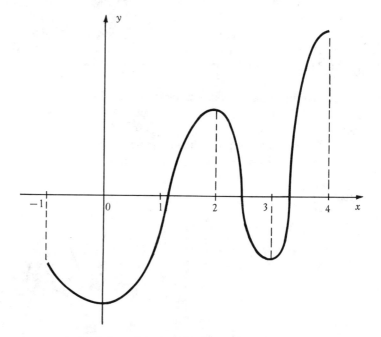

Example 3 Find all maxima and minima values of the function
$$f(x) = x^2 - 4x + 6,$$
where the domain of the function is $[-3, 4]$.

We can find the critical point by placing the derivative equal to 0:
$$f'(x) = 2x - 4 = 0,$$
from which $x = 2$. By the first derivative test, verify that $x = 2$ leads to a relative minimum. This relative minimum is given by $f(2) = 2^2 - 4(2) + 6 = 2$. By an inspection of the graph of the function (see Figure 5.9), it is seen that $x = -3$ leads to an absolute maximum, $f(-3) = 27$. Also from the graph, $x = 2$ yields the absolute minimum.

In summary, if a continuous function f has any absolute extrema, the absolute extrema can be found as follows:

1. Find $f'(x)$, and locate all critical points; that is, points for which $f'(x) = 0$ or for which $f'(x)$ does not exist.
2. Find any endpoints of the domain of the function.
3. Calculate the value of $f(x)$ for each critical point and for each endpoint. Select from these values the largest value, which is the absolute maximum. The smallest of these values is the absolute minimum.

5.1 Optimization Theory—The First Derivative Test

Figure 5.9

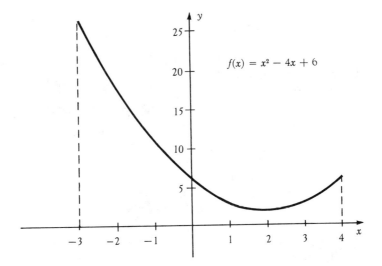

5.1 Exercises

Find all points at which the following functions have any relative maxima or relative minima. Find the value of any maxima or minima.

1. $f(x) = x^2 - 4x + 6$
2. $f(x) = x^2 - 8x + 12$
3. $f(x) = 2 + 8x - x^2$
4. $f(x) = 9 - 12x - x^2$
5. $f(x) = x^3 + 3x^2 - 24x + 2$
6. $f(x) = x^3 + 6x^2 + 9x - 8$
7. $f(x) = 2x^3 + 15x^2 + 36x - 4$
8. $f(x) = 2x^3 - 21x^2 + 60x + 5$

Find all points at which the following functions, having domains as specified, have any relative maxima or minima. Also, find any absolute extrema. Find the value of any extrema.

9. $f(x) = x^2 - 4x + 1$; $[-6, 5]$
10. $f(x) = x^2 + 6x + 2$; $[-6, 0]$
11. $f(x) = -x^2 + 8x - 2$; $[-5, 5]$
12. $f(x) = -x^2 + 2x - 3$; $[-4, 12]$
13. $f(x) = x^3 + 5x - 4$; $[-3, 2]$
14. $f(x) = x^3 - \frac{5}{2}x^2 - 2x + 1$; $[-1, 3]$
15. $f(x) = \frac{1}{3}x^3 + \frac{3}{2}x^2 - 4x + 1$; $[-5, 2]$
16. $f(x) = \frac{1}{3}x^3 - \frac{1}{2}x^2 - 6x + 3$; $[-4, 4]$

Find the vertex of each of the following parabolas. (Hint: the vertex of a parabola is the highest or lowest point on the parabola.)

17. $y = x^2 - 8x + 4$
18. $y = x^2 + 12x - 6$
19. $y = 2x^2 - 8x + 3$
20. $y = 3x^2 - 12x + 2$

21. $y = -2x^2 + 8x - 1$ 22. $y = -x^2 - 2x + 1$
23. $y = 2x^2 - 5x + 2$ 24. $y = 3x^2 - 8x + 4$
25. Find two numbers whose sum is 20 and whose product is maximized.
26. Find two numbers whose sum is 20 such that the sum of their squares is minimized.
27. A professor has found that the number of biology students in her class is approximated by $S(x)$, where

$$S(x) = -x^2 + 20x + 80,$$

and x is the number of daily hours that the student union is open. Find the number of hours that the union should be open so that the number of biology majors is the maximum. Find the maximum number of such majors.

28. The number of people visiting Timberline Ski Lodge on Washington's Birthday is approximated by $W(x)$, where

$$W(x) = -x^2 + 60x + 180,$$

and x is the total snowfall in inches for the previous week. Find the snowfall that will produce the maximum number of visitors. Find the maximum number of visitors.

29. The total cost to produce x units of rubber dog bones (where x is at least 5) is approximated by $B(x)$, where

$$B(x) = x^3 - 9x^2 - 120x + 1200.$$

Find the number of units of bones that must be sold in order to produce the minimum cost.

30. The number of salmon swimming upstream to spawn is approximated by $S(x)$, where

$$S(x) = -x^3 + 3x^2 + 360x + 5000,$$

with x representing the temperature of the water in degrees Centigrade. (This function is valid only if $6 \le x \le 20$.) Find the water temperature that produces the maximum number of salmon swimming upstream.

5.2 The Second Derivative Test

In this section we shall discuss another useful way to determine whether a given critical point x, where $f'(x) = 0$, leads to a maximum or a minimum. First, let us define a second derivative. The **second derivative** of a function f is the derivative of f'. We shall use f'', $f''(x)$, or y'' to represent the second derivative. To find the second derivative of $f(x) = 3x^2$, for example, first write $f'(x) = 6x$. Then take the derivative of this derivative to get $f''(x) = 6$.

5.2 The Second Derivative Test

Example 4 Let $f(x) = 6x^3 - 4x^2 + 8x - 5$, and find $f''(x)$. Then find $f''(0)$ and $f''(2)$.

First,
$$f'(x) = 18x^2 - 8x + 8,$$
so that
$$f''(x) = 36x - 8.$$
Using $f''(x) = 36x - 8$, we have $f''(0) = 36(0) - 8 = -8$, and $f''(2) = 64$.

The derivative at a critical point, if it exists, gives the slope of the tangent line, while the derivative of the derivative (the second derivative) gives the rate of change of the slope. As we have seen, a point just to the left of a relative minimum yields a tangent line with a negative slope, while a point just to the right of a relative minimum yields a tangent line with a positive slope. Thus, moving from left to right along the graph, past a relative minimum where the derivative exists, we see that the slope of the tangent line is increasing, from negative, to zero, to positive. Hence, the rate of change of the slope (the derivative of the slope) should be positive. An argument based on this discussion leads to the *second derivative test* for maxima and minima, as given in the following theorem.

Theorem 5.3 *(Second Derivative Test)* Let f be a function. Let x_0 be a number for which $f'(x_0) = 0$, and for which $f''(x_0)$ exists.
(a) If $f''(x_0) < 0$, then $f(x_0)$ is a relative maximum.
(b) If $f''(x_0) > 0$, then $f(x_0)$ is a relative minimum.
(c) If $f''(x_0) = 0$, then this test gives no information about relative maxima or minima.

⌒ max −
⌣ min +

Example 5 Find all relative maxima and minima of $f(x) = x^3 - 3x^2 - 9x + 5$.

Here we have
$$f'(x) = 3x^2 - 6x - 9,$$
and
$$f''(x) = 6x - 6.$$
If we set the first derivative equal to 0, we have
$$3x^2 - 6x - 9 = 0$$
$$x^2 - 2x - 3 = 0$$
$$(x - 3)(x + 1) = 0$$
$$x = 3 \quad \text{or} \quad x = -1.$$

Thus, $f'(3) = 0$ and $f'(-1) = 0$, and so both 3 and -1 are critical points of the function. Now we can apply the second derivative test at both $x = 3$ and $x = -1$. Calculate $f''(3)$ and verify that
$$f''(3) = 6(3) - 6 = 12.$$

Since $12 > 0$, part (b) of Theorem 5.3 holds, so that the function has a relative minimum at $x = 3$. This minimum is given by $f(3) = -22$. For $x = -1$, we have

$$f''(-1) = 6(-1) - 6 = -12.$$

Since $-12 < 0$, part (a) of the theorem holds, which means that the function has a relative maximum at $x = -1$. The maximum here is $f(-1) = 10$.

The second derivative test is easier to use than the first derivative test of the previous section, and normally should be tried first. However, there are some disadvantages to the second derivative test. For example, $f''(x)$ is very difficult to find for some functions. Also, the second derivative test does not always yield the desired result. By Theorem 5.3 above, if $f''(x_0) = 0$, we have no information. For example, if $f(x) = x^3$, then $f'(x) = 3x^2$. Note that $f'(0) = 0$, so that $x = 0$ is a critical point. Here $f''(x) = 6x$, and $f''(0) = 0$. However, as can be verified by a graph of $f(x) = x^3$ (see Figure 5.10(a)), the point $x = 0$ leads to neither a maximum nor a minimum for $f(x) = x^3$. On the other hand, it can be shown that $f(x) = (x - 1)^4$ has a relative minimum at $x = 1$ even though $f''(1) = 0$. See Figure 5.10(b). When the second derivative test does not supply the desired information, we must use the first derivative test to identify relative maxima and minima.

Figure 5.10

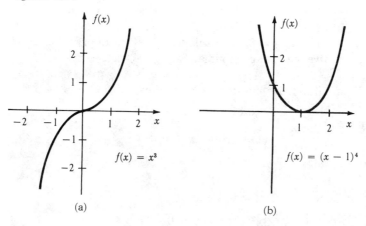

(a)   (b)

5.2 Exercises

For each of the following functions, find f''. Then find $f''(0)$, $f''(2)$, and $f''(-3)$.

1. $f(x) = x^3 + 4x^2 + 2$
2. $f(x) = 3x^3 - 4x + 5$
3. $f(x) = -x^4 + 2x^3 - x^2$
4. $f(x) = 3x^4 - 5x^3 + 2x^2$

5. $f(x) = 8x^2 + 6x + 5$
6. $f(x) = 3x^2 - 4x + 8$
7. $f(x) = (x - 2)^3$
8. $f(x) = (x + 4)^3$
9. $f(x) = \dfrac{x + 1}{x - 1}$
10. $f(x) = \dfrac{2x + 1}{x - 2}$

Find any critical points for the following functions, and use the second derivative test, if it works, to find whether the points are maxima or minima. If the second derivative test does not work, use the first derivative test.

11. $f(x) = x^2 - 12x + 36$
12. $f(x) = -x^2 - 10x - 25$
13. $f(x) = -9x^2 - 30x - 5$
14. $f(x) = 16x^2 - 24x + 9$
15. $f(x) = 12 - 8x + 4x^2$
16. $f(x) = 6 + 4x + x^2$
17. $f(x) = x^3 - 3x^2$
18. $f(x) = 2x^3 - 4x^2 + 2$
19. $f(x) = 3x^3 - 3x^2 + 1$
20. $f(x) = 2x^3 - x^2 + 4$
21. $f(x) = 2x^3 + 9x^2 - 60x + 8$
22. $f(x) = 2x^3 + 9x^2 - 108x + 10$
23. $f(x) = x^3 - \dfrac{15}{2}x^2 - 18x - \dfrac{1}{2}$
24. $f(x) = 2x^3 + \dfrac{7}{2}x^2 - 5x + \dfrac{1}{2}$
25. $f(x) = x^3 + x^2 - 8x + 3$
26. $f(x) = x^3 + \dfrac{x^2}{2} - 2x + 3$
27. $f(x) = x^3$
28. $f(x) = (x - 1)^4$
29. $f(x) = x^4 - 8x^2$
30. $f(x) = x + 1/x$
31. $f(x) = 2x - 8/x$
32. $f(x) = x^4 - 3x^3$
33. $f(x) = \dfrac{x^2 + 4}{x}$
34. $f(x) = \dfrac{x^2 + 25}{x}$
35. $f(x) = \dfrac{x}{1 + x}$
36. $f(x) = \dfrac{x - 1}{x + 1}$

37. Find two numbers x and y which have a sum of 40, such that the product x^2y is maximized.
38. Find two numbers x and y which have a sum of 20, such that the sum $x^2 + 2y$ is minimized.
39. The number of mosquitos, $M(x)$, in millions, in a certain area depends on the January rainfall, x, measured in inches, approximately as follows:

$$M(x) = 50 - 32x + 14x^2 - x^3.$$

Find the rainfall that will produce the maximum and the minimum number of mosquitos.

40. Because of raw material shortages, it is increasingly expensive to produce color print film. In fact, the profit, in thousands of dollars, from producing x million rolls is approximated by $F(x)$, where

$$F(x) = x^3 - 23x^2 + 120x + 60.$$

Find the level of production that will yield maximum profit and minimum profit.

5.3 Applications of the Theory of Extrema

To apply the theory of extrema to practical situations, it is first necessary to produce a mathematical description of the given situation. For example, suppose we are told that the number of units of a product that can be produced on a production line is approximated by

$$T(x) = 8x^{1/2} + 2x^{3/2} + 50,$$

where x is the number of employees on the production line. Once this equation has been obtained, we can use it to produce information about the production line. For example, we could take the derivative of this function and use it to estimate the marginal production resulting from additions of extra workers to the line.

However, in writing this equation, a mathematical model of a real world situation, we must be aware of certain restrictions on possible values of the variables involved. For example, since x represents the number of employees on a production line, x must certainly be restricted to the positive integers, or perhaps to a few common fractional values (we can conceive of half-time employees, but probably not $\frac{1}{16}$-time employees). Certainly, in this example we cannot have $x = \sqrt{2}$, or $x = \sqrt{3}$, or $x = \pi + 6$.

On the other hand, if we wish to apply the tools of calculus to obtain a maximum value or a minimum value for some function which might involve $T(x) = 8x^{1/2} + 2x^{3/2} + 50$, it is necessary that the function be defined and meaningful at every real number point in some interval. Because of this, the answer obtained from a mathematical formulation or model of a practical problem might be a real number that is not feasible in the setting of the problem.

Usually, the requirement of using a continuous function instead of a function which can take on only certain selected values is of theoretical interest only. In most cases, calculus gives results which *are* acceptable in a given situation. And if the methods of calculus should be used on a function f and lead to the conclusion that $80\sqrt{2}$ units should be produced in order to achieve the lowest possible cost, it is only necessary to calculate $f(80\sqrt{2})$, and then compare this to various values of $f(x)$, where x is an acceptable number. The lowest of these values of $f(x)$ then leads to x, the number that will minimize cost. In most cases, the result obtained will be very close to the theoretical minimum.

In the remainder of this section we shall discuss several examples showing applications of calculus to problems of finding maxima and minima.

Example 6 Suppose that the number of roosters, $r(x)$, sold in a farming community is given by

$$r(x) = 100 - 25x,$$

where x is the price, in dollars, of the roosters. Find the price that will lead to the maximum total revenue.

5.3 Applications of the Theory of Extrema

If $T(x)$ represents the total revenue, then $T(x)$ is given by the product of the number of roosters sold, $100 - 25x$, and the price per rooster, x. That is,

$$T(x) = (100 - 25x)x$$
$$= 100x - 25x^2.$$

To find the maximum revenue, calculate $T'(x)$, and then place $T'(x)$ equal to 0. This gives

$$T'(x) = 100 - 50x$$
$$0 = 100 - 50x$$
$$x = 2.$$

We must decide now whether $x = 2$ leads to the maximum revenue or the minimum revenue. The second derivative test is helpful here. Since $T''(x) = -50$, which is always negative, $x = 2$ must lead to the maximum revenue, as required. The maximum revenue is thus given by

$$T(2) = 2[100 - 25(2)]$$
$$= 100.$$

The function $T(x) = 100x - 25x^2$ is valid in this example only for values of x in the interval $[0, 4]$, since $r(x)$ is meaningful only for these values. The endpoints of this interval, $x = 0$ and $x = 4$, might lead to an absolute extremum for this function. Verify that both of these values lead to *minimum* revenue of 0.

Example 7 When Alfred E. Newman University charges $600 for a class that includes tours of several secondary and tertiary sewer plants in New Jersey, it attracts 1000 students. For each $20 decrease in the charge, an additional 100 students attend the class. What price should the university charge in order to receive the maximum revenue?

Let x represent the number of $20 decreases in the price. Then the price charged by the university is $600 - 20x$ dollars, and the number of students taking the class will be $1000 + 100x$. If $R(x)$ represents the revenue from x of the $20 decreases in price, then

$$R(x) = (600 - 20x)(1000 + 100x)$$
$$= 600,000 + 40,000x - 2000x^2.$$

Now calculate $R'(x)$ and solve the equation $R'(x) = 0$. Doing this, we have

$$R'(x) = 40,000 - 4000x$$
$$0 = 40,000 - 4000x$$
$$x = 10.$$

Here $R''(x) = -4000$, so that $R''(10) = -4000 < 0$. Thus, revenue is maximized with 10 of the $20 decreases in price, or a price of $600 - 20(10) = 400$ dollars. This price will attract $1000 + 100(10) = 2000$ students. The maximum total revenue is

$$R(10) = 400(2000) = 800,000$$

dollars.

Example 8 An open box is to be made by cutting squares from each corner of a 12-inch by 12-inch piece of tin and folding up the sides. What size square should be cut from each corner in order to produce the box of maximum volume?

Let x represent the side of the square cut from the corners of the sheet of tin, as shown in Figure 5.11. The width of the resulting box is $12 - 2x$, and the

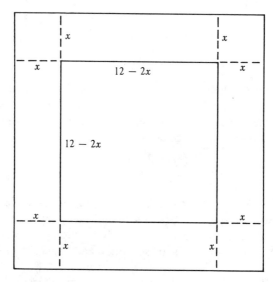

Figure 5.11

length is $12 - 2x$. Since the volume of a box is given by the product of the length, width, and height, we have

$$V = (12 - 2x)(12 - 2x)x$$
$$= 144x - 48x^2 + 4x^3.$$

If we find the derivative of this function and set it equal to 0, we can find the critical values of the volume function. We have

$$144 - 96x + 12x^2 = 0$$
$$x^2 - 8x + 12 = 0$$
$$(x - 6)(x - 2) = 0$$
$$x = 6 \quad \text{or} \quad x = 2.$$

By the second derivative test, $x = 2$ leads to the maximum volume, $V = 144(2) - 48(2^2) + 4(2^3) = 128$. Verify that x has domain $[0, 6]$. Both $x = 0$ and $x = 6$ lead to absolute minimum values for the volume, namely $V = 0$.

Example 9 A truck burns fuel at the rate of

$$G(x) = \frac{1}{200}\left(\frac{800}{x} + x\right)$$

gallons per mile when traveling x miles per hour on a straight, level road. If fuel costs 50¢ per gallon, find the speed that will produce the minimum total cost for a 1000-mile trip. Find the minimum total cost.

The total cost in cents, $C(x)$, is the product of the number of gallons per mile, the number of miles, and the cost per gallon, or

$$C(x) = 50(1000)\left[\frac{1}{200}\left(\frac{800}{x} + x\right)\right]$$

$$= \frac{200{,}000}{x} + 250x.$$

To find the minimum cost, we begin by calculating the first and second derivatives:

$$C'(x) = \frac{-200{,}000}{x^2} + 250$$

$$C''(x) = \frac{400{,}000}{x^3}.$$

If we now set the first derivative equal to 0, we have

$$\frac{-200{,}000}{x^2} + 250 = 0$$

$$\frac{200{,}000}{x^2} = 250$$

$$200{,}000 = 250x^2$$

$$800 = x^2$$

$$x = \pm 28.2 \text{ mph}.$$

We must reject $x = -28.2$ mph (why?), leaving $x = 28.2$ as a critical value. Since $C''(28.2) > 0$, we see that 28.2 leads to a minimum. Note that it is not necessary to actually calculate $C''(28.2)$—it is only necessary to know that $C''(28.2) > 0$.

The minimum total cost occurs at a speed of about 28.2 mph. Thus, the minimum total cost is about

$$C(28.2) = \frac{200{,}000}{28.2} + 250(28.2)$$

$$= 7092 + 7050$$

$$= 14{,}142.$$

Thus, the minimum total cost for the trip is about 14,000 cents, or $140.

5.3 Exercises

1. Find two numbers whose sum is 100 and whose product is maximized.
2. Find two numbers whose sum is 8 and whose product is maximized.
3. Find two numbers whose sum is 100 such that the sum of the squares of the two numbers is minimized.

4. Find two numbers whose sum is 8 such that the sum of the squares of the two numbers is minimized.
5. A farmer has 1200 feet of fencing. He wants to enclose a rectangular field that borders a river, so that he needs no fencing along the river's edge. Find the dimensions of the field that will produce the maximum possible area. Find the maximum area.
6. A company has found through experience that increasing its advertising also increases its sales, up to a point. The company believes that the relationship between profit, $P(x)$, and expenditure on advertising, x, is given by

$$P(x) = 80 + 108x - x^3.$$

What expenditure on advertising will lead to maximum profit? What is the maximum profit?

7. The total profit from the sale of x units of a certain prescription drug is given by $P(x)$, where

$$P(x) = -x^3 + 3x^2 + 72x - 280.$$

Find the number of units that should be sold in order to maximize the total profit. What is the maximum total profit?

8. The total profit from the sale of x units of Extra Crispy fried chicken is approximated by $C(x)$, where

$$C(x) = -x^3 + 6x^2 + 288x + 500.$$

Find the number of units that should be sold in order to maximize profit. Find the maximum profit.

9. The microbe concentration, $B(x)$ (in appropriate units), of Lake Gitchegoome depends on the oxygen concentration, x, again in appropriate units, approximately according to the function

$$B(x) = x^3 - 7x^2 - 160x + 800.$$

Find the oxygen concentration that will lead to the minimum microbe concentration.

10. The number of Chihuahuas in a certain city is approximated by $C(x)$, where

$$C(x) = x^3 + \frac{7}{2}x^2 - 40x + 100.$$

Here x represents the number, in hundreds, of German shepherds in the city. Find the population of German shepherds that will lead to the minimum Chihuahua population.

11. The manager of an 80-unit apartment complex is trying to decide on the rent to charge. It is known from experience that at a rent of $100, all the units will be full. However, on the average, one additional unit will remain vacant for each $5 increase in rent. Find the rent that should be charged to maximize total rental income. Find the maximum income.

12. The manager of a peach field is trying to decide when to arrange for the picking of his peaches. If they are picked now, the average yield per tree will be 100 pounds, which can be sold for 40¢ per pound. Past experience shows that the yield per tree will increase about 5 pounds per week, while the price will decrease about 2¢ per pound per week. When should the peaches be picked in order to produce maximum income? What is the maximum income?

13. In planning a small restaurant, it is estimated that a profit per seat of $5 per week will be made if the number of seats is between 60 and 80. On the other hand, if the number of seats is above 80, the profit on each seat will decrease by 5¢ for each seat above 80. Find the number of seats that will produce the maximum profit. What is the maximum profit?

14. The local hamburger fan club is arranging a charter flight to Key West, Florida, to see the southernmost McDonald's in the United States. The cost of the flight is $58 each for 75 passengers, with a refund of 40¢ per passenger for each passenger in excess of 75. Find the number of passengers that will maximize the revenue received by the organizers of the flight. Find the maximum revenue.

15. A closed box with a square base is to have a volume of 16,000 cubic inches. The material for the top and bottom of the box costs $3 per square inch, while the material for the sides costs $1.50 per square inch. Find the dimensions of the box that will lead to minimum total cost. Find the minimum total cost.

16. A farmer has 1800 yards of fencing. He wants to enclose a rectangular area on four sides, and then run an additional fence down the center (see the figure). What should the dimensions of the pasture be in order to enclose the maximum area?

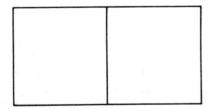

17. A mathematics book is to contain 36 square inches of printed matter per page, with margins of 1 inch along the sides, and $1\frac{1}{2}$ inches along the top and bottom. Find the dimensions of the page that will lead to the minimum amount of paper being used for a page.

18. In Example 9 we found the speed in miles per hour that minimized cost when we considered only the cost of the fuel. Rework the problem, assuming that the driver is paid $6 per hour. (Hint: if the trip is 1000 miles at x miles per hour, the driver will be paid for $1000/x$ hours.)

19. A company wishes to manufacture a box with volume 24 cubic feet which is open on top and which is twice as long as it is wide. Find the dimensions of the box produced from the minimum amount of material.

152 Further Applications of the Derivative

20. Assume that the number of bacteria, in thousands, present in a certain culture at time t is given by $N(t)$, where

$$N(t) = 24t(10 - 2t) + 15.$$

At what time will the population be at a maximum? Find the maximum population.

21. Assume that the number of bacteria, in thousands, present in a certain culture at time t is given by $R(t)$, where

$$R(t) = t^2(t - 18) + 96t + 1000.$$

At what time before 8 hours will the population be maximized? Find the maximum population.

22. An auto-catalytic chemical reaction is one in which the product being formed causes the rate of its formation to increase. The rate of reaction, V, of a certain auto-catalytic reaction is given by

$$V = 12x(100 - x),$$

where x is the quantity of the product present, and 100 represents the quantity of chemical present initially. For what value of x is the rate of the reaction a maximum?

23. A rectangular field is to be enclosed with a fence. One side of the field is against an existing fence, so that no fence is needed on that side. If material for the fence costs $2 per foot for the two ends, and $4 per foot for the side parallel to the existing fence, find the dimensions of the field of largest area that can be enclosed for $1000.

24. Decide what you would do if your assistant brought you a contract like this for your signature:

Your firm offers to deliver to a dealer 300 tables at $90 per table, and to reduce the price per table on the entire order by 25¢ for each additional table over 300.

Find the dollar total involved in the largest possible transaction between the manufacturer and the dealer under these conditions.

25. A piece of wire of length 12 feet is cut into two pieces. One piece is made into a circle and the other piece is made into a square. (See the figure.)

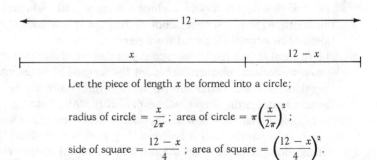

Let the piece of length x be formed into a circle;

$$\text{radius of circle} = \frac{x}{2\pi}; \quad \text{area of circle} = \pi\left(\frac{x}{2\pi}\right)^2;$$

$$\text{side of square} = \frac{12 - x}{4}; \quad \text{area of square} = \left(\frac{12 - x}{4}\right)^2.$$

(a) Where should the cut be made in order to make the sum of the areas enclosed by both figures a minimum?
(b) Where should the cut be made in order to make the sum of the areas enclosed by both figures a maximum? (Hint: recall how we find absolute extrema.)

26. A company wishes to run a utility cable from point A on the shore (see the figure) to an installation on the island (point B). The island is 6 miles from the shore. It costs \$400 per mile to run the cable on land, and \$500 per mile underwater. Assume that the cable starts from A and runs along the shoreline, then angles and runs underwater to the island. Find the point at which the line should begin to angle in order to yield the minimum total cost. (Hint: the length of the line underwater is $\sqrt{x^2 + 36}$.)

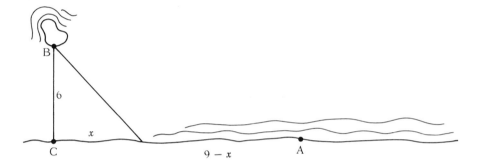

27. Use calculus to work Exercise 36 of Section 2.3.
28. Use calculus to work Exercise 37 of Section 2.3.

5.4 Curve Sketching

In several sections of this book we have discussed methods of graphing functions. For example, in Section 2.4 we discussed the graph of polynomial functions such as $f(x) = x^3 + x^2 - x + 1$. In that section we could only suggest that the graphs be found by selecting various values of x and finding the corresponding values of $f(x)$. Such a method is crude and time consuming, and especially so if the function does not have carefully selected "nice" values.

In this section we shall see how the idea of a derivative can be used as an aid in graphing such functions. Using techniques from calculus, it will be possible to graph a function more efficiently than by the method of plotting points.

To see how the derivative can be used in graphing functions, first glance at the curve of Figure 5.12. Note that as x moves from x_1 toward x_2, the height or y-value of a point on the graph is continually increasing. As x moves from x_2 to x_3, the function is decreasing, while as x moves from x_3 to x_4, the function is again increasing. In general, a function f is **increasing** on the interval (a, b) if

Figure 5.12

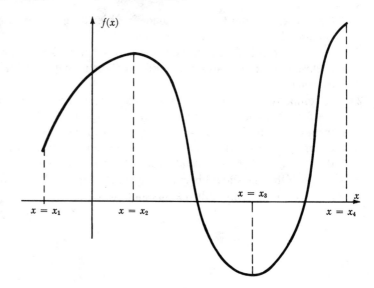

$f(m) < f(n)$ whenever m and n are any two points in (a, b) with $m < n$. Also, f is **decreasing** on the interval (a, b) in case $f(m) > f(n)$ whenever $m < n$.

Figure 5.13 shows the graph of a function which is increasing on an interval (a, b). The figure also shows several tangents to the graph of the function. Verify

Figure 5.13

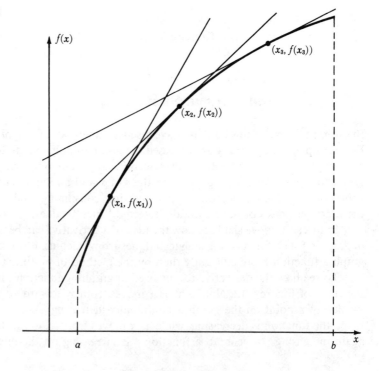

that each of these tangents has a positive slope, or, saying the same thing in other words, the derivative at the point of tangency is positive. Also, if $f'(x_0) < 0$ for every point x_0 in an interval (a, b), then f is a decreasing function on the interval (a, b). These results lead to the following theorem, which gives a method for identifying intervals on which functions are increasing or decreasing.

Theorem 5.4 Suppose a function f has a derivative at every point on an interval (a, b).
(a) If $f'(x) > 0$ for all values of x in (a, b), then the function f is increasing on (a, b).
(b) If $f'(x) < 0$ for all values of x in (a, b), then the function f is decreasing on (a, b).

Example 10 Let $f(x) = x^2 + 2x - 15$. Find all intervals on which the function is increasing or decreasing.

Here we have

$$f'(x) = 2x + 2.$$

Verify that $f'(x) = 0$ when $x = -1$. Also verify that for all values of x such that $x < -1$, we have $f'(x) < 0$. Thus, the function is decreasing on the interval $(-\infty, -1)$. For all values of x such that $x > -1$, we have $f'(x) > 0$. This means that the function is increasing on the interval $(-1, \infty)$. By using this information and plotting a few points, we get the graph of Figure 5.14.

Figure 5.14

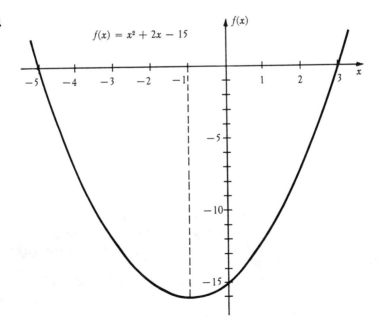

Example 11 Find all intervals over which the function $f(x) = x^3 + x^2 - x + 1$ is increasing or decreasing.

Here we have
$$f'(x) = 3x^2 + 2x - 1,$$
or
$$f'(x) = (3x - 1)(x + 1).$$

From this last result, we see that $f'(x) = 0$ when $x = 1/3$ or when $x = -1$. To find the interval or intervals for which $f'(x) > 0$, it is necessary to solve the inequality
$$(3x - 1)(x + 1) > 0.$$

This inequality can be solved by methods presented in Section 1.3. By the methods presented there, we see that $(3x - 1)(x + 1) > 0$ for all x in the intervals $(-\infty, -1)$ and $(1/3, +\infty)$. Also, $f'(x) < 0$ for all x in the interval $(-1, 1/3)$. Therefore, the function f is increasing on the interval $(-\infty, -1)$, decreasing on $(-1, 1/3)$, and increasing on $(1/3, +\infty)$. Using this information and plotting a few points, we obtain the graph of $f(x) = x^3 + x^2 - x + 1$ as shown in Figure 5.15.

Figure 5.15

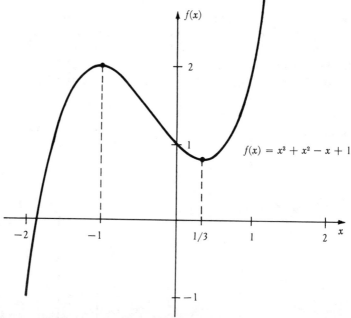

By using the first derivative of a function, we can determine intervals over which the function is increasing or decreasing. We can get further information about the graph of a function by using the second derivative. To do this, we must introduce the idea of the concavity of a curve. A curve is *concave upward* over an interval (a, b) if the curve looks like one of the curves of Figure 5.16. On the other hand, a curve is *concave downward* over an interval (a, b) if the curve looks like one of the curves of Figure 5.17.

Figure 5.16

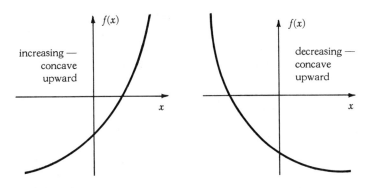

Figure 5.17

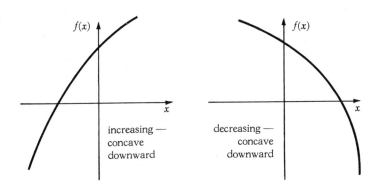

Figure 5.18

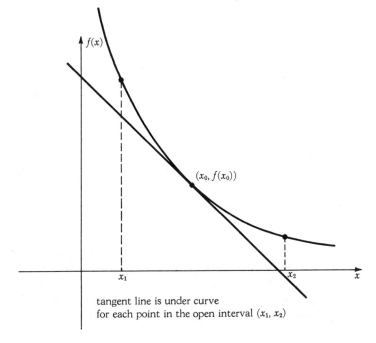

tangent line is under curve
for each point in the open interval (x_1, x_2)

Further Applications of the Derivative

To be a little more precise, we can say that the graph of f is **concave upward** at the point $(x_0, f(x_0))$ if there exists a neighborhood of x_0 such that the tangent line to the graph at the point $(x_0, f(x_0))$ is under the graph for every point in the neighborhood (except x_0). See Figure 5.18. The term **concave downward** can be defined in a similar manner.

If a function has a graph which is concave upward, then the slopes of the tangents to the graph increase as we move from left to right along the curve. Thus, the rate of change of the slopes of the tangents is positive. The rate of change of the slopes of the tangents is given by the derivative of the first derivative, which is the second derivative. Therefore, a graph is concave upward on any interval for which the second derivative is positive. Similarly, a curve is concave downward on any interval for which the second derivative is negative. We can summarize these statements in the following theorem.

> **Theorem 5.5** Suppose a function f has a second derivative $f''(x)$ for each point of the interval (a, b).
> (a) If $f''(x) > 0$ for each x in the interval (a, b), then the graph of f is concave upward on (a, b).
> (b) If $f''(x) < 0$ for each x in the interval (a, b), then the graph of f is concave downward on (a, b).

The use of this theorem as an aid in sketching a graph is illustrated in the next example.

Example 12 Sketch the graph of $f(x) = x^3 - 2x^2 - 4x + 3$.

To find all intervals for which the function is increasing or decreasing, we find $f'(x)$. Here we have

$$f'(x) = 3x^2 - 4x - 4$$
$$= (3x + 2)(x - 2).$$

Verify that $f'(x) = 0$ when $x = -2/3$ or when $x = 2$. We can use the second derivative test to check for maxima or minima. Here we have

$$f''(x) = 6x - 4,$$

which is positive when $x = 2$ and negative when $x = -2/3$. Thus, $f(2)$ is a relative minimum, and $f(-2/3)$ is a relative maximum.

The first derivative is positive on the intervals $(-\infty, -2/3)$ and $(2, +\infty)$, so that the function is increasing on these intervals. The first derivative is negative on the interval $(-2/3, 2)$, which shows that the function is decreasing on this interval.

The second derivative, $f''(x) = 6x - 4$, is 0 when $x = 2/3$. Using this fact, verify that the second derivative is negative for all values of x in the interval $(-\infty, 2/3)$ and positive for all values of x in the interval $(2/3, +\infty)$. By this result, the graph is concave downward on $(-\infty, 2/3)$ and concave upward on $(2/3, +\infty)$. A summary of this information is given in the chart below. Using these facts, and plotting a few points, we get the graph shown in Figure 5.19.

Interval	f'	f''	
$(-\infty, -2/3)$	+		increasing
$(-2/3, 2)$	−		decreasing
$(2, +\infty)$	+		increasing
$(-\infty, 2/3)$		−	concave downward
$(2/3, +\infty)$		+	concave upward

Figure 5.19

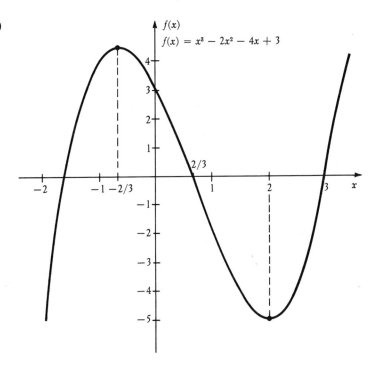

In the example above, we found that the function f was concave upward on the interval $(2/3, +\infty)$ and concave downward on $(-\infty, 2/3)$. At the point $x = 2/3$ the concavity changes. A point on a graph where the graph changes its concavity is called a **point of inflection**. As shown above, if $(x_0, f(x_0))$ is a point of inflection, then $f''(x_0) = 0$. However, just because $f''(x_0) = 0$, we have no assurance that a point of inflection has been located. For example, if $f(x) = (x - 1)^4$, then $f''(x) = 12(x - 1)^2$, which is 0 at $x = 1$. However, the graph of $f(x) = (x - 1)^4$ is always concave upward, and thus has no point of inflection. A sketch of the graph of $f(x) = (x - 1)^4$ is shown in Figure 5.20.

The steps involved in sketching the graph of a function are as summarized below:
1. Find f' and f''.
2. Find all values of x such that $f'(x) = 0$. Check these points to identify maxima or minima.
3. Find any intervals for which $f'(x) < 0$ or $f'(x) > 0$.

Figure 5.20

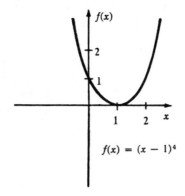

$f(x) = (x - 1)^4$

4. Find any values of x for which $f''(x) = 0$.
5. Find any intervals for which $f''(x) < 0$ or $f''(x) > 0$.
6. Locate as many points of the graph as needed, and sketch the graph, using the facts summarized in the following table.

$f'(x) > 0, f$ is increasing	$f''(x) > 0, f$ is concave upward
$f'(x) < 0, f$ is decreasing	$f''(x) < 0, f$ is concave downward

Example 13 When learning a new skill, it is common for learning to be slow at first, then increase, and finally level off at a fairly high degree of proficiency. A typical function showing such growth of learning is graphed in Figure 5.21. As shown by the graph, the derivative of the function P is increasing, up to the point $(t_0, P(t_0))$, and decreasing afterward. Thus, $P'(t)$ is a maximum at $(t_0, P(t_0))$. Because of this, $P''(t_0) = 0$, so that $(t_0, P(t_0))$ is a point of inflection for the function.

Figure 5.21

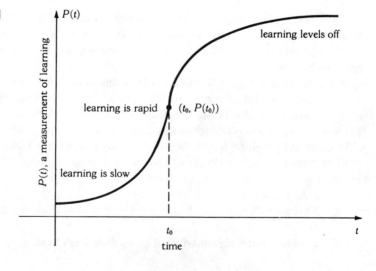

5.4 Exercises

Identify any intervals where the following functions are increasing or decreasing.

1. $f(x) = x^2 - 4x + 1$
2. $f(x) = x^2 + 8x - 2$
3. $f(x) = -4x^2 + 9x$
4. $f(x) = -3x^2 + 4x$
5. $f(x) = 2x^3 - 3x^2 - 12x + 2$
6. $f(x) = 2x^3 - 3x^2 - 72x - 4$
7. $f(x) = 4x^3 - 9x^2 - 30x + 6$
8. $f(x) = 4x^3 - 15x^2 - 72x + 5$
9. $f(x) = 6x + 3$
10. $f(x) = 4x + 8$
11. $f(x) = -4x + 1$
12. $f(x) = -8x + 9$

Find any intervals where the following functions are increasing or decreasing. Find any intervals where the functions are concave upward or concave downward. Find any points of inflection. Sketch the graph of the function.

13. $f(x) = x^3$
14. $f(x) = -x^3 + 6$
15. $f(x) = x^2 + 4x + 6$
16. $f(x) = x^2 - 8x + 10$
17. $f(x) = 3x^{2/3}$
18. $f(x) = 3x^{5/3}$
19. $f(x) = x^3 - 3x^2$
20. $f(x) = x^3 + 9x^2$
21. $f(x) = 2x^3 - 15x^2 - 36x + 8$
22. $f(x) = 4x^3 + 7x^2 - 10x - 3$
23. $f(x) = x^3 - 3x^2 + 3x - 7$
24. $f(x) = 2x^3 + 6x^2 - 48x + 9$
25. $f(x) = \dfrac{1}{x + 1}$
26. $f(x) = \dfrac{-4}{x + 3}$
27. $f(x) = x + 1/x$
28. $f(x) = -4x + 4/x$

5.5* Newton's Method

In looking for maxima and minima for a given function, it is necessary to find the first derivative and set it equal to 0. The resulting equation is then solved, giving critical points which can then be checked to see if they lead to maxima or minima, or neither. If the equation that results from placing the first derivative equal to 0 is of the first degree, such as $ax + b = 0$ $(a \neq 0)$, then the equation can be readily solved: $x = -b/a$. An equation of degree two, $ax^2 + bx + c = 0$ $(a \neq 0)$, can be solved by the quadratic formula. However, if the equation is of degree three or more, or if it involves complicated radicals or logarithms, then, in general, it is not easy to solve the equation. Because such complicated equations occur often in practice, a branch of mathematics called *numerical analysis* has been developed. Over the years, many methods for finding approximate numerical solutions for complicated equations have been developed. One of the most common methods of numerical analysis uses derivatives in an attempt to approximate the solutions of equations to any desired degree of accuracy. To see how this is done, let f be a continuous function defined on a closed interval $[a, b]$, with a derivative at each point in the interval (a, b). Assume that $f(a)$ and

---

* Material in this section will not be used in later chapters.

$f(b)$ are of opposite sign. Then, there must be a value c in the interval $[a, b]$ such that $f(c) = 0$. See Figure 5.22. To find an approximation to c, first make a guess for c, say c_1, in $[a, b]$. Then locate the point $(c_1, f(c_1))$ on the graph of $y = f(x)$ and draw the tangent line at this point. This tangent line will cut the y-axis at a point c_2. The number c_2 is often a better approximation to c than was c_1. It can be shown that

$$c_2 = c_1 - \frac{f(c_1)}{f'(c_1)}.$$

To try to improve the approximation to c, locate the point $(c_2, f(c_2))$ on the curve $y = f(x)$, and draw the tangent line at this point. This tangent line will cut the x-axis at a point c_3, which is usually a better approximation to c than was c_2. Also, c_3 is given by

$$c_3 = c_2 - \frac{f(c_2)}{f'(c_2)}.$$

Figure 5.22

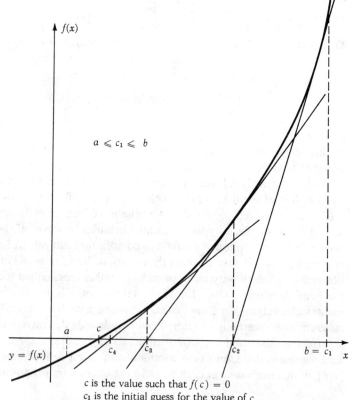

c is the value such that $f(c) = 0$
c_1 is the initial guess for the value of c
c_2, c_3, c_4 are successively better approximations

5.5 Newton's Method

We can often improve the approximation to c by repeating this process as many times as desired. If we have a value c_n which is an approximation to c, we can often find a better approximation c_{n+1} by the relationship

$$c_{n+1} = c_n - \frac{f(c_n)}{f'(c_n)}.$$

This process of obtaining a rough approximation to c, and replacing it by approximations that are often better and better, is called **Newton's method**, after the co-discoverer of the calculus, Sir Isaac Newton.

Example 14 Find a solution of the equation

$$3x^3 - x^2 + 5x - 12 = 0.$$

Approximate the solution to the nearest hundredth.

Here we have $f(x) = 3x^3 - x^2 + 5x - 12$, so that $f'(x) = 9x^2 - 2x + 5$. By trial, first verify that $f(1) = -5$, while $f(2) = 18$. Since $f(1)$ and $f(2)$ have opposite signs, there is a solution for the equation in the interval $[1, 2]$. As our first approximation to the solution, let $c_1 = 1$. A second approximation, c_2, can now be found as follows:

$$c_2 = c_1 - \frac{f(c_1)}{f'(c_1)}$$

$$c_2 = 1 - \frac{-5}{12}$$

$$= \frac{17}{12}$$

$$= 1.42.$$

A third approximation, c_3, can now be found:

$$c_3 = c_2 - \frac{f(c_2)}{f'(c_2)}$$

$$= 1.42 - \frac{1.67}{20.3}$$

$$= 1.34.$$

This result can be used to find c_4:

$$c_4 = 1.34 - \frac{0.122}{18.5}$$

$$= 1.33.$$

Now we can find c_5:

$$c_5 = 1.33 - \frac{-0.06}{14.3}$$

$$= 1.33.$$

Since $c_4 = c_5 = 1.33$, we can conclude that to two decimal places, $x = 1.33$ is a solution of $3x^3 - x^2 + 5x - 12 = 0$.

In this example, we had to go through 5 steps in order to get the desired degree of accuracy. The solutions of similar polynomial equations can usually be found in approximately as many steps, while other types of equations might require more steps.

In any case, if a solution can be found by Newton's method, it can usually be found by a computer in a few seconds of running time. Some functions, such as the one of Figure 5.23, have solutions which cannot be found by Newton's method. Because of the symmetry of the graph of Figure 5.23, c_1 will always equal c_3 and c_5, while c_2 will always equal c_4 and c_6. In such a situation, Newton's method provides no help in finding solutions. Luckily, such situations are rare in practice.

Figure 5.23

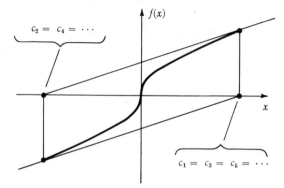

Newton's method can be used to approximate roots, as shown by the next example.

Example 15 Approximate $\sqrt{12}$ to the nearest thousandth.

First, note that $\sqrt{12}$ is a solution of the equation $x^2 - 12 = 0$. Hence, $f(x) = x^2 - 12$, and $f'(x) = 2x$. We know $3 < \sqrt{12} < 4$, so let us use $c_1 = 3$ as the first approximation to $\sqrt{12}$. Doing this, we have

$$c_2 = 3 - \frac{-3}{6} = 3.5.$$

Now we can find c_3:

$$c_3 = 3.5 - \frac{0.25}{7} = 3.464.$$

To find c_4, we write

$$c_4 = 3.464 - \frac{0.0007}{6.928} = 3.464.$$

Since $c_3 = c_4 = 3.464$, we can say that, to the nearest thousandth, $\sqrt{12} = 3.464$.

5.5 Exercises

Use Newton's method to find all real solutions of each of the following equations. Find all solutions to the nearest hundredth. Use a calculator or computer, if possible.

1. $3x^3 - 5x^2 + 8x - 4 = 0$
2. $5x^3 - 2x^2 - 2x - 7 = 0$
3. $3x^3 - 4x^2 - 4x - 7 = 0$
4. $3x^3 - 14x^2 + 17x - 22 = 0$
5. $5x^3 + 6x^2 + 4x + 33 = 0$
6. $7x^3 + 2x^2 - 30x - 27 = 0$
7. $x^3 - 3x^2 - 1 = 0$
8. $x^3 + 4x^2 + 6 = 0$
9. $6x^3 - 13x^2 - 36x + 28 = 0$
10. $6x^3 + x^2 - 46x + 15 = 0$
11. $3x^4 - 5x^3 - 16x^2 - 19x - 11 = 0$
12. $2x^4 + 7x^3 + 6x^2 + 7x - 6 = 0$

Use Newton's method to find each of the following roots to the nearest thousandth.

13. $\sqrt{2}$
14. $\sqrt{3}$
15. $\sqrt{11}$
16. $\sqrt{15}$
17. $\sqrt{250}$
18. $\sqrt{300}$
19. $\sqrt[3]{9}$
20. $\sqrt[3]{15}$
21. $\sqrt[3]{100}$
22. $\sqrt[3]{121}$

5.6* L'Hopital's Rule

In Chapter 3 we spent much time discussing limits. In many of the limits that we considered there, it was necessary to perform various algebraic manipulations in order to be able to evaluate the limit. In this section we shall discuss a method which uses derivatives to evaluate certain limits in an easier way. For example, derivatives can be used to evaluate a limit such as

$$\lim_{x \to 3} \frac{2x^2 - x - 15}{x - 3}.$$

If we substitute 3 for x in both numerator and denominator, we find that both the numerator and denominator equal 0, giving the fraction 0/0. The expression 0/0 is called an *indeterminate form*. When an indeterminate form such as this is obtained, it is often possible to evaluate the limit using L'Hopital's Rule. The first form of L'Hopital's Rule is given here, with an extended version given below.

L'Hopital's First Rule Let f and g be two functions with the following properties:
(a) $\lim_{x \to c} f(x) = 0$ and $\lim_{x \to c} g(x) = 0$ for c in some open interval (a, b).

* Material in this section will not be used in later chapters.

166 *Further Applications of the Derivative*

(b) f' and g' both exist on $[a, b]$, and $g'(x) \neq 0$ for all x ($x \neq c$) in (a, b).

(c) $\lim_{x \to c} \dfrac{f'(x)}{g'(x)}$ exists.

If the above conditions are satisfied, then

$$\lim_{x \to c} \frac{f(x)}{g(x)} = \lim_{x \to c} \frac{f'(x)}{g'(x)}.$$

There are many conditions required before this rule can be applied. However, these conditions are satisfied by most of the common functions, so that the result of the rule can be applied to most all problems that occur in practice.

Example 16 Find $\lim_{x \to 3} \dfrac{2x^2 - x - 15}{x - 3}$.

Here we can let $f(x) = 2x^2 - x - 15$ and $g(x) = x - 3$. We have $\lim_{x \to 3} f(x) = 2 \cdot 3^2 - 3 - 15 = 0$ and $\lim_{x \to 3} g(x) = 3 - 3 = 0$. Thus, we can use L'Hopital's rule to write $f'(x) = 4x - 1$ and $g'(x) = 1$. Thus,

$$\lim_{x \to 3} \frac{2x^2 - x - 15}{x - 3} = \lim_{x \to 3} \frac{4x - 1}{1} = 11.$$

Example 17 Find $\lim_{x \to -5} \dfrac{x^3 + x^2 - 14x + 30}{x + 5}$.

Let $f(x) = x^3 + x^2 - 14x + 30$ and $g(x) = x + 5$. Then $\lim_{x \to -5} f(x) = 0$ and $\lim_{x \to -5} g(x) = 0$. Since $f'(x) = 3x^2 + 2x - 14$, and $g'(x) = 1$, we have

$$\lim_{x \to -5} \frac{x^3 + x^2 - 14x + 30}{x + 5} = \lim_{x \to -5} \frac{3x^2 + 2x - 14}{1} = 51.$$

Example 18 Find $\lim_{x \to 4} \dfrac{x^4 - 10x^3 + 35x^2 - 56x + 48}{x^2 - 8x + 16}$.

Here we can let $f(x) = x^4 - 10x^3 + 35x^2 - 56x + 48$ and $g(x) = x^2 - 8x + 16$. Verify that $\lim_{x \to 4} f(x) = 0$ and $\lim_{x \to 4} g(x) = 0$. If we calculate $f'(x)$ and $g'(x)$, we get $f'(x) = 4x^3 - 30x^2 + 70x - 56$, and $g'(x) = 2x - 8$. Again, we have $\lim_{x \to 4} f'(x) = 0$ and $\lim_{x \to 4} g'(x) = 0$. To get around this problem, we use L'Hopital's rule on the quotient of the derivatives: we find that $f''(x) = 12x^2 - 60x + 70$ and $g''(x) = 2$. Therefore,

$$\lim_{x \to 4} \frac{x^4 - 10x^3 + 35x^2 - 56x + 48}{x^2 - 8x + 16} = \lim_{x \to 4} \frac{12x^2 - 60x + 70}{2} = 11.$$

5.6 L'Hopital's Rule

A word of warning: the limit $\lim_{x \to c} f(x)/g(x)$ cannot be found by calculating $\lim_{x \to c} f'(x)/g'(x)$ unless it is true that $\lim_{x \to c} f(x) = 0$ and $\lim_{x \to c} g(x) = 0$. For example, verify that

$$\lim_{x \to 4} \frac{x^2 + 16}{2x + 1} = \frac{32}{9}.$$

If we tried L'Hopital's rule here, we would end up with

$$\lim_{x \to 4} \frac{2x}{2} = 4,$$

which is not correct. The difficulty is that $\lim_{x \to 4} x^2 + 16 = 32 \neq 0$, and $\lim_{x \to 4} 2x + 1 = 9 \neq 0$.

Another form of L'Hopital's rule* is applicable when $\lim_{x \to c} f(x) = +\infty$ and $\lim_{x \to c} g(x) = \pm \infty$, where c is either a real number or $\pm \infty$. This version can be stated as follows.

> **Extended L'Hopital's Rule** Let $y = f(x)$ and $y = g(x)$ be two functions with the following properties.
> (a) $\lim_{x \to c} f(x) = \lim_{x \to c} g(x) = A$, where $A = 0$, $A = +\infty$, or $A = -\infty$, and where c is a real number, $c = +\infty$, or $c = -\infty$.
> (b) $f'(x)$ and $g'(x)$ exist, and $g'(x) \neq 0$ for an appropriate interval,
> (c) $\lim_{x \to c} \dfrac{f'(x)}{g'(x)}$ exists.

If the conditions above are satisfied, then

$$\lim_{x \to c} \frac{f(x)}{g(x)} = \lim_{x \to c} \frac{f'(x)}{g'(x)}.$$

Example 19 Evaluate $\lim_{x \to +\infty} \dfrac{2x + 5}{3x - 6}$.

If $f(x) = 2x + 5$ and $g(x) = 3x - 6$, then $\lim_{x \to +\infty} f(x) = +\infty$ and $\lim_{x \to +\infty} g(x) = +\infty$. Thus, we can apply the extended L'Hopital's rule. Here $f'(x) = 2$ and $g'(x) = 3$. We have $\lim_{x \to +\infty} f'(x)/g'(x) = 2/3$. Thus,

$$\lim_{x \to +\infty} \frac{2x + 5}{3x - 6} = \frac{2}{3}.$$

---

* A proof of L'Hopital's rule can be found in T. M. Apostol, *Mathematical Analysis*, second edition, Addison-Wesley, Reading, Massachusetts, 1974.

5.6 Exercises

Use L'Hopital's rule, where it applies, to evaluate each of the following limits.

1. $\lim\limits_{x \to 1} \dfrac{x^2 + 5x - 6}{x - 1}$

2. $\lim\limits_{x \to 2} \dfrac{x^2 - 7x + 10}{x - 2}$

3. $\lim\limits_{x \to -4} \dfrac{x^2 + 12x + 32}{x + 4}$

4. $\lim\limits_{x \to -3} \dfrac{x^2 - 2x - 15}{x + 3}$

5. $\lim\limits_{x \to 6} \dfrac{2x^2 - 9x - 18}{x - 6}$

6. $\lim\limits_{x \to -8} \dfrac{3x^2 + 22x - 16}{x + 8}$

7. $\lim\limits_{x \to 4} \dfrac{2x^3 - 11x^2 + 13x - 4}{x - 4}$

8. $\lim\limits_{x \to 6} \dfrac{3x^3 - 20x^2 + 13x - 6}{x - 6}$

9. $\lim\limits_{x \to -5} \dfrac{3x^3 + 19x^2 + 25x + 25}{x + 5}$

10. $\lim\limits_{x \to -3} \dfrac{2x^3 + 3x^2 - 8x + 3}{x + 3}$

11. $\lim\limits_{x \to 2} \dfrac{x^3 - 5x^2 + 8x - 4}{x^2 - x - 2}$

12. $\lim\limits_{x \to -3} \dfrac{x^3 - 2x^2 - 9x + 18}{x^2 + x - 6}$

13. $\lim\limits_{x \to -4} \dfrac{2x^3 + 7x^2 - x + 12}{2x^2 + 7x - 4}$

14. $\lim\limits_{x \to -3} \dfrac{3x^3 + 4x^2 - 16x - 3}{4x^2 + 11x - 3}$

15. $\lim\limits_{x \to 4} \dfrac{x^3 - 6x^2 + 32}{x^2 - 8x + 16}$

16. $\lim\limits_{x \to -6} \dfrac{x^3 + 14x^2 + 60x + 72}{x^2 + 12x + 36}$

17. $\lim\limits_{x \to -3/2} \dfrac{4x^3 + 8x^2 - 3x - 9}{4x^3 + 16x^2 + 21x + 9}$

18. $\lim\limits_{x \to 1/4} \dfrac{16x^3 + 24x^2 - 15x + 2}{16x^3 - 40x^2 + 17x - 2}$

19. $\lim\limits_{x \to +\infty} \dfrac{3x^2 - 4x + 1}{2x^2 + 3x - 1}$

20. $\lim\limits_{x \to +\infty} \dfrac{4x^2 + 2x - 3}{2x^2 - 3x + 6}$

21. $\lim\limits_{x \to +\infty} \dfrac{3x^5 - 2x^2 - 4x + 1}{2x^5 + 3x^4 + 8x^2 - 2}$

22. $\lim\limits_{x \to +\infty} \dfrac{8x^4 - 3x^2 + 5x - 8}{3x^4 + 2x^2 + 15x - 3}$

23. $\lim\limits_{x \to +\infty} \dfrac{4x^2 - 6x + 2}{8x^3 + 3x^2 - 4x + 1}$

24. $\lim\limits_{x \to +\infty} \dfrac{9x^2 + 6x - 5}{2x^4 - 3x^3 + x^2}$

25. $\lim\limits_{x \to +\infty} \dfrac{8x^5 - 2x^2 + 6x}{3x^4 - 5x + 2}$

26. $\lim\limits_{x \to +\infty} \dfrac{3x^4 + 2x^3 - 8x + 1}{2x^5 - 3x^3 + 4x^2 - 5x}$

27. $\lim\limits_{x \to -1} \dfrac{1 + 1/x}{x + 1}$

28. $\lim\limits_{x \to 2} \dfrac{1 - 2/x}{x - 2}$

29. $\lim\limits_{x \to 5} \dfrac{1 - 5/x}{x^2 - 10x + 25}$

30. $\lim\limits_{x \to -3} \dfrac{1 + 3/x}{x^2 + 6x + 9}$

● Case 5 Minimizing Manufacturing Costs

Suppose a company manufactures a certain number of units of a product per year. Let us assume that the product is such that it can be manufactured in a number of equal batches during the year, and that the demand for the product is constant throughout the year. In this case we shall find the number of batches of this product that the company should manufacture annually in order to minimize its total cost for producing the product. This case uses the following variables:

x = the number of batches of the product to be manufactured annually
k = cost of storing one unit of the product for a year
f = fixed cost of setting up the factory to manufacture the product
g = variable cost of manufacturing a single unit of the product
M = total number of units produced annually.

The company has two types of cost associated with producing its product: a cost associated with manufacturing the item, and a cost associated with storing the finished product.

Let us first consider manufacturing costs. During a year the company will produce x batches of the product. The company will thus produce M/x units of product per batch. Each batch has a fixed cost f, and a variable cost g per unit, so that the manufacturing cost per unit is

$$f + g\left(\frac{M}{x}\right).$$

Since there are x batches per year, the total annual manufacturing cost is

$$\left[f + g\left(\frac{M}{x}\right)\right]x. \tag{1}$$

Now we must determine the storage cost. Since each batch consists of M/x units, and since we have assumed demand is constant, it is customary to assume an average inventory of

$$\frac{1}{2}\left(\frac{M}{x}\right) = \frac{M}{2x}$$

units per year. It costs k to store one unit of the product for a year, making a total storage cost of

$$k\left(\frac{M}{2x}\right) = \frac{kM}{2x}. \tag{2}$$

The total production cost is the sum of the manufacturing and storage costs. If $T(x)$ is the total cost of producing x batches, we have

$$T(x) = \left[f + g\left(\frac{M}{x}\right)\right]x + \frac{kM}{2x}. \qquad (1) + (2)$$

To find the minimum value of $T(x)$, we take the derivative of $T(x)$ and set it equal to 0. Since f, g, k, and M are constants, we have

$$T'(x) = f - \frac{kM}{2}x^{-2}.$$

Setting this equal to 0 gives

$$f - \frac{kM}{2}x^{-2} = 0, \quad \text{or} \quad f = \frac{kM}{2x^2},$$

170 *Further Applications of the Derivative*

from which

$$2fx^2 = kM, \quad \text{or} \quad x^2 = \frac{kM}{2f},$$

which yields

$$x = \sqrt{\frac{kM}{2f}} \tag{3}$$

as the annual number of batches which will entail minimum total production cost.

Case 5 Exercises

1. Find the approximate number of batches to be produced annually if it costs $1 to store a unit for a year, if 100,000 units are to be manufactured, if the fixed factory set-up cost is $500, and if it costs $4 in variable costs to manufacture a unit.
2. How many units per batch will be manufactured in the example of Exercise 1?
3. Use the second derivative to show that the value of x obtained in the text [equation (3)] really leads to a minimum cost.
4. Why do you think the variable cost g does not appear in the answer for x [equation (3)]?

● Case 6 Minimum Warehouse Cost

If a company serves an area of relatively uniform population density, we can use the calculus we have developed to find an estimate of the number of warehouses that will serve the area at minimum cost.

First note that warehouse cost per unit goes down as the number of units in the warehouse increases. One large warehouse is cheaper than several smaller ones. However, transportation costs to the customers then increase. The company must find a balance between these two variables that will lead to the minimum total cost.

Let us assume that the cost of transporting the goods to the customers varies directly as the distance. If we assume, for simplicity, that the area served by a warehouse is a circle whose center is the warehouse, then the area served by the warehouse is given by πr^2, where r is the radius of the circle. From this, we conclude that the distance traveled, and thus the transportation cost, varies as the square root of the area served. Let

> K = number of units of sales per square mile per year (Here we use the assumption that the population is uniformly distributed.)
> x = area served by one warehouse
> b = annual fixed cost of one warehouse
> c = variable cost of transportation per unit of goods (c varies as the square root of the area)

Case 6 Minimum Warehouse Cost

A given warehouse serves x square miles, and distributes K items per square mile per year, for a total annual volume of Kx units. Since b is the annual fixed cost, we have

$$\frac{b}{Kx}$$

as the fixed cost per unit. The total cost of warehousing and transporting a unit of goods, $c(x)$, is given by

$$c(x) = \text{warehousing cost} + \text{transportation cost}$$

$$c(x) = \frac{b}{Kx} + c\sqrt{x}.$$

If we now take the derivative and set it equal to 0, we will be able to find the area that will produce the minimum total cost. Here b, K, and c are constants. Thus,

$$c'(x) = \frac{-b}{Kx^2} + \frac{c}{2\sqrt{x}}.$$

If we equate this to 0, we have

$$0 = \frac{-b}{Kx^2} + \frac{c}{2\sqrt{x}},$$

or

$$\frac{b}{Kx^2} = \frac{c}{2\sqrt{x}}.$$

Multiplying both sides by x^2 gives

$$\frac{b}{K} = \frac{c}{2} x^{3/2}$$

or

$$\frac{2b}{cK} = x^{3/2}$$

from which

$$x = \left(\frac{2b}{cK}\right)^{2/3}.$$

Case 6 Exercise

1. Find the area served by each warehouse if the company sells about 2 units of goods per square mile per year, if it costs about $320,000 annually to run a warehouse, and if $c = \$5$.

● Case 7 Pricing an Airliner—The Boeing Company*

The Boeing Company is one of the United States largest producers of civilian airliners. It has developed a series of successful jet aircraft, starting with the 707 in 1955. In this case we discuss the mathematics involved in determining the optimum price for a new model jet airliner. It is helpful to summarize here the variables to be used in this case.

p = price per airliner, in millions of dollars
$N(p)$ = total number of airliners that will be sold by the industry at a price of p
x = total number of airliners to be produced by Boeing
$C(x)$ = total cost to manufacture x airliners
h = share of the total market to be won by Boeing
P^* = total profit for Boeing

For the airliner in question, Boeing had only one competitor. The price charged by both Boeing and its competitor would have to be the same—any attempt by one firm to lower the price would of necessity be met by the other firm. Thus, the price charged by Boeing would have no effect on the share of the total market to be won by Boeing. However, the price charged by Boeing (and the competitor) would have considerable effect on the *size* of the total market.

Figure 1

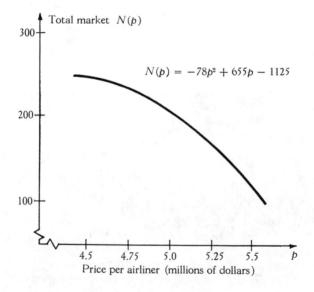

---

* Based on Brigham, Georges, "Pricing, Investment, and Games in Strategy," in *Management Sciences, Models and Techniques*, edited by C. W. Churchman and M. Verhulst, v. 1, pp. 271–87 (Pergamon Press, 1960).

Case 7 Pricing an Airliner

In fact, Boeing sales analysts made predictions of the total market at various price levels, and found that the function

$$N(p) = -78p^2 + 655p - 1125$$

gave a reasonable estimate of the total market; that is, $N(p)$ is the total number of planes that will be sold, by both Boeing and its competitor, at a price p, in millions of dollars per plane. A graph of $N(p)$ is shown in Figure 1.

Production analysts at Boeing estimated that if $C(x)$ is the total cost to manufacture x airplanes, then

$$C(x) = 50 + 1.5x + 8x^{3/4},$$

where $C(x)$ is measured in millions of dollars.

The company desires to know the price, p, it should charge per plane so that the total profit, P^*, will be a maximum. Profit is given as the numerical product of the price per plane, p, and the total number of planes sold by Boeing, x, minus the cost to manufacture x planes, $C(x)$. Thus, the profit function is

$$P^* = p \cdot x - C(x). \tag{1}$$

If h is the fractional share of the total market for this plane that will be won by Boeing (note: $0 \le h \le 1$), then

$$x = h \cdot N(p).$$

(The number of planes sold by Boeing equals its share of the market times the total market.) Substituting $h \cdot N(p)$ for x in the profit function, equation (1), gives

$$P^* = p \cdot h \cdot N(p) - C[h \cdot N(p)].$$

To find the maximum profit, we must take the derivative of this function with respect to p. Assume h is a constant; treat $p \cdot h \cdot N(p)$ as the product $(ph) \cdot N(p)$, and use the chain rule on $C[h \cdot N(p)]$. This gives

$$(P^*)' = (ph) \cdot N'(p) + N(p) \cdot h - C'[h \cdot N(p)] \cdot h \cdot N'(p).$$

For convenience, replace $h \cdot N(p)$ by x in the expression $C'[h \cdot N(p)]$. Then put the derivative equal to 0 and simplify:

$$(ph) \cdot N'(p) + N(p) \cdot h - C'(x) \cdot h \cdot N'(p) = 0,$$

$$p \cdot N'(p) + N(p) = C'(x) \cdot N'(p),$$

from which

$$p + \frac{N(p)}{N'(p)} = C'(x). \tag{2}$$

Thus, the optimum price, p, must satisfy equation (2). (It is necessary to verify that this value of p leads to a *maximum* profit and not a minimum. This calculation, using the second derivative test, is left for the energetic reader.)

Returning to the functions $N(p)$ and $C(x)$ given above, verify that

$$N'(p) = -156p + 655 \quad \text{and} \quad C'(x) = 1.5 + 6x^{-1/4}.$$

Using p, $N(p)$, and $N'(p)$, we can sketch a graph of the left-hand side of equation (2), as shown by the left-hand graph of Figure 2. On the right in Figure 2 is the graph of the right-hand side of equation (2), $C'(x) = 1.5 + 6x^{-1/4}$, for various values of x considered feasible by the company. We know that the maximum profit is produced when the left-hand and right-hand sides of equation (2) are equal. Using this fact, we can read the price that leads to the maximum profit from Figure 2. If $x = 60$ (the company sells a total of 60 airplanes), the price per plane will be about $5.1 million, while if $x = 120$, the price should be a little less than $5 million.

Figure 2

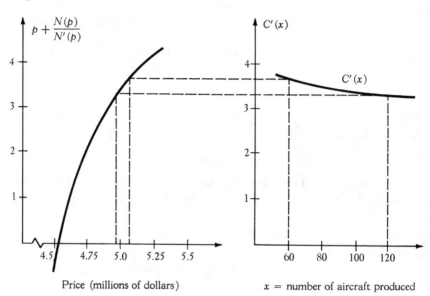

Price (millions of dollars) $x = $ number of aircraft produced

Case 7   Exercises

1. (a) Find the total cost of manufacturing 120 planes. (Hint: $8 \cdot 120^{3/4} \approx 290$)
   (b) Assume each plane sells for $5 million, and find the total profit from the sale of 120 planes, using equation (1).
2. (a) Find the total market at a price of $5 million per plane.
   (b) If Boeing sells 120 planes at a price of $5 million each, how many will be sold by the competitor?
3. (a) Find $C(60)$. (Hint: $8 \cdot 60^{3/4} \approx 170$)
   (b) Assume each plane is sold for $5.1 million, and find the total profit from the sale of 60 planes, using equation (1).

Chapter 6 Pretest

1. Find (a) $|2|$; (b) $|0|$; (c) $|-4|$. [1.1]
2. Find (a) 2^0; (b) $(\frac{1}{2})^{-1}$; (c) $(\frac{1}{2})^3$; (d) 2^{-2}; (e) $2^{1/2}$. [1.5]
3. Find (a) 3^{-2^2}; (b) $(\frac{1}{2})^{-1^2}$; (c) $2^{-(-2)^2}$. [1.5]
4. Find (a) 10^{-1}; (b) 10^0; (c) $10^{-1/2}$; (d) 10^3; (e) 10^{-3}. [1.5]
5. Find x: (a) $3^x = 27$; (b) $2^x = \frac{1}{4}$; (c) $4^x = 2$. [1.5]
6. Find the derivative of each of the following. [4.2]
    (a) $f(x) = -0.8x$
    (b) $f(x) = 1.4x^2$
    (c) $f(x) = 4x^2 - 2x$
    (d) $f(x) = \sqrt{2x - 5}$

6
Exponential and Logarithmic Functions

In Chapter 2 we studied several types of functions that are useful in setting up mathematical descriptions or models of practical situations. While the functions that we have already studied are certainly important, it is probably a safe generalization to say that exponential functions are the single most important type of function in practical applications. *Exponential functions*, and the closely related *logarithmic functions*, are often used in management, social science, and biology to describe growth and decay. In this chapter we shall see exponential functions used to describe growth of populations, increase in sales, growth of money with time, as well as decay of radioactive samples and decay of sales in the absence of advertising.

6.1 Exponential Functions

We looked at integer and rational number exponents in Section 1.5 and gave meaning to exponentials such as 2^x for any integer or rational number value of x. However, if we were to try to graph the function $y = 2^x$, we would have a problem. The domain of this function would not be the set of all real numbers, but only the set of rational numbers. The graph of $y = 2^x$ would not be a smooth curve, but rather would be a series of points, representing only the rational values of x. To take care of this problem, we must define 2^x for *all* real numbers x, and not just for the rational values of x.

6.1 Exponential Functions

In general, for a real number $a > 0$ and $a \neq 1$, and for any real numbers x, y, and z, we will define a^x so that the following properties are satisfied.

1. a^x is a unique real number.
2. If $x = y$, then $a^x = a^y$.
3. If $a^x = a^y$, then $x = y$.
4. If $a > 1$, and if $x < y < z$, then $a^x < a^y < a^z$.
   If $0 < a < 1$, and if $x < y < z$, then $a^x > a^y > a^z$.

It is necessary to assume $a > 0$ so that property (1) will be valid. For example, $a^{1/2}$ is not a real number if $a < 0$. We must assume $a \neq 1$ for property (3) to hold. For example, $1^4 = 1^5$, even though $4 \neq 5$, so that property (3) would not hold if $a = 1$.

By property (1), we know that 2^x represents a single unique real number for any real number value of x. For example, the symbol $2^{\sqrt{3}}$ represents a real number. To find a decimal value for the symbol $2^{\sqrt{3}}$, we can use property (4). It can be shown that $1.73 < \sqrt{3} < 1.74$; thus, by property (4),

$$2^{1.73} < 2^{\sqrt{3}} < 2^{1.74}.$$

Both 1.73 and 1.74 are rational numbers, so that $2^{1.73}$ and $2^{1.74}$ could be evaluated (at least in theory) by the methods of Chapter 1. Common logarithms (see the next section) could also be used to evaluate these numbers, as could a computer. By any of these methods, it can be shown that $2^{1.73} \approx 3.317$ and $2^{1.74} \approx 3.340$, and

$$3.317 < 2^{\sqrt{3}} < 3.340,$$

so that to one decimal place, $2^{\sqrt{3}} = 3.3$. If we were to evaluate $2^{1.73}$ and $2^{1.74}$ to more and more decimal places, then we could get better and better approximations to $2^{\sqrt{3}}$. For example, our school calculator gives $2^{\sqrt{3}}$ to six decimal places as $2^{\sqrt{3}} = 3.321997$.

We know that $2^{1.73} < 2^{\sqrt{3}} < 2^{1.74}$. Thus, the point on the graph of $y = 2^x$ for $x = \sqrt{3}$ will be higher than the point for $x = 1.73$ and lower than the point for $x = 1.74$. This means that the graph will be smooth, with no gaps or sudden jumps, and makes plausible the continuity of the function.

The four properties above give meaning to a^x for all real numbers x, where a is a positive real number and $a \neq 1$. It can be shown that with this definition of a^x, all the properties of exponents from Section 1.5 are still valid. In particular, if a and b are positive real numbers, $a \neq 1$, $b \neq 1$, and x and y are any real numbers, then:

$$a^x \cdot a^y = a^{x+y} \qquad (a^x)^y = a^{xy} \qquad a^0 = 1$$

$$\frac{a^x}{a^y} = a^{x-y} \qquad (ab)^x = a^x b^x \qquad a^{-x} = \frac{1}{a^x}.$$

$$\left(\frac{a}{b}\right)^x = \frac{a^x}{b^x}$$

178 *Exponential and Logarithmic Functions*

Now that we have defined a^x for any real number x and positive real number a (with $a \neq 1$), we can define an **exponential function**, $y = a^x$, or $f(x) = a^x$. For example, to graph the exponential function $y = 2^x$ we can make a table of pairs of values of x and y, as shown in Figure 6.1. If we then plot these points and draw a smooth curve through them, we get the graph shown in Figure 6.1. This graph is typical of the graphs of exponential functions of the form $y = a^x$, where $a > 1$.

Figure 6.1

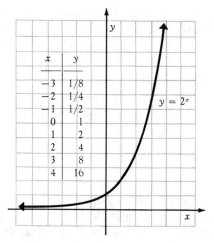

Possibly the single most useful exponential function is the function $y = e^x$, where e is an irrational number that occurs often in practical applications, as we shall see later in this chapter. It is shown by more advanced methods that, to seven decimal places of accuracy, $e = 2.7182818$. To graph the function $y = e^x$, we must find a number of ordered pairs belonging to the graph. Table 4, in the Appendix, gives values of e^x for various values of x. Use the table to verify the ordered pairs shown in the chart. The graph of $y = e^x$ is shown in Figure 6.2.

Figure 6.2

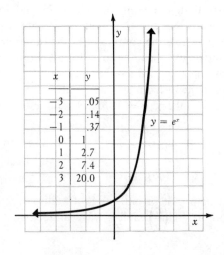

6.1 Exponential Functions

Example 1 Graph $y = 2^{-x} = 1/2^x = (1/2)^x$.

Once again we construct a table of values and draw a smooth curve through the resulting points (see Figure 6.3). This graph is typical of the graphs of exponential functions of the form $y = a^x$, where $0 < a < 1$.

Figure 6.3

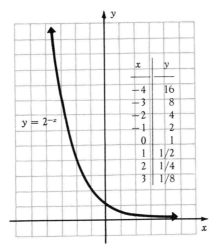

Example 2 Graph $y = 2^{|x|}$.

By completing a table of ordered pairs, we can sketch the graph of $y = 2^{|x|}$, as shown in Figure 6.4.

Figure 6.4

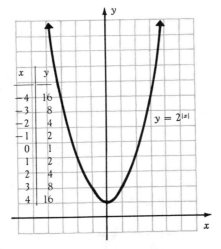

Example 3 Graph $y = e^{-x^2}$.

If we use Table 4 to help plot several pairs of values of x and y, we get the graph shown in Figure 6.5. This graph is very important in the study of

180 *Exponential and Logarithmic Functions*

probability. In fact, this graph, with a slight modification,* is the normal curve, one of the major topics of probability and statistics.

Figure 6.5

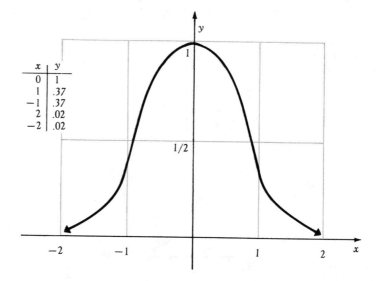

x	y
0	1
1	.37
−1	.37
2	.02
−2	.02

Several practical examples of exponential functions are shown in Figure 6.6.

Figure 6.6a

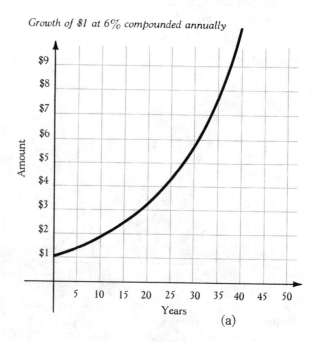

Growth of $1 at 6% compounded annually

Years	Amount
0	1.00
5	1.34
10	1.79
15	2.40
20	3.21
25	4.29
30	5.74
35	7.69
40	10.29
45	13.76
50	18.42

(a)

* The normal curve itself has equation $y = e^{-x^2}/\sqrt{2\pi}$.

Figure 6.6b

Selected year	Consumption
1915	43
1920	68
1925	110
1930	150
1935	170
1940	220
1945	260
1950	340
1955	480
1960	780
1965	1130
1970	1670
1971	1800
1973	2050
1974	1930

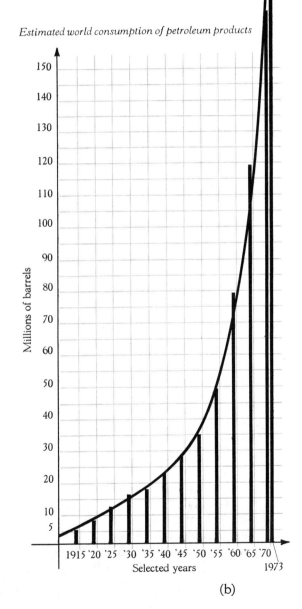

(b)

Figure 6.6c

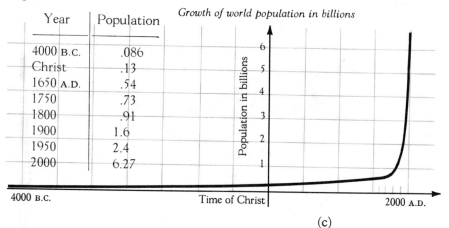

(c)

6.1 Exercises

Graph each of the following exponential functions.

1. $y = 3^x$
2. $y = 4^x$
3. $y = 3^{-x}$
4. $y = 4^{-x}$
5. $y = (1/4)^x$
6. $y = (1/3)^x$
7. $y = (1/3)^{-x}$
8. $y = (1/4)^{-x}$
9. $y = 3^{|x|}$
10. $y = 4^{|x|}$
11. $y = 2^{-|x|}$
12. $y = 3^{-|x|}$
13. $y = 2^{x^2}$
14. $y = 3^{x^2}$
15. $y = 3^{-x^2}$
16. $y = 2^{-x^2}$
17. $y = 2^{x+1}$
18. $y = 3^{x+1}$
19. $y = 2^{1-x}$
20. $y = 3^{1-x}$
21. $y = e^{x+1}$
22. $y = e^{2x}$
23. $y = e^{-x}$
24. $y = 2e^{-x}$

6.2 Logarithmic Functions

Exponential functions can be used as mathematical models in situations involving growth at an increasing rate as we shall see in the next section. Logarithmic functions can be used as mathematical models of growth involving a decreasing rate of growth. Logarithmic functions were originally studied because of their usefulness in arithmetic calculations. However, the growth of the availability of small and large computers has led to less need for logarithms as an aid in numerical calculation. Logarithmic functions are now studied primarily because they provide another function which can be used in making up a mathematical model of a real world situation.

Logarithmic functions can be obtained from exponential functions using the following definition. For $a > 0$, and $a \neq 1$,

$$x = a^y \quad \text{means the same as} \quad y = \log_a x$$

6.2 Logarithmic Functions

Read $y = \log_a x$ as "y is the logarithm of x to the base a." For example since $16 = 2^4$, we can write $4 = \log_2 16$. Also, since $10^3 = 1000$, we have $\log_{10} 1000 = 3$.

Example 4 In this example we show several pairs of equivalent statements. The same statement is written in both exponential and logarithmic forms.

	Exponential Form	Logarithmic Form
(a)	$3^2 = 9$	$\log_3 9 = 2$
(b)	$\left(\dfrac{1}{5}\right)^{-2} = 25$	$\log_{1/5} 25 = -2$
(c)	$10^5 = 100{,}000$	$\log_{10} 100{,}000 = 5$
(d)	$4^{-3} = 1/64$	$\log_4 1/64 = -3$

The exponential function $f(x) = 2^x$ and the logarithmic function $g(x) = \log_2 x$ are closely related. Note that $f(3) = 2^3 = 8$, while $g(8) = \log_2 8 = 3$. Hence, $f(3) = 8$ and $g(8) = 3$. Also, $f(5) = 32$ and $g(32) = 5$. In fact, for any number m, if $f(m) = n$, then $g(n) = m$. Functions related in this way are called **inverses** of each other. Verify that $f(x) = 3x + 4$ and $g(x) = (x - 4)/3$ are also inverses of each other. In general, $f(x) = a^x$ and $g(x) = \log_a x$ are inverses of each other. A general discussion of inverse functions would carry us too far afield; such a discussion can be found in books listed in the bibliography.

By finding a series of points that satisfy $y = \log_2 x$ and another series satisfying $y = \log_{1/2} x$, we get the graphs shown in Figure 6.7. Note that these graphs are the graphs of functions. The graph of Figure 6.7(a) is typical of the graphs of logarithmic functions of base $a > 1$, while the graph of Figure 6.7(b) is typical of the graphs of logarithmic functions of base a, where $0 < a < 1$.

Figure 6.7a

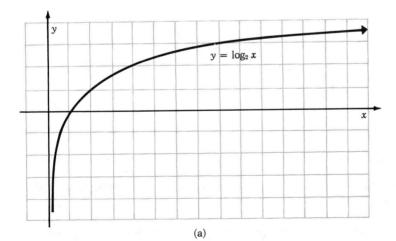

(a)

Figure 6.7b

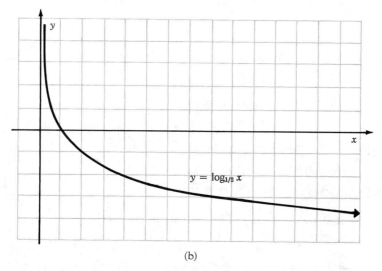

(b)

One of the main reasons for the usefulness of logarithmic functions comes from the properties of logarithms given in the following theorem.

Theorem 6.1 Let x and y be any positive real numbers, and let r be any real number. Let a be a positive real number, $a \neq 1$. Then

(a) $\log_a xy = \log_a x + \log_a y$
(b) $\log_a x/y = \log_a x - \log_a y$
(c) $\log_a x^r = r \log_a x$
(d) $\log_a a = 1$
(e) $\log_a 1 = 0$.

Example 5 Using the results of Theorem 6.1, we have

(a) $\log_6 7 \cdot 9 = \log_6 7 + \log_6 9$
(b) $\log_2 5/7 = \log_2 5 - \log_2 7$
(c) $\log_{10} 5^{3.2} = 3.2 \log_{10} 5$
(d) $\log_8 8 = 1$
(e) $\log_4 1 = 0$.

Historically, the main application of logarithms has been as an aid to numerical calculations. Using the properties above, and tables of logarithms, many numerical problems can be greatly simplified. Since our number system is base 10, logarithms to base 10 are the most convenient for numerical calculations.

Base 10 logarithms are called **common logarithms**. For simplicity, $\log_{10} x$ is abbreviated $\log x$. Using this notation, we have, for example

$$\log 1000 = 3$$
$$\log 100 = 2$$
$$\log 1 = 0$$
$$\log 0.01 = -2$$
$$\log 0.0001 = -4.$$

Other common logarithms can be found from Table 3 in the Appendix. The use of common logarithms to simplify numerical calculations is discussed in the appendix following this chapter. Common logarithms have few applications other than numerical calculations, and especially few applications involving calculus. In most other practical applications of logarithms, the number e is used as the base. (Recall: to seven decimal places, $e = 2.7182818$.) Logarithms to base e are called **natural logarithms**, and written $\ln x$. A table of natural logarithms is given as Table 5 in the Appendix. From this table, for example, we find that

$$\ln 55 = 4.0073$$
$$\ln 1.9 = 0.6419$$
$$\ln 0.4 = 9.0837 - 10,$$

and so on.

While common logarithms may seem more "natural" than logarithms to base e, there are several good reasons for using natural logarithms instead. In fact, many textbooks use "log" as an abbreviation for natural logarithm, with no abbreviation used for common logarithms. In this text, $\log x$ represents the common logarithm of x. A major reason for the importance of e and natural logarithms is given in Section 6.4.

Example 6 The cost in dollars, $C(x)$, of manufacturing x picture frames, where x is measured in thousands, is given by

$$C(x) = 5000 + 2000 \ln (x + 1).$$

(a) Find the fixed cost.
(b) Find the cost of manufacturing 19,000 frames.

We use the cost function given above to answer the questions presented.

(a) The fixed cost represents the cost of setting up the plant, designing the items, and so on. To find the fixed cost, let $x = 0$. This gives

$$C(0) = 5000 + 2000 \ln (0 + 1)$$
$$= 5000 + 2000 \ln 1$$
$$= 5000 + 2000 \cdot 0 \quad (\text{Recall: } \ln 1 = 0)$$
$$= 5000.$$

The fixed cost is $5000.

(b) To find the cost of producing 19,000 frames, let $x = 19$. This gives

$$C(19) = 5000 + 2000 \ln (19 + 1)$$
$$= 5000 + 2000(2.9957)$$

(From Table 5, ln 20 = 2.9957.)

$$= 5000 + 5991$$
$$= 10{,}991.$$

Thus, 19,000 frames cost a total of $10,991, or about $11,000.

Table 5 gives the natural logarithms of many numbers, but often it is necessary to find natural logarithms of numbers that are not found in this table. To do this, the following theorem may be used to convert common logarithms to natural logarithms.

Theorem 6.2 For any positive number x,

$$\ln x \approx 2.3026 \log x,$$

where $\log x$ is the common logarithm of x.*

Example 7 Find (a) ln 83; (b) ln 6000.

By Theorem 6.2 above, $\ln x \approx 2.3026 \log x$. Using Table 3, common logarithms, we have

(a) $\ln 83 \approx 2.3026 \log 83$
    $\approx 2.3026(1.9191)$
    ≈ 4.4189

(b) $\ln 6000 \approx 2.3026(3.7782)$
    $\approx 8.6997.$

Some practical examples of logarithm functions are shown in Figure 6.8.

Figure 6.8a

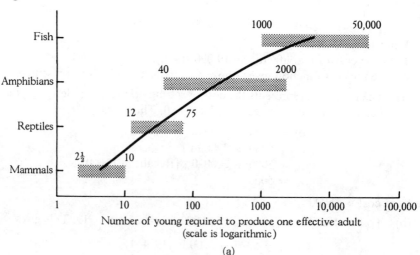

(a)

* $2.3026 \approx 1/\log e.$

Figure 6.8b

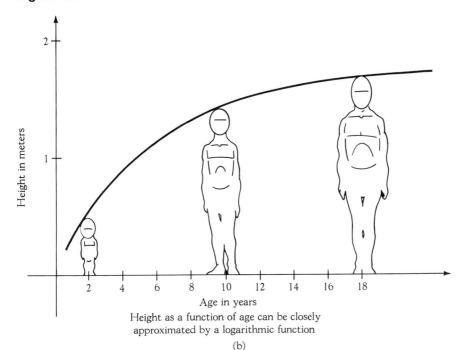

Height as a function of age can be closely approximated by a logarithmic function

(b)

6.2 Exercises

Express each of the following using logarithms.

1. $2^3 = 8$
2. $5^2 = 25$
3. $3^4 = 81$
4. $6^3 = 216$
5. $(1/3)^{-2} = 9$
6. $(3/4)^{-2} = 16/9$

Write each of the following using exponents.

7. $\log_2 8 = 3$
8. $\log_3 27 = 3$
9. $\log_{10} 100 = 2$
10. $\log_2 1/8 = -3$
11. $\log 100{,}000 = 5$
12. $\ln e^5 = 5$

Evaluate each of the following.

13. $\log_{10} 10{,}000$
14. $\log_9 81$
15. $\log_4 64$
16. $\log_2 1/4$
17. $\log 0.01$
18. $\log 0.00001$

Write each of the following as a single logarithm, or as a single number if possible.

19. $\log_4 8 + \log_4 2$
20. $\log_9 3 + \log_9 27$
21. $\log_7 15 - \log_7 11$
22. $\log_8 21 - \log_8 40$

188 *Exponential and Logarithmic Functions*

23. $0.2(\log_{10} 30)$
24. $-0.8(\log_5 12)$
25. $2 \ln e + \ln 1/e$
26. $-\ln e^{-1} - \ln e$
27. $0.3(\ln 5) - 0.4(\ln 6)$
28. $-0.2(\ln 12) + 0.5(\ln 8)$

Find each of the following natural logarithms.

29. ln 20
30. ln 35
31. ln 60
32. ln 50
33. ln 800
34. ln 920
35. ln 532
36. ln 255
37. ln 768
38. ln 324
39. ln 58,500
40. ln 12,400

Complete the following tables of ordered pairs and then graph the given functions.

41. $y = \log_3 x$

x	27	9	3	1	1/3	1/9	1/27
y							

42. $y = \log_4 x$

x	64	16	4	1	1/4	1/16	1/64
y							

43. $y = \log_{10} x$

x	100	10	1	1/10	1/100	1/1000
y						

44. $y = \ln x$ (Hint: use Table 5.)

x	100	10	e	1	0.1
y					

45. $y = \log_{1/4} x$

x	64	16	4	1	1/4	1/16	1/64
y							

46. $y = \log_{1/3} x$

x	27	9	3	1	1/3	1/9	1/27
y							

47. A company finds that its total sales, $T(x)$, from the distribution of x catalogs, measured in thousands, is approximated by

$$T(x) = 5000 \ln (x + 1).$$

Find the total sales resulting from the distribution of
(a) 0 catalogs
(b) 5000 catalogs
(c) 24,000 catalogs
(d) 49,000 catalogs.

48. The population of an animal species that is introduced in a certain area may grow rapidly at first, and then grow more slowly as time goes on. A logarithmic function can provide an excellent description of such growth. Suppose that the population of foxes, $F(t)$, in an area t months after the species is introduced in the area is given by

$$F(t) = 500 \ln (2t + 2).$$

Find the population of foxes
(a) When they are first released into the area (that is, when $t = 0$)
(b) After 3 months
(c) After 7 months
(d) After 10 months.

6.3 Applications of Exponential and Logarithmic Functions

In this section we shall discuss several examples which illustrate applications of exponential and logarithmic functions. One of the most common applications of exponential functions depends on the fact that in many situations involving growth or decay of a population, the amount or number present at time t can be closely approximated by a function of the form

$$y = y_0 e^{kt},$$

where y_0 is the amount or number present at time $t = 0$, and k is a constant. (The reason for using this particular function is given in Section 8.1.)

For example, suppose that the population, $P(t)$, of a certain city is given by

$$P(t) = 10{,}000 e^{0.04t},$$

where t represents time measured in years. The population at time $t = 0$ is given by

$$\begin{aligned} P(0) &= 10{,}000 e^{(0.04) \cdot 0} \\ &= 10{,}000 e^0 \\ &= 10{,}000 \cdot 1 \quad (\text{Recall: } e^0 = 1) \\ &= 10{,}000. \end{aligned}$$

Thus, the population of the city is 10,000 at time $t = 0$.

The number present initially is often expressed with the subscript 0. For example, here we could express the fact that the population is 10,000 at time $t = 0$ by saying that $P_0 = 10{,}000$. The population of the city in year $t = 5$ is given by

$$\begin{aligned} P(5) &= 10{,}000 e^{0.04(5)} \\ &= 10{,}000 e^{0.2}. \end{aligned}$$

The number $e^{0.2}$ can be found in Table 4. From this table, $e^{0.2} = 1.22140$, so that
$$P(5) = 10,000(1.22140) = 12214.$$
Thus, in 5 years the population of the city will be about 12,200.

Example 8 Sales of a new product often grow rapidly at first, and then begin to level off with time. For example, suppose that the sales, $S(x)$, in some appropriate units, of a new model typewriter are approximated by the exponential function
$$S(x) = 1000 - 800e^{-x},$$
where x represents the number of years that the typewriter has been on the market. Calculate S_0, $S(1)$, $S(2)$, and $S(4)$. Graph $y = S(x)$.

Verify that $S_0 = S(0) = 1000 - 800 \cdot 1 = 200$. Using Table 4, we have
$$\begin{aligned} S(1) &= 1000 - 800e^{-1} \\ &= 1000 - 800(0.36787) \\ &= 1000 - 294 \\ &= 706. \end{aligned}$$

In the same way, verify that $S(2) = 892$, and $S(4) = 986$. By plotting several such points we get the graph shown in Figure 6.9. Note that the line $y = 1000$ is an asymptote for the graph. This shows that in this case sales will tend to level off with time, and gradually approach a level of 1000 units.

Figure 6.9

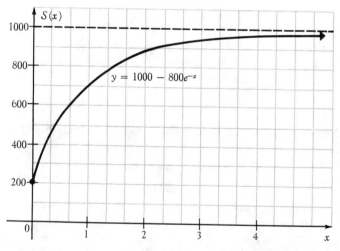

Example 9 For a variety of reasons, assembly line operations tend to have a high turnover of employees. Thus, the companies involved must spend much time and money in training new workers. It has been found that workers new to the operation of a certain machine (or a certain task on the assembly line) will produce items according to the function
$$P(x) = 25 - 25e^{-0.3x},$$

6.3 Applications of Exponential and Logarithmic Functions

where $P(x)$ items are produced on day x. Using this function, how many items will be produced by a new worker on day 8?

We must evaluate $P(8)$:

$$\begin{aligned} P(8) &= 25 - 25e^{-0.3(8)} \\ &= 25 - 25e^{-2.4} \\ &= 25 - 25(0.09071) \\ &= 25 - 2.3 \\ &= 22.7. \end{aligned}$$

Thus, on the eighth day, the worker can be expected to produce about 23 items. By plotting several such points, we can obtain the graph of $y = P(x)$, as shown in Figure 6.10. Such a graph is called a *learning curve*. According to the graph, a new worker tends to learn quickly at first, and then learning tapers off and approaches some upper limit. This is characteristic of the learning of certain skills involving the repetitive performance of the same task.

Figure 6.10

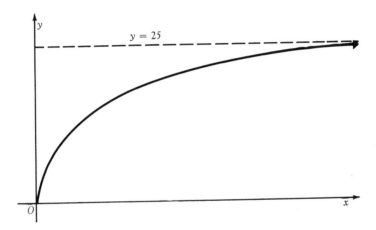

Carbon-14 is a radioactive isotope of carbon which has a half-life of about 5600 years (that is, of a given quantity of carbon-14 about half would decay in 5600 years). The Earth's atmosphere contains much carbon, mostly in the form of carbon dioxide gas, with small traces of carbon-14. Most atmospheric carbon is the nonradioactive isotope carbon-12. The ratio of carbon-14 to carbon-12 is virtually constant in the atmosphere. However, as a plant absorbs carbon dioxide from the air in the process of photosynthesis, the carbon-12 stays in the plant while the carbon-14 is converted into nitrogen. Thus, in a plant, the ratio of carbon-14 to carbon-12 is smaller than it is in the atmosphere. Even when the plant is eaten by an animal, this ratio will continue to decrease. Based on these facts, a method of dating objects, called *carbon-14 dating* has been developed. This method is explained in the following examples.

Example 10 Suppose an Egyptian mummy has been discovered in which the ratio of carbon-14 to carbon-12 is only about half the ratio found in the atmosphere. How long ago did the individual die whose body was mummified?

To solve this, note that in 5600 years half the carbon-14 in a specimen will decay. Thus, the individual died about 5600 years ago. (If the ratio in an object is 1/4 the atmospheric ratio, then the object is about $2 \cdot 5600 = 11{,}200$ years old.)

Example 11 Let R be the (nearly constant) ratio of carbon-14 to carbon-12 found in the atmosphere, and let r be the ratio of carbon-14 to carbon-12 as found in an observed specimen. It can then be shown that the relationship between R and r is given by

$$R = r \cdot e^{(t \ln 2)/5600},$$

where t is the age of the specimen in years. Verify the formula for $t = 0$.

To do this, substitute 0 for t in the formula. This gives

$$R = r \cdot e^{(0 \cdot \ln 2)/5600}$$
$$= r \cdot e^0$$
$$= r \cdot 1$$
$$= r.$$

This result is correct—when $t = 0$ the specimen has just died, so that R and r should be the same.

Example 12 Suppose a specimen has been found in which $r = \dfrac{2}{3} R$. Estimate the age of the specimen.

Here we use the formula given in Example 11 above.

$$R = r \cdot e^{(t \ln 2)/5600}$$
$$= \frac{2}{3} R \cdot e^{(t \ln 2)/5600}.$$

Dividing through by R, and multiplying through by 3/2, we have

$$\frac{3}{2} = e^{(t \ln 2)/5600}.$$

Taking natural logarithms of both sides of this last result gives

$$\ln \frac{3}{2} = \ln e^{(t \ln 2)/5600}.$$

Using properties of logarithms, we have

$$\ln \frac{3}{2} = \left[\frac{t \ln 2}{5600} \right] \ln e.$$

Since $\ln e = 1$, we have

$$\ln \frac{3}{2} = \frac{t \ln 2}{5600}.$$

6.3 Applications of Exponential and Logarithmic Functions

To solve this equation for t, the age of the specimen, multiply both sides by $5600/\ln 2$. This gives

$$\frac{5600 \ln (3/2)}{\ln 2} = t.$$

Using Table 5, we have $\ln (3/2) = \ln 1.5 = 0.4055$ and $\ln 2 = 0.6931$. Thus,

$$t = \frac{5600(0.4055)}{0.6931} = 3276.$$

The specimen is about 3276 years old.

6.3 Exercises

Suppose that the population, $P(t)$, of a city is given by

$$P(t) = 1{,}000{,}000 e^{0.02t},$$

where t represents time measured in years. Find each of the following values.

1. P_0
2. $P(2)$
3. $P(4)$
4. $P(10)$

Suppose the quantity, $Q(t)$, measured in grams, of a radioactive substance present at time t is given by

$$Q(t) = 500 e^{-0.05t},$$

where t is measured in days. Find the quantity present at each of the following times.

5. $t = 0$
6. $t = 4$
7. $t = 8$
8. $t = 20$

Let the number of bacteria, $B(t)$, present in a certain culture be given by

$$B(t) = 25{,}000 e^{0.2t},$$

where t is measured in hours, and $t = 0$ corresponds to noon. Find the number of bacteria present at

9. noon
10. 1 PM
11. 2 PM
12. 5 PM

When a bactericide is introduced into a certain culture, the number of bacteria present, $D(t)$, is given by

$$D(t) = 50{,}000 e^{-0.01t},$$

where t is time measured in hours. Find the number of bacteria present at time

13. $t = 0$
14. $t = 5$
15. $t = 20$
16. $t = 50$

Exponential and Logarithmic Functions

Sales of a new model can opener are approximated by

$$S(x) = 5000 - 4000e^{-x},$$

where x represents the number of years that the can opener has been on the market, and $S(x)$ represents the sales in thousands. Find each of the following.

17. $S(0)$
18. $S(1)$
19. $S(2)$
20. $S(5)$
21. $S(10)$
22. Graph $y = S(x)$

Assume that a person new to an assembly line will produce

$$P(x) = 500 - 500e^{-x}$$

items per day, where x is measured in days. Find

23. $P(0)$
24. $P(1)$
25. $P(2)$
26. $P(5)$
27. $P(10)$
28. Graph $y = P(x)$.

The number of years, $N(r)$, since two independently evolving languages split off from a common ancestral language is approximated by

$$N(r) = -5000 \ln r,$$

where r is the proportion of the words from the ancestral language that are common to both languages now.

29. Find $N(0.9)$
30. Find $N(0.5)$
31. Find $N(0.3)$
32. How many years have elapsed since the split if 80 percent of the words of the ancestral language are common to both languages today?

In Example 11, find the age of a specimen in which

33. $r = 0.8R$
34. $r = 0.4R$
35. $r = 0.1R$
36. $r = 0.01R$

Under certain conditions, the total number of facts of a certain kind that are remembered is approximated by

$$N(t) = 1000 \left(\frac{1 + e}{1 + e^{t+1}} \right)$$

where $N(t)$ is the number of facts remembered at time t measured in days.

37. Calculate $N(0)$.
38. Calculate $N(1)$.
39. Calculate $N(5)$.
40. Graph $y = N(t)$. This graph is called a *forgetting curve*.

A sociologist has shown that the fraction, $y(t)$, of a population who have heard a rumor after t days, is approximated by

$$y(t) = \frac{y_0 e^{kt}}{1 - y_0(1 - e^{kt})}$$

6.4 Derivatives of Exponential and Logarithmic Functions

where y_0 is the fraction of people who heard the rumor at time $t = 0$, and k is a constant. A graph of $y(t)$ for a particular value of k is shown in the figure.

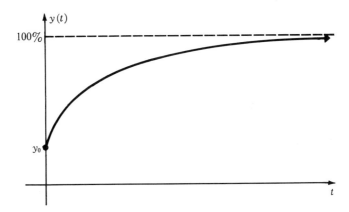

41. If $k = 0.1$, and $y_0 = 0.05$, find $y(10)$.
42. If $k = 0.2$, and $y_0 = 0.10$, find $y(5)$.

(A hand calculator will be helpful for the following problem.) Helms* says that one measure of the evaporation, W, of water from Ponderosa pines is approximated by

$$W \approx 4.6e^z,$$

where $z = \dfrac{17.3C}{C + 237}$. C is the temperature of the air surrounding the tree, in degrees Centigrade. Estimate W, using Table 4, if

43. $C = 15°$ 44. $C = 20°$
45. $C = 22°$ 46. $C = 25°$

6.4 Derivatives of Exponential and Logarithmic Functions

We shall complete our discussion of exponential and logarithmic functions for this chapter by discussing methods of finding derivatives of these functions.

Almost all our work with exponential functions has involved the number e and most results for logarithmic functions have been given for natural logarithms, which are logarithms to base e. We first mentioned the number e in Section 6.1; at that time we promised that the main reason for the importance of the number e would become clear when this section was reached. This reason can now be given: the exponential function with e as base, $f(x) = e^x$, has the simplest derivative of any exponential function. In fact, this function is its own derivative. Moreover, e^x is the only useful function which is its own derivative. In summary,

* Helms, John A., Environmental Control of Net Photosynthesis in Naturally Growing *Pinus Ponderosa* Nets, *Ecology*, Winter, 1972, page 92.

Theorem 6.3 If $f(x) = e^x$, then $f'(x) = e^x$.

The proof of this theorem, which is somewhat complex, can be found in any of the calculus books listed in the bibliography. If some other base were to be used for the exponential function, the derivative would be more complicated. For example, if $f(x) = 2^x$, it can be shown that $f'(x) = (2^x)(\ln 2)$, which is not as simple as the derivative of $f(x) = e^x$.

Example 13 If $f(x) = 3e^{-4x}$, find $f'(x)$.

We can use the chain rule, together with Theorem 6.3, to find $f'(x)$. Since $f(x) = 3e^{-4x}$, let $f(u) = 3e^u$, where $u = g(x) = -4x$. Therefore $f'(u) = 3e^u$, and $g'(x) = -4$, so that

$$f'(x) = (3e^{-4x})(-4) = -12e^{-4x}.$$

In general, if $f(x) = ae^{kx}$, for any real numbers a and k, then

$$f'(x) = ake^{kx}.$$

Example 14 Let $f(x) = \dfrac{100{,}000}{1 + 100e^{-0.3x}}$. Find $f'(x)$.

If we use the quotient rule, we have

$$f'(x) = \frac{(1 + 100e^{-0.3x})(0) - 100{,}000(-30e^{-0.3x})}{(1 + 100e^{-0.3x})^2}$$

$$= \frac{3{,}000{,}000e^{-0.3x}}{(1 + 100e^{-0.3x})^2}.$$

Example 15 Often a population, or the sales of a certain product, will start growing slowly, then grow more rapidly, and then gradually level off. Such growth can often be approximated by a function of the form

$$f(x) = \frac{b}{1 + ae^{kx}}$$

for appropriate constants a, b and k. For example, suppose that population planners predict that the population, $P(x)$, of a certain city will be approximated for the next few years by the function

$$P(x) = \frac{100{,}000}{1 + 100e^{-0.3x}},$$

where x is time in years. Find the rate of change of the population at time $x = 4$.

We use $f'(x)$ from the previous example.

$$f'(4) = \frac{3{,}000{,}000e^{-0.3(4)}}{[1 + 100e^{-0.3(4)}]^2}$$

$$= \frac{3{,}000{,}000e^{-1.2}}{(1 + 100e^{-1.2})^2}.$$

6.4 Derivatives of Exponential and Logarithmic Functions

From Table 4, we see that $e^{-1.2} = 0.301$. Using this value, we have

$$f'(4) = \frac{3{,}000{,}000(0.301)}{[1 + 100(0.301)]^2}$$

$$= \frac{903{,}000}{(1 + 30.1)^2}$$

$$= \frac{903{,}000}{967}$$

$$= 934.$$

The rate of change of the population at time $x = 4$ is an increase of 934 persons per year.

The next theorem shows how to find the derivative of the natural logarithm function.

Theorem 6.4 If $f(x) = \ln x$, then $f'(x) = 1/x$.

Example 16 Let $f(x) = \ln(3x^2 - 4x)$ and find $f'(x)$.
We have, by the chain rule,

$$f'(x) = \frac{1}{(3x^2 - 4x)}(6x - 4)$$

$$= \frac{6x - 4}{3x^2 - 4x}.$$

Example 17 Find $f'(x)$ if $f(x) = e^x \ln x$.
Using the product rule, we have

$$f'(x) = e^x(1/x) + (\ln x)(e^x)$$
$$= e^x(1/x + \ln x).$$

6.4 Exercises

Find the derivative of each of the following functions.

1. $f(x) = e^{4x}$
2. $f(x) = e^{-2x}$
3. $f(x) = -6e^{-2x}$
4. $f(x) = 8e^{4x}$
5. $f(x) = -8e^{2x}$
6. $f(x) = 0.2e^{5x}$
7. $f(x) = -16e^{x+1}$
8. $f(x) = -4e^{-0.1x}$
9. $f(x) = e^{x^2}$
10. $f(x) = e^{-x^2}$
11. $f(x) = 4e^{2x^2-4}$
12. $f(x) = -3e^{3x^2+5}$
13. $f(x) = xe^x$
14. $f(x) = x^2 e^{-2x}$
15. $f(x) = (x - 3)^2 e^{2x}$
16. $f(x) = (3x^2 - 4x)e^{-3x}$
17. $f(x) = \ln(2x^2 - 4x)$
18. $f(x) = \ln(-8x^2 + 6x)$
19. $f(x) = \ln e^{x^2}$
20. $f(x) = \ln e^{-3x}$
21. $f(x) = \ln \sqrt{x + 5}$
22. $f(x) = \ln \sqrt{2x + 1}$

23. $f(x) = \dfrac{e^x}{\ln x}$

24. $f(x) = \dfrac{e^x - 1}{\ln x}$

25. $f(x) = \dfrac{e^x + e^{-x}}{x}$

26. $f(x) = \dfrac{e^x - e^{-x}}{x}$

27. $f(x) = \dfrac{5000}{1 + 10e^{0.4x}}$

28. $f(x) = \dfrac{600}{1 - 50e^{0.2x}}$

29. $f(x) = \dfrac{10{,}000}{8 + 4e^{-0.2x}}$

30. $f(x) = \dfrac{500}{10 + 5e^{-0.5x}}$

31. Suppose that the population, $P(t)$, of a certain collection of rare Brazilian ants is given by
$$P(t) = 1000e^{0.2t},$$
where t represents the time in days. Find the rate of change of the population when $t = 2$; when $t = 8$. (Hint: Use Table 4 of the Appendix.)

32. Often, sales of a new product grow rapidly, and then level off with time. Such a situation can be represented by an equation of the form
$$S(t) = 100 - 90e^{-0.3t},$$
where t represents time in years, and $S(t)$ represents sales. Find the rate of change of sales when $t = 1$; when $t = 10$.

33. Suppose $P(x) = e^{-0.02x}$ represents the proportion of shoes manufactured by a given company that are still wearable after x days of use. Find the proportion of shoes wearable after
   (a) 1 day
   (b) 10 days
   (c) 100 days
   (d) Calculate and interpret $P'(100)$.

34. Consider an experiment in which equal numbers of male and female insects of a certain species are permitted to intermingle in a controlled experiment. Assume $M(t)$, given by
$$M(t) = (e^{0.1t} + 1)[\ln \sqrt{1000(t + 15)}],$$
represents the number of matings observed among the insects in an hour, where t is the temperature in degrees Centigrade. (Note: the formula is an approximation at best, and holds only for specific temperature intervals.)
   (a) Find $M(15)$.
   (b) Find $M(25)$.
   (c) Find the rate of change of $M(t)$ when $t = 15$.

35. The concentration, $P(x)$, of pollutants in grams per liter in the east fork of the Big Weasel River is approximated by
$$P(x) = 0.04e^{-4x},$$
where x is the number of miles downstream from a paper mill that the measurement is taken. Find
   (a) $P(0.5)$
   (b) $P(1)$
   (c) $P(2)$.

Find the rate of change of the concentration with respect to distance at
(d) $x = 0.5$
(e) $x = 1$
(f) $x = 2$.

Appendix Common Logarithms

Common logarithms are logarithms to base 10. In this appendix we shall discuss the use of common logarithms as an aid in numerical calculations. Recall from Section 6.2 that the common logarithm of a number x is written $\log x$. Also recall from Section 6.2 that we can use the definition of logarithm to write

$$\log 1000 = 3$$
$$\log 100 = 2$$
$$\log 10 = 1$$
$$\log 1 = 0$$
$$\log 0.1 = -1$$
$$\log 0.01 = -2,$$

and so on. In order to find the common logarithm of a number that is not a power of 10, special logarithm tables must be used. An excerpt from such a table is included below; for the complete table, see Table 3 in the Appendix at the back of the book. Let us use the excerpt printed below to find $\log 4230$. We can write 4230 as 4.23×1000, so that

$$\log 4230 = \log (4.23 \times 1000).$$

Using Theorem 6.1(a), we can write

$$\log 4230 = \log 4.23 + \log 1000.$$

Since $\log 1000 = 3$, we have

$$\log 4230 = \log 4.23 + 3.$$

We can find $\log 4.23$ from the excerpt of the logarithm table below. To find $\log 4.23$ in the table, find 4.2 down the left column, and find 3 across the top row. This leads to the table entry 0.6263, so that $\log 4.23 = 0.6263$. We can now finish finding the common logarithm of 4230:

$$\log 4230 = \log 4.23 + 3$$
$$= 0.6263 + 3$$
$$= 3.6263.$$

x	0	1	2	3	4	5	6	7	8	9
4.0	0.6021	0.6031	0.6042	0.6053	0.6064	0.6075	0.6085	0.6096	0.6107	0.6117
4.1	0.6128	0.6138	0.6149	0.6160	0.6170	0.6180	0.6191	0.6201	0.6212	0.6222
4.2	0.6232	0.6243	0.6253	0.6263	0.6274	0.6284	0.6294	0.6304	0.6314	0.6325
4.3	0.6335	0.6345	0.6355	0.6365	0.6375	0.6385	0.6395	0.6405	0.6415	0.6425
4.4	0.6435	0.6444	0.6454	0.6464	0.6474	0.6484	0.6493	0.6503	0.6513	0.6522

200 Exponential and Logarithmic Functions

The integer part of the logarithm (3 in the example above) is called the **characteristic**, while the decimal part (0.6263 in the example) is called the **mantissa**. The mantissa identifies the digits, while the characteristic is used to locate the decimal point.

Example 1 Find each of the following common logarithms.

(a) log 418,000.

Since $418{,}000 = 4.18 \times 100{,}000$, we have

$$\begin{aligned}\log 418{,}000 &= \log(4.18 \times 100{,}000)\\ &= \log 4.18 + \log 100{,}000\\ &= 0.6212 + 5\\ &= 5.6212.\end{aligned}$$

(b) $\log 4.47 = 0.6503$.

Here the characteristic is 0.

(c) log 0.000435.

Note that $0.000435 = 4.35 \times 0.0001$. Thus,

$$\begin{aligned}\log 0.000435 &= \log 4.35 + \log 0.0001\\ &= 0.6385 + (-4).\end{aligned}$$

The characteristic here is -4. It is customary, and useful, to write the characteristic as $-4 = 6 - 10$, so that log 0.000435 becomes

$$\log 0.000435 = 6.6385 - 10.$$

Example 2 Find the numbers having each of the following logarithms.

(a) 4.6503.

As shown in the logarithm table, $0.6503 = \log 4.47$. We also know that $4 = \log 10{,}000$. Thus,

$$\begin{aligned}4.6503 &= 4 + 0.6503\\ &= \log 10{,}000 + \log 4.47\\ &= \log[(10{,}000)(4.47)]\\ &= \log 44{,}700.\end{aligned}$$

(b) $8.6096 - 10$.

We can rewrite $8.6096 - 10$ as $-2 + 0.6096$, since $8 - 10 = -2$. This gives

$$\begin{aligned}-2 + 0.6096 &= \log 0.01 + \log 4.07\\ &= \log[(0.01)(4.07)]\\ &= \log 0.0407.\end{aligned}$$

We can now use properties of logarithms from Section 6.1 and Table 3 to simplify numerical calculations. We know that $\log xy = \log x + \log y$, and $\log x/y = \log x - \log y$. Therefore, we can evaluate

$$\frac{(4350)(5.86)}{28.3}$$

by writing

$$\log\left[\frac{(4350)(5.86)}{28.3}\right] = \log(4350)(5.86) - \log 28.3$$
$$= \log 4350 + \log 5.86 - \log 28.3.$$

From Table 3, we find $\log 4350 = 3.6385$, $\log 5.86 = 0.7679$, and $\log 28.3 = 1.4518$. Thus,

$$\log 4350 + \log 5.86 - \log 28.3 = 3.6385 + 0.7679 - 1.4518$$
$$= 2.9546.$$

From Table 3, we see that 9.01 is the number whose mantissa is closest to 0.9546. We thus have

$$2.9546 = 2 + 0.9546$$
$$= \log 100 + \log 9.01$$
$$= \log 901.$$

Therefore,

$$\frac{(4350)(5.86)}{28.3} \approx 901.$$

Example 3 Use logarithms to evaluate $21^{3.4}$.

Using Theorem 6.1(c), we have

$$\log 21^{3.4} = 3.4 \log 21.$$

From Table 3, $\log 21 = 1.3222$. Thus,

$$\log 21^{3.4} = 3.4 \log 21$$
$$= 3.4 (1.3222)$$
$$= 4.4955.$$

Again from Table 3, we see that

$$21^{3.4} \approx 31{,}300.$$

Example 4 The **scrap value** of a machine is the value of the machine at the end of its useful life. By one method of calculating scrap value, where it is assumed a constant percentage of value is lost annually, the scrap value, S, is given by

$$S = C(1 - r)^n,$$

where C is the original cost, n is the useful life of the machine in years, and r is the constant percentage of value lost. Find the scrap value of a machine costing $30,000, having a useful life of 12 years, and a constant annual rate of value lost of 15%.

Here $C = \$30{,}000$, $n = 12$, and $r = 15\% = 0.15$. The scrap value is given by

$$S = C(1 - r)^n$$
$$= 30{,}000(1 - 0.15)^{12}$$
$$= 30{,}000(0.85)^{12}.$$

We can use common logarithms to evaluate this last quantity:

$$\log 30{,}000(0.85)^{12} = \log 30{,}000 + \log (0.85)^{12}$$
$$= \log 30{,}000 + 12 \log 0.85.$$

The evaluation of $12 \log 0.85$ can be performed as follows:

$$\log 0.85 = 9.9294 - 10$$
$$\underline{ \times 12}$$
$$119.1528 - 120.$$

Rewrite $119.1528 - 120$ as $9.1528 - 10$. We then have

$$\log 30{,}000 + 12 \log 0.85 = 4.4771 + (9.1528 - 10)$$
$$= 13.6299 - 10,$$

or simply 3.6299. From Table 3, we finally have

$$S = 30{,}000(0.85)^{12} \approx \$4260.$$

The scrap value of the machine is about \$4260.

Appendix Exercises

Find each of the following common logarithms.

1. $\log 749$
2. $\log 8560$
3. $\log 9.71$
4. $\log 86.4$
5. $\log 7.13$
6. $\log 912$
7. $\log 0.810$
8. $\log 0.0712$
9. $\log 0.00348$
10. $\log 0.0000426$

Find the numbers which have each of the following common logarithms.

11. 0.3502
12. 0.6675
13. 2.9484
14. 3.8470
15. 1.4886
16. 1.1847
17. $8.9355 - 10$
18. $7.7210 - 10$
19. $6.9294 - 10$
20. $19.4771 - 20$

Use logarithms to evaluate each of the following.

21. $\dfrac{(71.4)(23.6)}{48.2}$
22. $\dfrac{(7150)(9830)}{4280}$
23. $12.8^{2.2}$
24. $3.67^{2.7}$
25. $\sqrt{34500}$ (Hint: $\sqrt{a} = a^{1/2}$)
26. $\sqrt{23.7}$
27. $\sqrt[3]{689}$
28. $\sqrt[3]{1.93}$
29. $67.8^{3/2}$
30. $26700^{5/4}$

Use the formula for scrap value presented in Example 4 and find the scrap value for each of the following machines.

31. Original cost, $54,000; life, 8 years; annual rate of value loss, 12%.
32. Original cost, $178,000; life, 11 years; annual rate of value loss, 14%.

When fed a certain nutrient, the number of fish, $F(t)$, present in a tank at time t, measured in days, is about

$$F(t) = 4750t^{1.3}.$$

Find the number of fish present at time

33. $t = 8$
34. $t = 12$
35. $t = 30$

A common problem in archaeology is to determine estimates of populations. Several methods have been used to calculate the number of people who occupied a site. One method relates the total surface area of a site to the number of occupants. If P represents the population of a site which covers an area of a square units, then

$$\log P = k \log a,$$

where k is an appropriate constant which varies for hilly, coastal, or desert environments, or for sites with single family dwellings or multiple family dwellings.*

For example, for a type of site where $k = 0.8$, let us find the estimated population of a site which covers 1050 square meters. In the formula, $k = 0.8$ and $a = 1050$. We have

$$\log P = 0.8 \log 1050$$
$$= (0.8)(3.0212)$$
$$= 2.4170.$$

From the table, we find that $P = 261.19$, so that the population of the site was about 260 persons.

Find the populations of sites with the following areas using $k = 0.8$.

36. $a = 230$
37. $a = 95$
38. $a = 20,000$

● Case 8 Compound Interest

Compound interest is interest paid on interest. For example, suppose $1000 is deposited in a bank paying 5% interest compounded annually. At the end of the first year that the money is on deposit, the bank calculates interest: Interest = Principal × Rate × Time, or

$$\text{Interest} = \$1000 \times 0.05 \times 1$$
$$= \$50.$$

This interest of $50 will be deposited to the account, giving a total of $1000 + $50 = $1050 on deposit after one year. Interest for the second year is calculated

* Cook, S. F., 1950, "The Quantitative Investigation of Indian Mounds." Berkeley: University of California Publications in American Archaeology and Ethnology, 40: 231–233.

on the total of $1050 on deposit:

$$\text{Interest} = \$1050 \times 0.05 \times 1$$
$$= \$52.50.$$

At the end of the second year, the account will contain $1050 + \$52.50 = \$1102.50$. This process can be continued indefinitely, for as many years as the money is on deposit.

In general, if P dollars is deposited into an account paying interest at the rate r compounded annually, then the total amount on deposit after n years, A, is given by

$$A = P(1 + r)^n.$$

For example, if $1000 is deposited for 20 years in an account paying 5% interest compounded annually, the total amount on deposit will be given by

$$A = 1000(1 + 0.05)^{20}$$
$$= 1000(1.05)^{20}.$$

The number $(1.05)^{20}$ can be found on a hand calculator, or by the use of common logarithms. If much work must be done with compound interest, it would be a good idea to buy a book of tables that give such numbers. A good source for these tables is the CRC Publishing Company's *Standard Mathematical Tables*, any edition. This company seems to have a perpetual sale on this book of tables, so that you should be able to pick up a copy for $5 or so. (Tell them that Marge and Chuck sent you.) From any of these sources, you will find that to five decimal places, $(1.05)^{20} \approx 2.65330$, so that the amount on deposit will be given by

$$A = 1000(1.05)^{20}$$
$$\approx 1000(2.65330)$$
$$= \$2653.30.$$

Because of competitive pressures, most banks now compound interest more often than annually. In this case, it is necessary to use the following generalization of the formula given above:

$$A = P\left(1 + \frac{r}{f}\right)^{nf},$$

where P is the amount of principal deposited initially, r is the annual rate of interest, f is the number of times that interest is compounded annually (for semi-annual compounding, $f = 2$ for example), n is the number of years that the money is left on deposit, and A is the final amount (principal and interest) on deposit.

For example, suppose $1000 is deposited for 20 years in a bank paying 5% interest compounded quarterly. Then we have $P = \$1000$, $r = 0.05$, $n = 20$, and $f = 4$. Using the second formula from above, we have

$$A = 1000\left(1 + \frac{0.05}{4}\right)^{20(4)}$$
$$= 1000(1 + 0.0125)^{80}$$
$$= 1000(1.0125)^{80}.$$

Case 8 Compound Interest

From a table, we find that $(1.0125)^{80} \approx 2.70148$, so that

$$A \approx 1000(2.70148)$$
$$\approx \$2701.48.$$

By performing similar calculations for other periods of compounding, we get the results shown in the table below.

Deposit of $1000 at 5% per year for 20 years

Frequency of Compounding	Amount on Deposit After 20 Years
annual	$2653.30
semi-annual	$2685.06
quarterly	$2701.48
monthly	$2712.65
daily	$2718.09

As the frequency of compounding increases, the final amount on deposit increases, but at a decreasing rate. In fact, even if interest were to be compounded *every instant*, the final amount on deposit would only be 19¢ more than the amount for daily compounding shown in the chart above. In fact, it can be shown that the final amount on deposit after n years at a rate of interest r compounded every instant is given by

$$A = Pe^{nr},$$

which shows yet another application of the number e. In the example considered in the table above, we have

$$A = Pe^{nr}$$
$$= 1000e^{20(0.05)}$$
$$= 1000e^{1}$$
$$\approx 1000(2.71828)$$
$$= \$2718.28.$$

As another example, $2500 deposited at 4% interest compounded every instant for 10 years would become

$$A = 2500e^{(0.04)10}$$
$$= 2500e^{0.4}$$
$$\approx 2500(1.49182) \quad \text{(from Table 4)}$$
$$= 3729.56$$

Case 8 Exercises

Find the final amount on deposit for each of the following initial deposits.

1. $1000 at 6% compounded annually for 2 years
2. $4000 at 5% compounded annually for 3 years

3. $2000 at 4% compounded quarterly for 2 years
4. $1000 at 8% compounded semi-annually for 4 years
5. $2000 at 6% compounded every instant for 5 years
6. $6000 at 5% compounded every instant for 8 years
7. $2500 at 3% compounded every instant for 20 years
8. $8000 at 4% compounded every instant for 10 years

Find the total interest earned by each of the following deposits. (Hint: interest = total on deposit − original deposit.)

9. $1000 at 5% compounded annually for 5 years
10. $2000 at 6% compounded semi-annually for 3 years
11. $10,000 at 4% compounded every instant for 5 years
12. $18,000 at 5% compounded every instant for 10 years
13. Tom Smith deposits $2000 in an account paying 5% per year, compounded annually. The rate of inflation in the economy is 6% per year. Smith must pay tax on the interest earned by his account at the rate of 25%. After all this goes on, find the net value of Smith's money at the end of one year.
14. Joann Friedlander deposits $10,000 in an account paying 4% per year interest. The rate of inflation in the economy is about 8% per year. She pays income tax on the interest earned by her account at the rate of 25%. Find the net value of Friedlander's money if interest is compounded annually.

● **Case 9 Allocating Catalog Advertising— Montgomery Ward***

Catalog advertising is expensive. Thus, companies that advertise heavily by catalog must try to find ways to maximize sales and profits from a given expenditure on catalogs. The company must decide how many catalog pages to allocate to each of its different product lines, whether or not the pages for a particular line should be in color, and the number of catalogs that should be distributed, all within the financial and other limitations imposed by management.

An increase in the number of catalogs distributed, or in the number of pages allocated to a particular line of merchandise, is likely to increase sales, but not linearly. That is, increasing the space allocated to an item from a half-page to a full page is not likely to double the sales, while the first thousand catalogs distributed produce more sales than the last thousand, and so on. (In the language of economics: the marginal productivity of either variable decreases with increases in the variable.)

Several functions could be found to describe such a situation—appropriate quadratic, exponential, or logarithmic functions for example. The function finally chosen could depend on several factors, such as the degree of accuracy desired, the time available for the analyst to use in considering possible functions,

---

* Based on a paper by W. D. Sokolick and Philip H. Hartung, of the Montgomery Ward Division of Marcor Corporation. See *Management Science*, v. 15, No. 10, June 1969, pp. B521–529.

the number of variables involved, the amount of computer time available, and so on. After considering all these factors, the company analysts here finally decided that the best function to use for predicting catalog sales of a line of merchandise is

$$\ln S = \ln k + 0.56 \ln x_1 + 0.56 \ln x_2 + 0.23 \ln x_3 + 0.46 \ln x_4,$$

where

S = total sales of the line of merchandise being considered, in thousands of dollars
x_1 = number of catalogs distributed (in thousands)
x_2 = number of pages devoted to this line
x_3 = number of items in the line
x_4 = color factor (a variable describing color printing and paper quality)
k = a constant, determined from past records.

By studying past company records, and by applying certain advanced statistical tests, the company found that the coefficients 0.56, 0.56, 0.23, 0.46 of the equation above were relatively constant for all lines of merchandise in a particular catalog. The number of variables to be included in an equation such as the one above is also arbitrary, and depends on many of the same factors as does the choice of the type of equation itself.

To see how well sales are predicted by the equation above, the analysts took past records and used the data from these records to predict sales. For example, for one line of merchandise it was found that $x_1 = 570$, $x_2 = 12$, $x_3 = 400$, $x_4 = 1.8$, and $k = 1.1$. Substituting these values into the equation above, we have

$$\ln S = \ln 1.1 + 0.56(\ln 570) + 0.56(\ln 12) + 0.23(\ln 400) + 0.46(\ln 1.8).$$

Some of the logarithms are given directly in Table 5, while others must be calculated by the formula $\ln x = 2.3026 \log x$, given in Section 3.4. In either case, we have

$$\ln S = 0.0953 + 0.56(6.3456) + 0.56(2.4849) + 0.23(5.9915) + 0.46(0.5878)$$
$$= 0.0953 + 3.5535 + 1.3915 + 1.3780 + 0.2709$$
$$= 6.6892.$$

We can now evaluate S by consulting a large table of natural logarithms, or by using the relationship

$$\log x = \frac{\ln x}{2.3026}.$$

Using this relationship, we compute

$$\log S = \frac{6.6892}{2.3026} = 2.9051.$$

Using Table 5, we finally obtain

$$S = 804,$$

so that predicted sales for this line of merchandise is $804,000.

This same process was performed by the analysts for several different lines of merchandise, with the results as displayed below.

Line Number	Predicted Sales	Actual Sales	Percent Error
1	6,359	6,516	−2.4
2	9,919	9,292	6.7
3	1,403	1,571	−10.7
4	4,712	3,999	17.8
5	3,187	3,459	−7.9

When these results are tabulated, it is then necessary to make a decision about the percent of error—is it likely to be so great as to make the results meaningless? If this is the case, then it will be necessary to go back and do more work—either increase the number of variables in the equation used, or else scrap the equation completely and try another type.

Case 9 Exercises

Given the data below, and assuming $k = 1.1$, *estimate sales for the following lines of merchandise, based on the information given.*

Line Number		x_1	x_2	x_3	x_4	Predicted Sales
1.	A	562	10	514	1.6	
2.	B	578	12	605	1.4	
3.	C	581	14	558	1.8	
4.	D	610	7	435	1.7	

Chapter 7   Pretest

1. Given $f(x) = 1 - x^2$, find $f(1); f(1\tfrac{1}{2}); f(1 + n); f\left(1 + \dfrac{1}{n}\right)$.   [2.1]

2. If $x = -1$ when $y = 2$, find C given $y = 50x - \dfrac{x}{2} + C$.   [1.2]

Solve.   [1.3]

3. $2m^2 - 2m = 15 - m$
4. $x^2 - 9 = 0$
5. $3x^2 - x = 4x - 1$

Find the derivative.

6. $f(x) = 2x - 3$ [4.2]
7. $f(x) = 4x^2 - 3x + 2$ [4.2]
8. $f(x) = \sqrt{x} - x^{2/3}$ [4.2]
9. $f(x) = \sqrt{2x - x^3}$ [4.5]
10. $f(x) = e^x$ [6.4]
11. $f(x) = 2e^{3x}$ [6.4]
12. $f(x) = \dfrac{2}{x^2 - 1}$ [4.4]
13. $f(x) = 3 \ln (x^2 + 1)$ [6.4]
14. Find $\lim\limits_{n \to \infty} \left[\dfrac{1}{2} - \dfrac{1}{n} + \dfrac{1}{n^2}\right]$ [3.4]

7
Integration

Calculus is divided into two broad areas—*differential calculus*, which we discussed in Chapters 4 and 5, and *integral calculus*, which we consider in this chapter. Like the derivative of a function, the definite integral of a function is a special limit with many applications. Geometrically, the definite integral is related to the area under a curve, as we shall see.

7.1 The Antiderivative

In Chapter 4 we saw that the marginal cost to produce x items was given by the derivative $f'(x)$ of the cost function $f(x)$. Now suppose we know the marginal cost function and wish to find the cost function. For example, suppose the marginal cost at a level of production of x units is

$$f'(x) = 2x - 10.$$

Can we find the cost function? In other words, can we find a function whose derivative is $2x - 10$?

By trial and error, with a little thought on how the derivative of $f(x)$ is found, we find

$$f(x) = x^2 - 10x,$$

which has $2x - 10$ as its derivative. Is $x^2 - 10x$ the only function with derivative $2x - 10$? No, there are many, for example,

$$f(x) = x^2 - 10x + 2,$$
$$f(x) = x^2 - 10x + 5,$$
$$f(x) = x^2 - 10x - 8,$$

and so on. (Verify that each of these functions has $2x - 10$ as its derivative.) In fact, if we add any real number to $x^2 - 10x$ we have a function whose derivative is $2x - 10$. To express this fact, we write

$$f(x) = x^2 - 10x + C,$$

where C represents any constant. The function $f(x)$ is called an **antiderivative** of $f'(x)$. It can be shown that every antiderivative of $f'(x)$ is of the form $f(x) = x^2 - 10x + C$.

The trial and error method of finding an antiderivative used in the example above is not very satisfactory. We need some rules for finding antiderivatives. First, recall that to take the derivative of x^n we multiply by n and reduce the power on x by 1:

$$\frac{d}{dx} x^n = nx^{n-1}.$$

Thus, to antidifferentiate, that is, to undo what was done, we should increase the power by 1 and divide by the new power $n + 1$. In general, we have the following rule.

Theorem 7.1 If $n \neq -1$, then an antiderivative of $f(x) = x^n$ is

$$F(x) = \frac{1}{n+1} x^{n+1} + C.$$

Note that we are labeling the antiderivative of a function $f(x)$ as $F(x)$. We can check that $F(x)$ is the antiderivative of $f(x)$ by showing that the derivative of $F(x)$ is $f(x)$. Here we would have

$$\frac{d}{dx}\left(\frac{1}{n+1} x^{n+1} + C\right) = \frac{n+1}{n+1} x^n + 0 = x^n.$$

Example 1 Find an antiderivative for each of the following.
(a) $f(x) = x^3$
By Theorem 7.1, an antiderivative of $f(x) = x^3$ is

$$F(x) = \frac{1}{3+1} x^{3+1} + C = \frac{1}{4} x^4 + C = \frac{x^4}{4} + C.$$

(b) $f(x) = \sqrt{x}$
First write $\sqrt{x}$ as $x^{1/2}$. Then we use the theorem to write

$$F(x) = \frac{1}{1/2 + 1} x^{3/2} + C = \frac{2}{3} x^{3/2} + C.$$

To check this result, take the derivative of F.

(c) $f(x) = 1$
To use Theorem 7.1 here, write 1 as x^0, so that the antiderivative of $f(x) = 1 = 1 \cdot x^0$ is given by

$$F(x) = \frac{1}{1} x^1 + C = x + C.$$

In Theorem 7.1 we made the restriction $n \neq -1$. This is necessary because $n = -1$ makes the denominator of the antiderivative equal 0, which is not possible. Thus, Theorem 7.1 cannot be used to find the antiderivative of $f(x) = x^{-1} = 1/x$. The antiderivative of $f(x) = 1/x$ is $\ln x + C$. (See Section 7.6.)

In Chapter 4 we saw that the derivative of the product of a constant and a function is the product of the constant and the derivative of the function. We could expect a similar rule to apply to antidifferentiation and it does. Also, since we differentiate term by term, it seems reasonable to antidifferentiate term by term. The next two theorems state these properties of antiderivatives.

Theorem 7.2 If $F(x)$ is an antiderivative of $f(x)$ and if k is any constant, then the antiderivative of $k \cdot f(x)$ is $k \cdot F(x) + C$, where C is a constant.

Example 2 Find an antiderivative of $f(x) = 2x^3$.

By Theorems 7.1 and 7.2, we have the antiderivative given by

$$F(x) = 2\left(\frac{1}{4}x^4\right) + C = \frac{x^4}{2} + C.$$

Theorem 7.3 If $F(x)$ is an antiderivative of $f(x)$ and $G(x)$ is an antiderivative of $g(x)$, then an antiderivative of $f(x) \pm g(x)$ is $F(x) \pm G(x) + C$.

Example 3 Find an antiderivative of $f(x) = 3x^2 - 4x + 5$.

Using all the theorems of this section, we have

$$F(x) = 3\left(\frac{1}{3}x^3\right) - 4\left(\frac{1}{2}x^2\right) + 5x + C$$

$$= x^3 - 2x^2 + 5x + C,$$

as an antiderivative of $f(x) = 3x^2 - 4x + 5$.

Example 4 Suppose a company has found that the marginal cost at a level of production of x thousand books is given by

$$C'(x) = \frac{50}{\sqrt{x}},$$

and the fixed cost (the cost to produce 0 books) is $25,000. Find the cost function $C(x)$.

Writing $50/\sqrt{x}$ as $50x^{-1/2}$, we find that an antiderivative of $C'(x)$ is

$$C(x) = 50(2x^{1/2}) + k = 100x^{1/2} + k,$$

where k is a constant. We can find k by using the fact that when $x = 0$, $C(x)$ is 25,000.

$$25,000 = 100\sqrt{0} + k$$
$$k = 25,000.$$

Therefore, the cost function is $C(x) = 100x^{1/2} + 25,000$.

7.1 Exercises

Find an antiderivative for each of the following.

1. $f(x) = 4x$
2. $f(x) = -8x$
3. $f(x) = 5x^2$
4. $f(x) = 6x^3$
5. $f(x) = 6$
6. $f(x) = 2$
7. $f(x) = 2x + 3$
8. $f(x) = 3x - 5$
9. $f(x) = x^2 + 6x$
10. $f(x) = x^2 - 2x$
11. $f(x) = x^2 - 4x + 5$
12. $f(x) = 5x^2 - 6x + 3$
13. $f(x) = \sqrt{x}$
14. $f(x) = x^{1/3}$
15. $f(x) = x^{1/2} + x^{3/2} + x^{5/2}$
16. $f(x) = \sqrt{x} + x\sqrt{x} - x^2\sqrt{x}$
17. $f(x) = x^2\sqrt{x} + 3x\sqrt{x} + 2\sqrt{x}$
18. $f(x) = x^{3/2} + x^{1/2} - x^{-1/2}$
19. $f(x) = 10x^{3/2} - 14x^{5/2}$
20. $f(x) = 56x^{5/2} + 18x^{7/2}$
21. $f(x) = 9x^{1/2} + 4$
22. $f(x) = -18x^{1/2} - 6$
23. $f(x) = 1/x^2$
24. $f(x) = 4/x^3$
25. $f(x) = 1/x^3 - 1/\sqrt{x}$
26. $f(x) = \sqrt{x} + (1/x^2)$
27. $f(x) = 2x^{-1}$
28. $f(x) = -4x^{-1}$
29. $f(x) = -9x^{-2} - 2x^{-1}$
30. $f(x) = 8x^{-3} + 4x^{-1}$

31. The marginal profit of Henrietta's Hamburgers is

$$P'(x) = -2x + 20,$$

where x is the sales volume in thousands of hamburgers. Henrietta knows that her profit is -50 when she sells no hamburgers. What is her profit function?

32. Suppose the marginal cost function for producing x items is

$$C'(x) = 3x^2 + 6x + 4,$$

and the cost to produce 0 items is $40. Find the cost function.

33. The slope of the tangent line to a curve is given by

$$f'(x) = 6x^2 - 4x + 3.$$

If the point $(0, 1)$ is on the curve, find the equation of the curve.

34. Find the equation of a curve whose tangent line has a slope of

$$f'(x) = x^{2/3},$$

if the point $(1, 3/5)$ is on the curve.

7.2 Area and the Definite Integral

In this section, we consider a method for finding the area of a region bounded

Figure 7.1

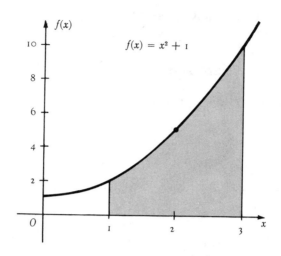

on one side by a curve. For example, Figure 7.1 shows the region between the lines $x = 1$, $x = 3$, the x-axis, and the graph of the function defined by

$$f(x) = x^2 + 1.$$

To get an approximation of this area, we could use two rectangles as shown in Figure 7.2. We use $f(1) = 2$ as the height of the rectangle on the left and $f(2) = 5$ as the height of the rectangle on the right. The width of each rectangle is 1; thus, the total area of the two rectangles is

$$1 \cdot f(1) + 1 \cdot f(2) = 2 + 5 = 7 \text{ square units.}$$

Figure 7.2

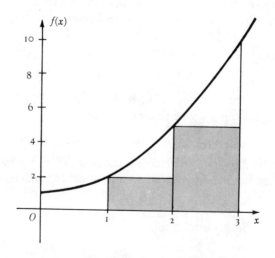

7.2 Area and the Definite Integral

As shown in Figure 7.2, this approximation is smaller than the actual area. To improve the accuracy of the approximation, we can divide the interval from 1 to 3 into four equal parts of width 1/2, as shown in Figure 7.3. The total area of the four rectangles is

$$\frac{1}{2}\cdot f(1) + \frac{1}{2}\cdot f\left(1\frac{1}{2}\right) + \frac{1}{2}\cdot f(2) + \frac{1}{2}\cdot f\left(2\frac{1}{2}\right)$$

$$= \frac{1}{2}(2) + \frac{1}{2}\left(\frac{13}{4}\right) + \frac{1}{2}(5) + \frac{1}{2}\left(\frac{29}{4}\right)$$

$$= 1 + \frac{13}{8} + \frac{5}{2} + \frac{29}{8}$$

$$= 8.75 \text{ square units.}$$

Figure 7.3

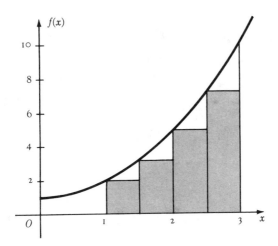

This approximation is better, but still less, than the actual area we seek. To improve the approximation, we could divide the interval from 1 to 3 into 8 parts with equal widths of $\frac{1}{4}$. The total area of the rectangles formed would then be

$$\frac{1}{4}\cdot f(1) + \frac{1}{4}\cdot f\left(1\frac{1}{4}\right) + \frac{1}{4}\cdot f\left(1\frac{1}{2}\right) + \frac{1}{4}\cdot f\left(1\frac{3}{4}\right)$$
$$+ \frac{1}{4}\cdot f(2) + \frac{1}{4}\cdot f\left(2\frac{1}{4}\right) + \frac{1}{4}\cdot f\left(2\frac{1}{2}\right) + \frac{1}{4}\cdot f\left(2\frac{3}{4}\right)$$

$$= 9.69 \text{ square units.}$$

The process we have been using here, of approximating the area under a curve by using more and more rectangles to get a better and better approximation can be generalized. To do this, we divide the interval from 1 to 3 into n equal parts, as shown in Figure 7.4. Each of these n intervals will have width

$$\frac{3-1}{n} = \frac{2}{n}.$$

Figure 7.4

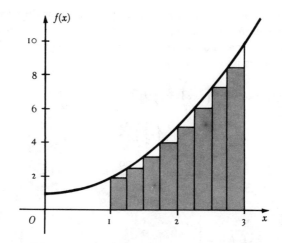

The heights and areas of the n rectangles are shown in the following chart.

Rectangle	Height	Area
1	$f(1)$	$\frac{2}{n}\cdot f(1)$
2	$f\left(1+\frac{2}{n}\right)$	$\frac{2}{n}\cdot f\left(1+\frac{2}{n}\right)$
3	$f\left(1+2\cdot\frac{2}{n}\right)$	$\frac{2}{n}\cdot f\left(1+2\cdot\frac{2}{n}\right)$
4	$f\left(1+3\cdot\frac{2}{n}\right)$	$\frac{2}{n}\cdot f\left(1+3\cdot\frac{2}{n}\right)$
⋮	⋮	⋮
n	$f\left(1+(n-1)\frac{2}{n}\right)$	$\frac{2}{n}\cdot f\left(1+(n-1)\frac{2}{n}\right)$

The total of the areas listed in the last column approximates the area under the curve, above the x-axis, and between the lines $x = 1$ and $x = 3$. As n becomes larger and larger the approximation here is better and better.

The simplification of this expression is not easy, and fortunately, it is not necessary for us to go through all the steps. It is sufficient to say that the sum of the areas of the n rectangles simplifies to

$$\frac{32}{3} - \frac{8}{n} + \frac{4}{3n^2}.$$

Since n is the number of rectangles, this result can be used to evaluate the sum of the areas of any number of rectangles. For example, if $n = 20$, the total area of the rectangles is

$$\frac{32}{3} - \frac{8}{20} + \frac{4}{3(20)^2} = 10.27,$$

a result very close to the actual area under the curve which we are trying to find. If we find the limit of the expression

$$\frac{32}{3} - \frac{8}{n} + \frac{4}{3n^2}$$

as the number of rectangles is increased without bound, that is as $n \to \infty$, we have

$$\lim_{n \to \infty} \left(\frac{32}{3} - \frac{8}{n} + \frac{4}{3n^2} \right) = \frac{32}{3} = 10\frac{2}{3}.$$

This limit is the area of the region under the graph of $f(x) = x^2 + 1$ and above the x-axis, between the lines $x = 1$ and $x = 3$. This number is called the **definite integral** of $f(x) = x^2 + 1$ from $x = 1$ to $x = 3$, written

$$\int_1^3 (x^2 + 1)\, dx = \frac{32}{3}.$$

The symbol $\int$ is called an **integral sign**, with 3 called the **upper limit** of integration and 1 the **lower limit** of integration. In the next section we shall see how the antiderivative is used in finding the definite integral and thus the area under a curve.

7.2 Exercises

In each of the following exercises, first approximate the area under the given curve and above the x-axis by using two rectangles. Let the height of the rectangle be given by the value of the function at the left side of the rectangle. Then repeat the process and approximate the area using four rectangles.

1. $f(x) = 3x + 2$ from $x = 1$ to $x = 5$
2. $f(x) = -2x + 1$ from $x = -4$ to $x = 0$
3. $f(x) = x + 5$ from $x = 2$ to $x = 4$
4. $f(x) = 3 + x$ from $x = 1$ to $x = 3$
5. $f(x) = x^2$ from $x = 1$ to $x = 5$
6. $f(x) = x^2$ from $x = 0$ to $x = 4$
7. $f(x) = x^2 + 2$ from $x = -2$ to $x = 2$
8. $f(x) = -x^2 + 4$ from $x = -2$ to $x = 2$
9. Let $f(x) = x^2$. If the region above the x-axis, between $x = 1$ and $x = 4$, is divided into n rectangles, as in the text, and the sum of the areas of the n rectangles evaluated, the result simplifies to

$$21 - \frac{45}{n} + \frac{9}{2n^2}.$$

   (a) Approximate the area of the region described above using four rectangles.
   (b) Approximate the area of the region using eight rectangles.
   (c) Find $\int_1^4 x^2\, dx$. This definite integral gives the exact area.

7.3 The Fundamental Theorem of Calculus

In the previous section we found that

$$\int_1^3 (x^2 + 1)\, dx = \frac{32}{3}.$$

This definite integral was found by a long calculation using rectangles and a limit. In this section we shall develop a simpler procedure for finding this number. An antiderivative of $f(x) = x^2 + 1$ is $F(x) = \frac{1}{3}x^3 + x + C$. If we calculate $F(3) - F(1)$ we get

$$F(3) - F(1) = \left(\frac{27}{3} + 3 + C\right) - \left(\frac{1}{3} + 1 + C\right)$$
$$= \frac{32}{3},$$

which is the value of the definite integral

$$\int_1^3 (x^2 + 1)\, dx,$$

which we found in the last section. This amazing coincidence is expressed as the *fundamental theorem of calculus*.

Theorem 7.4 *(Fundamental Theorem of Calculus)* Let $f(x)$ be the derivative of $F(x)$ and let $f(x)$ be continuous in the interval $[a, b]$. Then

$$\int_a^b f(x)\, dx = F(b) - F(a).$$

To represent $F(b) - F(a)$ we shall often use

$$F(x)\Big]_a^b$$

The fundamental theorem of calculus is really quite remarkable. It deserves its name which sets it apart as the most important theorem of calculus. It is the key connection between the differential calculus and the integral calculus, which originally were developed separately without knowledge of this connection between them.

Because of this relationship between the definite integral and the antiderivative given in the fundamental theorem, it is customary to call an antiderivative an indefinite integral (or sometimes just an integral). Indefinite integrals are indicated with the $\int$ symbol, but with no limits of integration. For

7.3 The Fundamental Theorem of Calculus

example, $F(x) = x^3 - 4x^2 + 5x$ is an antiderivative of $f(x) = 3x^2 - 8x + 5$. This fact can be expressed by writing

$$\int (3x^2 - 8x + 5)\, dx = x^3 - 4x^2 + 5x + C.$$

This process of finding an antiderivative is called **integrating**; we integrate $3x^2 - 8x + 5$ to obtain $x^3 - 4x^2 + 5x + C$.

Example 5 First find $\int 4x^3\, dx$ and then find $\int_1^2 4x^3\, dx$.

By the results given before, we have

$$\int 4x^3\, dx = x^4 + C.$$

Thus, by the fundamental theorem, we have

$$\int_1^2 4x^3\, dx = x^4 + C \Big]_1^2$$
$$= (16 + C) - (1 + C)$$
$$= 15.$$

The number 15 represents the area under the curve $f(x) = 4x^3$ that is above the x-axis and between the lines $x = 1$ and $x = 2$. Since the constant C does not appear in the final answer, it can be omitted when evaluating the definite integral. It must be included, however, when finding an indefinite integral.

Theorems 7.2 and 7.3 given in Section 7.1 also apply to the definite integral and so are restated here.

Theorem 7.5 For any real constant k,

$$\int_a^b k \cdot f(x)\, dx = k \cdot \int_a^b f(x)\, dx.$$

Theorem 7.6 If all indicated definite integrals exist, then

$$\int_a^b [f(x) + g(x)]\, dx = \int_a^b f(x)\, dx + \int_a^b g(x)\, dx.$$

Example 6 Evaluate

$$\int_1^4 2x^3\, dx.$$

Using Theorems 7.1 and 7.2 and the fundamental theorem of calculus, we have

$$\int_1^4 2x^3\, dx = 2\int_1^4 x^3\, dx = 2\left(\frac{x^4}{4}\right)\Big]_1^4 = 2\left(\frac{256}{4} - \frac{1}{4}\right) = \frac{255}{2}.$$

Example 7 Find

$$\int_1^2 (3x^2 - 4x + 5)\, dx.$$

By using all the theorems of this section, we have

$$\int_1^2 (3x^2 - 4x + 5)\, dx = \int_1^2 3x^2\, dx - \int_1^2 4x\, dx + \int_1^2 5\, dx$$

$$= 3\int_1^2 x^2\, dx - 4\int_1^2 x\, dx + 5\int_1^2 dx$$

$$= 3\left(\frac{x^3}{3}\right)\bigg]_1^2 - 4\left(\frac{x^2}{2}\right)\bigg]_1^2 + 5(x)\bigg]_1^2$$

$$= x^3\bigg]_1^2 - 2x^2\bigg]_1^2 + 5x\bigg]_1^2$$

$$= (8 - 1) - (8 - 2) + (10 - 5)$$

$$= 6.$$

7.3 Exercises

Evaluate each of the following definite integrals.

1. $\int_1^6 x\, dx$
2. $\int_1^5 x\, dx$
3. $\int_0^4 (4x + 3)\, dx$
4. $\int_5^8 (6x - 5)\, dx$
5. $\int_4^8 6\, dx$
6. $\int_3^9 2\, dx$
7. $\int_8^{10} (2x + 3)\, dx$
8. $\int_5^{10} (4x - 7)\, dx$
9. $\int_{-1}^1 (x + 6)\, dx$
10. $\int_1^4 (8x + 3)\, dx$
11. $\int_1^5 3x^2\, dx$
12. $\int_0^3 -9x^2\, dx$
13. $\int_1^3 (6x^2 - 4x + 5)\, dx$
14. $\int_1^4 (15x^2 - 6x + 3)\, dx$
15. $\int_1^4 (3x^2 - 4x + 1)\, dx$
16. $\int_0^3 (3x^2 - 2x + 5)\, dx$
17. $\int_4^9 (10x^{3/2} - 6x^{1/2})\, dx$
18. $\int_4^9 (14x^{5/2} + 5x^{3/2} + 3x^{1/2})\, dx$
19. $\int_1^4 \sqrt{x}\, dx$
20. $\int_4^9 \sqrt{x}\, dx$

21. Let $f(x) = (x^2 + 2x)^2$. Verify that $f'(x) = 2(x^2 + 2x)(2x + 2) = 4(x + 1)(x^2 + 2x)$, and then find

$$\int_2^3 4(x + 1)(x^2 + 2x)\, dx.$$

22. Let $f(x) = \sqrt{x^2 - 4}$. Verify that $f'(x) = x/\sqrt{x^2 - 4}$, and then find

$$\int_3^4 \frac{x\, dx}{\sqrt{x^2 - 4}}.$$

23. Find the area of the region between the x-axis and the graph of $f(x) = x^2 + 4x + 4$ from $x = -1$ to $x = 2$.
24. Find the area of the region enclosed by the x-axis, $y = x$, $x = 0$, and $x = 4$.
25. Find the area of the region enclosed by the x-axis, $y = x^3$, $x = 0$, and $x = 3$.
26. Find the area of the region enclosed by the x-axis, $y = x^2 + x$, $x = 1$, and $x = 4$.

7.4 Some Applications of Integrals

In this section, we consider a few of the many applications of integrals. Since a marginal function $f(x)$ is the derivative of some function $F(x)$, the antiderivative, or indefinite integral, can be used to find $F(x)$ when the marginal function is known.

Example 8 An analyst tells a company that its marginal cost of production at its main plant is given by

$$C'(x) = 200 + x,$$

where x is the number of items produced. If the fixed cost of the plant is 30 dollars, find the cost function $C(x)$.

To find $C(x)$, we integrate the marginal cost function, as follows:

$$C(x) = \int (200 + x)\, dx = 200x + \frac{x^2}{2} + K.$$

To determine the constant K, we use the fact that the fixed cost is 30 dollars—that is, at a production of $x = 0$ units, the company has an expense of 30 dollars. Thus, $C(0) = 30$, which can be used to find K.

$$C(x) = 200x + \frac{x^2}{2} + K$$

$$C(0) = 200(0) + \frac{0^2}{2} + K$$

$$C(0) = K$$

$$K = 30.$$

The cost function, $C(x)$, is thus given by

$$C(x) = 200x + \frac{x^2}{2} + 30.$$

In Chapter 8, we discuss this use of integrals in more detail.

The definite integral can be used to express total value over a period of time as in the following example.

Suppose a leasing company wants to decide on the yearly lease fee for a certain new typewriter. The company expects to lease the typewriter for 5 years, and it expects the rate of maintenance, $M(t)$, in dollars, which it must supply, to approximate

$$M(t) = 10 + 2t + t^2,$$

where t is the number of years the typewriter has been used. The definite integral can be used to find the total maintenance charge the company can expect over the life of the typewriter. Figure 7.5 shows the graph of $M(t)$. The total maintenance charge for the 5-year period will be given by the shaded area of the figure, which can be found as follows:

$$\int_0^5 (10 + 2t + t^2)\, dt = \left(10t + t^2 + \frac{t^3}{3} \right)\Big]_0^5$$

$$= 50 + 25 + \frac{125}{3} - 0$$

$$\approx 116.67.$$

The company can expect the total maintenance charge for 5 years to be about $117. Hence, the company should add about

$$\frac{\$117}{5} = \$23.40$$

to its annual lease price.

Figure 7.5

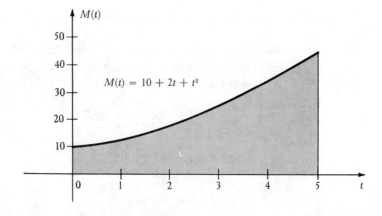

7.4 Some Applications of Integrals

Example 9 Elizabeth, who runs a factory that makes signs, has been shown a new machine to staple the signs to the handles. She estimates that the rate of savings, $S(x)$, from the machine will be approximated by

$$S(x) = 3 + 2x,$$

where x represents the number of years the stapler has been in use. If the machine costs $70, would it pay for itself in 5 years?

We need to find the area under the savings curve shown in Figure 7.6, between the lines $x = 0$ and $x = 5$, and the x-axis. Using definite integrals, we have

$$\int_0^5 (3 + 2x)\, dx = 3x + x^2 \Big]_0^5 = 40.$$

Figure 7.6

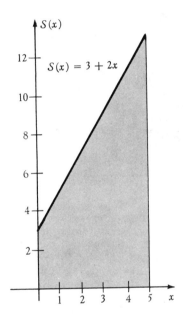

The total savings in five years are $40, and so the machine will not pay for itself in this time period.

To find the number of years it will take for the machine of Example 9 to pay for itself, note that, since the machine cost a total of $70, it will pay for itself when the area under the savings curve of Figure 7.7 equals 70, or at a time t such that

$$\int_0^t (3 + 2x)\, dx = 70.$$

Evaluating the definite integral we have

$$(3x + x^2)\Big]_0^t = (3t + t^2) - (3 \cdot 0 + 0^2) = 3t + t^2.$$

Since we want the total savings to equal 70, we have

$$3t + t^2 = 70.$$

Solve this quadratic equation to verify that the machine will pay for itself in seven years.

Figure 7.7

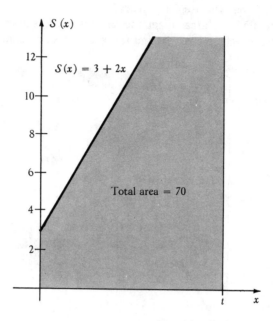

Example 10 Elizabeth, of the previous example, believes that the rate of sales of her signs, $T(x)$, is given by

$$T(x) = 15 + 10x,$$

where x is the number of years she has been in business. When can she expect to sell her 1000th sign?

Let t be the time at which Elizabeth will sell her 1000th sign. The area under the sales curve between $x = 0$ and $x = t$ gives the total sales during that period. Since we want the total sales to be 1000, we have

$$\int_0^t (15 + 10x)\, dx = 1000.$$

Since

$$\int_0^t (15 + 10x)\, dx = 15x + 5x^2 \Big]_0^t = 15t + 5t^2,$$

we have

$$15t + 5t^2 = 1000$$
$$5t^2 + 15t - 1000 = 0.$$

7.4 Some Applications of Integrals

If we divide both sides of this equation by 5, we have

$$t^2 + 3t - 200 = 0,$$

from which we find, using the quadratic formula,

$$t = 12.72 \text{ years.}$$

Hence, Elizabeth can expect to sell her 1000th sign about twelve years, nine months after she goes into business.

7.4 Exercises

Find the cost function for each of the following marginal cost functions.

1. $C'(x) = 4x - 5$, fixed cost is $8.
2. $C'(x) = 2x + 3x^2$, fixed cost is $15.
3. $C'(x) = 0.2x^2$, fixed cost is $10.
4. $C'(x) = x^2 + 3x$, fixed cost is $5.
5. $C'(x) = \sqrt{x}$, 16 units cost $40.
6. $C'(x) = x^{2/3} + 2$, 8 units cost $58.
7. $C'(x) = x^2 - 2x + 3$, 3 units cost $15.
8. $C'(x) = x + 1/x^2$, 2 units cost $5.50.

9. A car-leasing firm must decide how much to charge for maintenance on the cars it leases. After careful study, it is decided that the rate of maintenance, $M(x)$, on a new car will approximate

$$M(x) = 60(1 + x^2),$$

where x is the number of years the car has been in use. What total maintenance charge can the company expect for a two-year lease? What amount should it add to the monthly lease payments to pay for maintenance?

10. Using the function of Exercise 9, find the maintenance charge the company can expect during the third year. Find the total charge during the first three years. What monthly charge should be added to take care of a three-year lease?

11. A company is considering a new manufacturing process. It knows that the rate of savings, $S(t)$, from the process will be about

$$S(t) = 2t + 30,$$

where t is the number of years of use of the process. Find the total savings during the first year. Find the total savings during the first six years.

12. Assume that the new process in Exercise 11 costs $1000. About when will it pay for itself?

13. A company is introducing a new product. Production is expected to grow slowly, because of difficulties in the start-up process. It is expected that the rate of production, $P(x)$, will be approximated by

$$P(x) = 600x^{3/2},$$

where x is the number of years since the introduction of the product. Will the company be able to supply 10,000 units during the first four years?

14. Pollution from a factory is entering a lake. The rate of concentration, $P(t)$, of the pollutant at time t, is given by

$$P(t) = 140t^{5/2},$$

where t is the number of years since the factory started introducing pollutant into the lake. Ecologists estimate that the lake can accept a total level of pollution of 4850 units before all the fish life in the lake ends. Can the factory operate for 4 years without killing all the fish in the lake?

7.5 The Area Between Two Curves

In Section 7.2 we saw that the definite integral $\int_a^b f(x)\, dx$ can be used to find the area below the graph of the function $y = f(x)$, above the x-axis, and between the lines $x = a$ and $x = b$. In this section we extend this idea to find other areas.

Suppose we want to find the area between the x-axis and the graph of $f(x) = x^2 - 4$ from $x = 0$ to $x = 4$. As shown in Figure 7.8, part of the area lies above the x-axis and part lies below the x-axis. If we were to use the definite integral to evaluate the area below the x-axis, the result would be a negative number, since the function values there are all negative. Since area is a positive quantity, the correct value of that part of the area is given by the negative of the appropriate definite integral. In cases like this, the two parts of the area must be calculated separately, and then the two positive numbers added to give the total area.

To find the area between the x-axis and the graph of $f(x) = x^2 - 4$ from $x = 0$ to $x = 4$, we must first find the point where the graph crosses the x-axis. This is done by solving the equation

$$x^2 - 4 = 0.$$

The solutions of this equation are 2 and -2. We are only interested in values of x in the interval $[0, 4]$ so we discard the solution -2. Thus, the total area will be given by

$$-\int_0^2 (x^2 - 4)\, dx + \int_2^4 (x^2 - 4)\, dx = -\left(\frac{1}{3}x^3 - 4x\right)\Big]_0^2 + \left(\frac{1}{3}x^3 - 4x\right)\Big]_2^4$$

$$= -\left(\frac{8}{3} - 8\right) + \left(\frac{64}{3} - 16\right) - \left(\frac{8}{3} - 8\right)$$

$$= 16.$$

7.5 The Area Between Two Curves 227

Figure 7.8

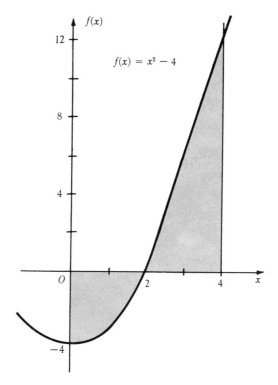

Figure 7.9 shows the curves $f(x) = x^{1/2}$ and $g(x) = x^3$. To find the area bounded by these two curves and the lines $x = 0$ and $x = 1$, we can use definite integrals. The area between $f(x) = x^{1/2}$, the lines $x = 0$ and $x = 1$, and the x-axis is

$$\int_0^1 x^{1/2}\, dx,$$

Figure 7.9

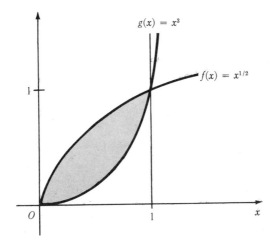

while the area between $g(x) = x^3$, the lines $x = 0$, $x = 1$, and the x-axis is

$$\int_0^1 x^3 \, dx.$$

As shown in Figure 7.9, the area between these two curves is given by the difference between these two integrals, or

$$\int_0^1 x^{1/2} \, dx - \int_0^1 x^3 \, dx,$$

which, by Theorem 7.6, we can also write as

$$\int_0^1 (x^{1/2} - x^3) \, dx.$$

Using the Fundamental Theorem of Calculus, we have

$$\int_0^1 (x^{1/2} - x^3) \, dx = \left(\frac{x^{3/2}}{\frac{3}{2}} - \frac{x^4}{4} \right) \Big]_0^1$$

$$= \frac{2}{3} x^{3/2} - \frac{x^4}{4} \Big]_0^1$$

$$= \left(\frac{2}{3} \cdot 1 - \frac{1}{4} \right)$$

$$= \frac{5}{12}.$$

Thus, the area between the two curves is 5/12.

The difference between two integrals can be used to find the area between the graphs of two functions even if one graph lies below the x-axis or if both graphs lie below the x-axis. In general, if $f(x) \le g(x)$ for all values of x in the interval $[a, b]$, then the area between the two graphs is

$$\int_a^b [g(x) - f(x)] \, dx.$$

Example 11 A company is considering the introduction of a new manufacturing process in one of its plants. The new process provides substantial initial savings, with the savings declining with time x according to the savings function

$$S(x) = 100 - x^2.$$

At the same time, the cost of operating the new process increases with time x, according to the cost function

$$C(x) = x^2 + \frac{14}{3} x.$$

(a) For how many years will the company realize savings?
(b) What will be the total savings during this period?

Figure 7.10

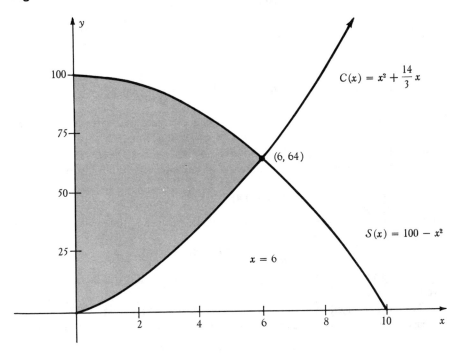

(a) Figure 7.10 shows the graphs of the savings and the cost functions. The company should use this new process until the time at which the savings and cost functions intersect. That is, the company should find the value of x such that

$$C(x) = S(x),$$

or such that

$$100 - x^2 = x^2 + \frac{14}{3} x.$$

Solving the equation, we find that the only valid solution is $x = 6$. Thus, the company should use the new process for 6 years.

(b) The total savings over the 6-year period are given by the area between the cost and savings curves and the lines $x = 0$ and $x = 6$, which we can evaluate with definite integrals as follows:

$$\text{total savings} = \int_0^6 (100 - x^2) \, dx - \int_0^6 \left(x^2 + \frac{14}{3} x \right) dx$$

$$= \int_0^6 \left[(100 - x^2) - \left(x^2 + \frac{14}{3} x \right) \right] dx$$

$$= \int_0^6 \left(100 - \frac{14}{3} x - 2x^2 \right) dx$$

$$= \left(100x - \frac{7}{3}x^2 - \frac{2}{3}x^3\right)\Big]_0^6$$

$$= 100(6) - \frac{7}{3}(36) - \frac{2}{3}(216)$$

$$= 372.$$

Thus, the company will save a total of $372 over the 6-year period.

Example 12 A farmer has been using a new fertilizer that gives him a better yield, but because it exhausts the soil of other nutrients, he must use other fertilizers in greater and greater amounts, so that his costs increase each year. The new fertilizer produces an increase in revenue (in hundreds of dollars) given by

$$R(t) = -0.4t^2 + 8t + 10,$$

where t is measured in years. The yearly costs due to use of the fertilizer increase according to the function

$$C(t) = 2t + 5.$$

How long can the farmer profitably use the fertilizer? What will be his increased revenue over this period?

The farmer should use the new fertilizer until the additional costs equal the increase in revenue. Thus, we need to solve the equation $R(t) = C(t)$, as follows.

$$-0.4t^2 + 8t + 10 = 2t + 5$$
$$-4t^2 + 80t + 100 = 20t + 50$$
$$-4t^2 + 60t + 50 = 0$$
$$t = 15.8.$$

The new fertilizer will be profitable for about 15.8 years.

To find the total amount of additional revenue over the 15.8-year period, we must find the area between the graphs of the revenue and cost functions, as shown in Figure 7.11. We have

$$\text{total savings} = \int_0^{15.8} [R(t) - C(t)]\, dt$$

$$= \int_0^{15.8} [(-0.4t^2 + 8t + 10) - (2t + 5)]\, dt$$

$$= \frac{-0.4t^3}{3} + \frac{6t^2}{2} + 5t\Big]_0^{15.8}$$

$$= 302.01.$$

The total savings will amount to about $30,000 over the 15.8-year period.

It is probably not realistic to say that the farmer will need to use the new process for 15.8 years—he will probably have to use it for 15 years or for 16 years. In this case, when the mathematical model produces results that are not in the domain of the function, it will be necessary to find the total savings after 15 years and after 16 years, and then select the best result.

Figure 7.11

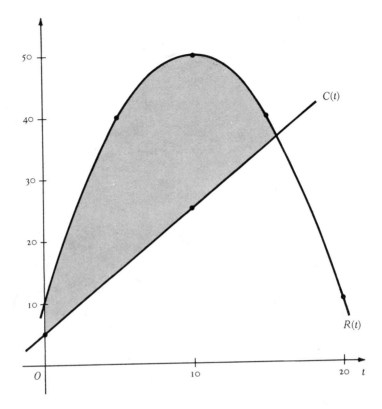

Example 13 Suppose the demand, $D(x)$, for a certain product is given by

$$D(x) = 900 - 20x - x^2,$$

where x is the price charged for the product, while the supply, $S(x)$, for this particular product is given by

$$S(x) = x^2 + 10x.$$

The graphs of both functions are shown in Figure 7.12, along with the equilibrium point at which supply and demand are equal. To find the equilibrium point x^*, we need to solve the equation

$$\begin{aligned} D(x^*) &= S(x^*) \\ 900 - 20x^* - (x^*)^2 &= (x^*)^2 + 10x^* \\ 0 &= 2(x^*)^2 + 30x^* - 900 \\ 0 &= (x^*)^2 + 15x^* - 450. \end{aligned}$$

Solving this last equation, we have

$$x^* = 15.$$

Figure 7.12

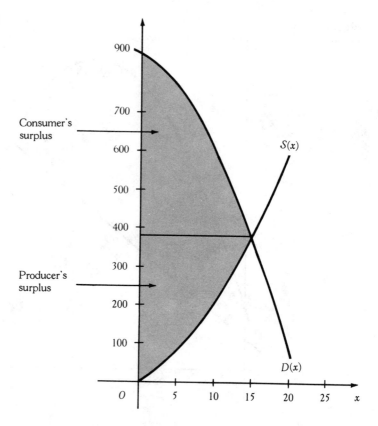

At a price of $x^* = 15$, the supply and demand will be in equilibrium. The demand (and supply) at that point is given by

$$D(15) = S(15) = 900 - 20(15) - 15^2 = 375.$$

Even though equilibrium is reached at a price of $x = 15$, the supply function shows that there are some producers who are able to supply the product at a lower price. At a price of only $x = 2$, for example, we still find $S(2) = 24$. Those producers who can sell at a lower price than the equilibrium price of $x^* = 15$, thus, will make additional profit when they sell at a price of $x^* = 15$. This surplus, shown by the indicated area in Figure 7.12, is called the **producer's surplus**. We can use a definite integral to find this area. We have

$$\int_0^{15} (375 - x^2 - 10x)\, dx = \left(375x - \frac{x^3}{3} - 5x^2 \right) \Big|_0^{15}$$

$$= 375(15) - \frac{15^3}{3} - 5(15)^2$$

$$= 3375.$$

The producer's surplus is $3375.

7.5 The Area Between Two Curves

When the consumer is willing and able to pay more than the equilibrium price, he has gained. This gain, called the **consumer's surplus**, can also be evaluated by a definite integral as follows:

$$\int_0^{15} (900 - 20x - x^2 - 375) \, dx = \int_0^{15} (525 - 20x - x^2) \, dx$$

$$= 525x - 10x^2 - \frac{x^3}{3} \Big]_0^{15}$$

$$= 525(15) - 10(15)^2 - \frac{15^3}{3}$$

$$= 4500.$$

The consumer's surplus is $4500.

7.5 Exercises

Find the areas between the following curves.

1. $x = -3, x = 1, f(x) = x + 1, y = 0$
2. $x = -2, x = 0, f(x) = 1 - x^2, y = 0$
3. $x = -2, x = 1, f(x) = (x + 1)^2 - 1, y = 0$
4. $x = 2, x = 1, f(x) = x^3, y = 0$
5. $x = 0, x = 5, f(x) = \frac{7}{5}x, g(x) = \frac{3}{5}x + 10$
6. $x = 0, x = 1, f(x) = x, S(x) = x^2$
7. $x = 0, x = 2, f(x) = 150 - x^2, S(x) = x^2 + \frac{11}{4}x$
8. $x = 0, x = 8, f(x) = 150 - x^2, g(x) = x^2 + \frac{11}{4}x$
9. $x = 0, x = 2, f(x) = x^2, g(x) = \frac{1}{2}x$
10. $x = 1, x = 4, f(x) = x^2, g(x) = x^3$
11. Suppose the supply function for a certain commodity is given by
    $$S(x) = 100 + 3x + x^2,$$
    where x is the price. Suppose that supply and demand are in equilibrium at a price of $x^* = 3$. Find the producer's surplus.
12. Find the consumer's surplus if the demand for an item is given by
    $$D(x) = 50 - x^2,$$
    at a price of x, assuming supply and demand are in equilibrium at a price of $x^* = 5$.
13. Find the consumer's surplus if the demand for an item is given by
    $$D(x) = -(x + 4)^2 + 66,$$
    where x is the price, if supply and demand are in equilibrium at a price of $x^* = 3$.

14. Suppose the supply of a certain item is given by

$$S(x) = \frac{7}{5}x,$$

and the demand is given by

$$D(x) = -\frac{3}{5}x + 10,$$

where x is the price.
(a) Graph the supply and demand curves.
(b) Find the price at which supply and demand are in equilibrium.
(c) Find the supply at the equilibrium price.
(d) Find the consumer's surplus.
(e) Find the producer's surplus.

15. Repeat the five steps in Exercise 14 for the supply function

$$S(x) = x^2 + \frac{11}{4}x$$

and the demand function

$$D(x) = 150 - x^2.$$

16. Suppose a company wants to introduce a new machine which will produce annual savings given by

$$S(x) = 150 - x^2,$$

where x is the number of years of operation of the machine, while producing costs of

$$C(x) = x^2 + \frac{11}{4}x.$$

(a) For how many years will it be profitable to use this new machine?
(b) What are the total savings during the first year of use of the machine?
(c) What are the total savings over the entire period of use of the machine?

17. A new smog control device will reduce the output of oxides of sulfur from automobile exhausts. It is estimated that the total savings to the community from the use of this device will be approximated by

$$S(x) = -x^2 + 4x + 8,$$

where $S(x)$ is the rate of savings in millions of dollars after x years of use of the device. The new device cuts down on the production of oxides of sulfur, but causes an increase in the production of oxides of nitrogen. The additional costs to the community are approximated by

$$C(x) = \frac{3}{25}x^2,$$

where $C(x)$ is the additional cost in millions after x years.
(a) For how many years will it pay to use the new device?
(b) What will be the total savings over this period of time?

7.6 Additional Techniques of Integration

In this section we discuss some additional techniques for integrating functions. Recall that the integral sign $\int$ is used to represent antiderivatives. For example,

$$\int (6x^2 - 8x + 3)\,dx = 2x^3 - 4x^2 + 3x + C,$$

since

$$\frac{d(2x^3 - 4x^2 + 3x + C)}{dx} = 6x^2 - 8x + 3.$$

Using this notation, Theorem 7.1 can be restated as

$$\int x^n\,dx = \frac{1}{n+1}x^{n+1} + C, \qquad n \neq -1.$$

Many integrals can be found by using an analogue of the chain rule for derivatives that was discussed in Chapter 4. However, before discussing this procedure, we need to define the differential. In Figure 7.13, PM is the line tangent to the graph of the function $y = f(x)$ at the point P. Line PM then has slope $f'(x)$. If we let $PR = dx = \Delta x$, and $MR = dy$, then by the definition of slope, the slope of line PM can be written as dy/dx. Since the slope of the tangent is given by the derivative, we have

$$f'(x) = dy/dx,$$

or

$$dy = f'(x)\,dx.$$

Figure 7.13

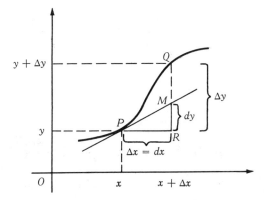

We call dy the **differential of y** and dx the **differential of x**. For small values of Δx and $\Delta x \neq 0$,

$$dy \approx \Delta y.$$

Example 14 Find dy if $y = x^2 + 3x$.

Since $dy = f'(x)\,dx$, we have
$$dy = (2x + 3)\,dx.$$

The differential can be used to extend our techniques of integration. For example, to integrate
$$\int 2x\sqrt{x^2 - 1}\,dx = \int 2x(x^2 - 1)^{1/2}\,dx,$$

which cannot be integrated by the methods presented so far, we shall use substitution to change the function to one which can be integrated by Theorem 7.1. We let $u = x^2 - 1$. The derivative of $x^2 - 1$ is $2x$. Then, by definition, $du = f'(x)\,dx = 2x\,dx$. By substituting u for $x^2 - 1$ and du for $2x\,dx$ in the integral above, we get
$$\int 2x(x^2 - 1)^{1/2}\,dx = \int u^{1/2}\,du, \quad \text{where } u = x^2 - 1.$$

We can find $\int u^{1/2}\,du$ by Theorem 7.1 as follows:
$$\int u^{1/2}\,du = \frac{u^{3/2}}{3/2} + C = \frac{2}{3}u^{3/2} + C.$$

We can now substitute $x^2 - 1$ for u:
$$\int 2x\sqrt{x^2 - 1}\,dx = \frac{2}{3}u^{3/2} + C = \frac{2(x^2 - 1)^{3/2}}{3} + C.$$

Example 15 Find $\int 6x(3x^2 + 4)^4\,dx$.

If we let $u = 3x^2 + 4$, then, since the derivative of $3x^2 + 4$ is $6x$, we have $du = 6x\,dx$. Thus,
$$\int 6x(3x^2 + 4)^4\,dx = \int (3x^2 + 4)^4(6x\,dx) = \int u^4\,du.$$

Since
$$\int u^4\,du = \frac{u^5}{5} + C,$$

and since $u = 3x^2 + 4$, we then have
$$\int 6x(3x^2 + 4)^4\,dx = \frac{u^5}{5} + C = \frac{(3x^2 + 4)^5}{5} + C,$$

as can be verified by using the chain rule to find the derivative of $(3x^2 + 4)^5/5 + C$.

Example 16 Find
$$\int x^2\sqrt{x^3 + 1}\,dx.$$

7.6 Additional Techniques of Integration

If we let $u = x^3 + 1$, then $du/dx = 3x^2$ and $du = 3x^2 \, dx$. The integral does not contain the constant 3, which is needed for du. To take care of this, we solve the equation $du = 3x^2 \, dx$ for dx as follows:

$$du = 3x^2 \, dx$$

$$dx = \frac{du}{3x^2}.$$

Now we substitute into the given integral u for $x^3 + 1$ and $du/3x^2$ for dx.

$$\int x^2 \sqrt{x^3 + 1} \, dx = \int x^2 u^{1/2} \cdot \frac{du}{3x^2}$$

$$= \int \frac{1}{3} u^{1/2} \, du$$

$$= \frac{1}{3} \int u^{1/2} \, du$$

$$= \frac{1}{3} \left(\frac{2}{3} u^{3/2} \right) + C$$

$$= \frac{2}{9} u^{3/2} + C.$$

Since $u = x^3 + 1$, we write

$$\int x^2 \sqrt{x^3 + 1} \, dx = \frac{2}{9} (x^3 + 1)^{3/2} + C.$$

The substitution method given in the examples above will not always work. For example, we might try to find

$$\int x^3 \sqrt{x^3 + 1} \, dx$$

by substituting $u = x^3 + 1$, so that $du/dx = 3x^2$ and $du = 3x^2 \, dx$. From this last result, we have $dx = du/3x^2$. If we substitute in the given integral, we get

$$\int x^3 \sqrt{x^3 + 1} \, dx = x^3 \cdot u^{1/2} \cdot \frac{du}{3x^2}$$

$$= xu^{1/2} \, du,$$

an expression which cannot be evaluated by the methods of this course. In general, to find $\int f(x) \, dx$, it must be the case that dx is the *exact* differential of $f(x)$, except for a nonzero constant.

Example 17 Find the value of

$$\int_0^4 (x^2 + 5x)^{3/2}(2x + 5) \, dx.$$

238 Integration

To evaluate this definite integral, we first find

$$\int (x^2 + 5x)^{3/2}(2x + 5)\, dx.$$

Let $u = (x^2 + 5x)$, so that $du = (2x + 5)\, dx$. This gives

$$\int (x^2 + 5x)^{3/2}(2x + 5)\, dx = \int u^{3/2}\, du$$

$$= \frac{2}{5} u^{5/2} + C.$$

We cannot use here the limits of integration given in the original problem, 0 and 4, since they are with reference to x, and this antiderivative contains u. Thus, we must substitute $x^2 + 5x$ for u, obtaining

$$\frac{2}{5} u^{5/2} + C = \frac{2}{5}(x^2 + 5x)^{5/2} + C.$$

This antiderivative can now be used to evaluate the given definite integral. Using the fundamental theorem of calculus, we have

$$\int_0^4 (x^2 + 5x)^{3/2}(2x + 5)\, dx = \frac{2}{5}(x^2 + 5x)^{5/2}\Big|_0^4$$

$$= \frac{2}{5}(4^2 + 5\cdot 4)^{5/2} - \frac{2}{5}(0^2 + 5\cdot 0)^{5/2}$$

$$= \frac{2}{5}(36)^{5/2} - 0$$

$$= \frac{2}{5}(6)^5.$$

Recall that it is not necessary to consider the constant C when evaluating a definite integral.

The formulas presented in Chapter 6 for derivatives of exponential and logarithmic functions have analogues for integrals which are given in the next three theorems.

Theorem 7.7 $\int e^x\, dx = e^x + C.$

Example 18 Find $\int e^{4x}\, dx.$

Let $u = 4x$, so that $du/dx = 4$ and $du = 4\, dx$. Thus, $dx = du/4$. Upon substituting these values into $\int e^{4x}\, dx$, we get

$$\int e^{4x}\, dx = \int e^u \cdot \frac{du}{4}$$

7.6 Additional Techniques of Integration

$$\int e^{4x}\,dx = \frac{1}{4}\int e^u\,du$$

$$= \frac{1}{4}e^u + C$$

$$= \frac{1}{4}e^{4x} + C.$$

Based on results similar to this one, we can state a more general form of Theorem 7.7, as follows.

Theorem 7.8 For any real number $a \neq 0$, we have

$$\int e^{ax}\,dx = \frac{1}{a}e^{ax} + C.$$

The following example shows how substitution can be used in integrals involving exponentials.

Example 19 Evaluate $\int xe^{x^2}\,dx$.

If we let $u = x^2$, then $du = 2x\,dx$, and $dx = du/(2x)$. Hence,

$$\int xe^{x^2}\,dx = \int xe^u \cdot \frac{du}{2x}$$

$$= \frac{1}{2}\int e^u\,du$$

$$= \frac{1}{2}e^u + C$$

$$= \frac{1}{2}e^{x^2} + C.$$

In Theorem 7.1, which showed how to find $\int x^n\,dx$, we said that $n \neq -1$. The case $n = -1$ is taken care of with the next theorem.

Theorem 7.9 $\int \frac{1}{x}\,dx = \int x^{-1}\,dx = \ln x + C.$

As we mentioned in Chapter 6, the domain of this logarithm function is the set of positive real numbers. Thus, when we write an expression such as $\ln x$, it is always understood that $x > 0$.

Example 20 $\int \frac{4}{x}\,dx = 4\int \frac{1}{x}\,dx = 4\ln x + C.$

240 Integration

Example 21 Find $\int \frac{(2x - 3) \, dx}{x^2 - 3x}$.

We can let $u = x^2 - 3x$, so that $du = (2x - 3) \, dx$. Then,

$$\int \frac{(2x - 3) \, dx}{x^2 - 3x} = \int \frac{du}{u}$$
$$= \ln u + C$$
$$= \ln (x^2 - 3x) + C.$$

The integral $\int \ln x \, dx$ is more complicated than those presented here and will be discussed in Section 7.7.

7.6 Exercises

Find the following antiderivatives.

1. $\int 5x^6 \, dx$
2. $\int 8x^{-2} \, dx$
3. $\int (4x^2 - 6x + 2) \, dx$
4. $\int (-3x^3 + 2x^2 - 4x + 1) \, dx$
5. $\int \frac{4}{x^3} \, dx$
6. $\int \frac{-3}{x^4} \, dx$
7. $\int \sqrt{x} \, dx$
8. $\int (\sqrt{x})^3 \, dx$
9. $\int 2x(x^2 + 1)^3 \, dx$
10. $\int 3x^2(x^3 - 4)^3 \, dx$
11. $\int x\sqrt{x^2 - 5} \, dx$
12. $\int x\sqrt{x^2 + 2} \, dx$
13. $\int (\sqrt{x^2 + 12x})(2x + 12) \, dx$
14. $\int (\sqrt{x^2 - 6x})(2x - 6) \, dx$
15. $\int x^2\sqrt{x^3 + 1} \, dx$
16. $\int x\sqrt{x^2 + 1} \, dx$
17. $\int e^{2x} \, dx$
18. $\int 3e^{-2x} \, dx$
19. $\int (-4e^{2x}) \, dx$
20. $\int 5e^{-0.3x} \, dx$
21. $\int \frac{-8}{x} \, dx$
22. $\int \frac{9}{x} \, dx$
23. $\int \frac{5}{x + 1} \, dx$
24. $\int \frac{4}{x - 4} \, dx$
25. $\int x^2 e^{x^3} \, dx$
26. $\int x e^{-x^2} \, dx$

27. $\int \dfrac{dx}{2x+1}$

28. $\int \dfrac{dx}{5x-2}$

29. $\int \dfrac{6x\,dx}{(3x^2+2)^4}$

30. $\int \dfrac{4x\,dx}{(2x^2-5)^3}$

31. $\int \dfrac{1-x}{(2x-x^2)^2}\,dx$

32. $\int \dfrac{4-6x}{(4x-3x^2)^3}\,dx$

33. $\int \dfrac{x-1}{(2x^2-4x)^2}\,dx$

34. $\int \dfrac{2x+1}{(x^2+x)^3}\,dx$

35. $\int \left(\dfrac{1}{x}+x\right)\left(1-\dfrac{1}{x^2}\right)dx$

36. $\int \left(\dfrac{2}{x}-x\right)\left(\dfrac{-2}{x^2}-1\right)dx$

37. $\int (2x^3-2x)(3x^2-1)\,dx$

38. $\int 5(x^4-x^2)(4x^3-2x)\,dx$

Find each of the following definite integrals.

39. $\displaystyle\int_1^4 \sqrt{x}\,dx$

40. $\displaystyle\int_4^9 2x^{1/2}\,dx$

41. $\displaystyle\int_0^1 2x(x^2+1)^3\,dx$

42. $\displaystyle\int_0^1 3x^2(x^3-4)^3\,dx$

43. $\displaystyle\int_0^4 (\sqrt{x^2+12x})(2x+12)\,dx$

44. $\displaystyle\int_0^8 (\sqrt{x^2-6x})(2x-6)\,dx$

45. $\displaystyle\int_1^e \dfrac{1}{x}\,dx$ (Hint: ln $e=1$)

46. $\displaystyle\int_1^e \dfrac{-4}{x}\,dx$

47. $\displaystyle\int_0^1 e^{2x}\,dx$

48. $\displaystyle\int_1^2 e^{-x}\,dx$

49. $\displaystyle\int_2^4 e^{-0.2x}\,dx$

50. $\displaystyle\int_1^4 e^{-0.3x}\,dx$

7.7 Tables of Integrals

It is a fairly straightforward task to find the derivative of most of the useful functions. The chain rule, together with the other formulas of Chapter 4, can be used to find derivatives of almost all functions that are useful in practical applications. However, this is not true for integration. There are many useful functions whose integrals cannot be found by the methods we have discussed. For example, $f(x)=e^{-x^2}$, used to obtain the normal curve of statistics, is one of many possible examples. Some calculus courses spend a considerable amount of time on techniques of integration. These techniques often depend on trigonometry and complicated algebraic substitution and manipulation.

Another, less time-consuming, approach is possible. This is to list all commonly needed integrals in a *table of integrals*. This table can then be referred to as needed. One such table is included as Table 6, in the Appendix. Larger and more complete tables are listed in the books in the bibliography. The remainder of this section will be devoted to examples showing how to use Table 6.

Example 22 Find $\int \frac{1}{\sqrt{x^2 + 16}} \, dx$.

By inspecting Table 6, we see that if $a = 4$, this antiderivative is the same as entry 5 of the table. Entry 5 of the table reads as follows:

$$\int \frac{1}{\sqrt{x^2 + a^2}} \, dx = \ln \left(\frac{x + \sqrt{x^2 + a^2}}{a} \right) + C.$$

By substituting 4 for a in this entry of the table, we get

$$\int \frac{1}{\sqrt{x^2 + 16}} \, dx = \ln \left(\frac{x + \sqrt{x^2 + 16}}{4} \right) + C.$$

This result could be verified by taking the derivative of the right-hand side of this last equation.

Example 23 Find $\int \frac{8}{16 - x^2} \, dx$.

We can convert this antiderivative into the one given in entry 7 of the table by writing the 8 in front of the integral sign (permissible only with constants) and by letting $a = 4$. Doing this gives

$$8 \int \frac{1}{16 - x^2} \, dx = 8 \left[\frac{1}{2 \cdot 4} \ln \left(\frac{4 + x}{4 - x} \right) \right] + C$$

$$= \ln \left(\frac{4 + x}{4 - x} \right) + C.$$

In entry 7 of the table, the condition $x^2 < a^2$ is given. Here $a = 4$. Hence, the result given above is valid only for $x^2 < 16$, so that the final answer should be written as follows:

$$\int \frac{8}{16 - x^2} \, dx = \ln \left(\frac{4 + x}{4 - x} \right) + C, \quad \text{for } x^2 < 16.$$

Example 24 Evaluate $\int \frac{12x}{2x + 1} \, dx$.

For $a = 2$ and $b = 1$, this antiderivative can be rewritten to match entry 11 of the table. Thus

$$\int \frac{12x}{2x + 1} \, dx = 12 \int \frac{x}{2x + 1} \, dx$$

$$= 12 \left[\frac{x}{2} - \frac{1}{4} \ln (2x + 1) \right] + C$$

$$= 6x - 3 \ln (2x + 1) + C.$$

Example 25 Find $\int \sqrt{9x^2 + 1} \, dx$.

This antiderivative seems most similar to entry 15 of the table. However, entry 15 requires that the coefficient of the x^2 term be 1. We can satisfy that requirement here by factoring out the 9, as follows:

$$\int \sqrt{9x^2 + 1} \, dx = \int \sqrt{9\left(x^2 + \frac{1}{9}\right)}$$

$$= \int 3\sqrt{x^2 + \frac{1}{9}} \, dx$$

$$= 3 \int \sqrt{x^2 + \frac{1}{9}} \, dx.$$

Now, using entry 15 with $a = 1/3$, we have

$$\int \sqrt{9x^2 + 1} \, dx = 3\left[\frac{x}{2}\sqrt{x^2 + \frac{1}{9}} + \frac{\left(\frac{1}{3}\right)^2}{2} \cdot \ln\left(x + \sqrt{x^2 + \frac{1}{9}}\right)\right] + C$$

$$= \frac{3x}{2}\sqrt{x^2 + \frac{1}{9}} + \frac{1}{6}\ln\left(x + \sqrt{x^2 + \frac{1}{9}}\right) + C.$$

(We might mention here that there is no table entry in our table for the antiderivative of $\sqrt{x^2 - a^2}$, since that result requires a trigonometric function.)

Example 26 Find $\int x^2 e^x \, dx$.

Using entry 17, with $n = 2$ and $a = 1$, we have

$$\int x^2 e^x \, dx = \frac{x^2 e^x}{1} - \frac{2}{1}\int xe^x \, dx + C. \tag{1}$$

We must now use entry 17 again, this time to find $\int xe^x \, dx$. Here $n = 1$ and $a = 1$. Thus,

$$\int xe^x \, dx = \frac{xe^x}{1} - \frac{1}{1}\int x^0 e^x \, dx + K$$

$$= xe^x - \int e^x \, dx + K \quad \text{(Recall: } x^0 = 1\text{)}$$

$$= xe^x - e^x + K.$$

Substituting this result back into equation **(1)** gives

$$\int x^2 e^x \, dx = x^2 e^x - 2(xe^x - e^x + K) + C$$

$$= x^2 e^x - 2xe^x + 2e^x + M$$

$$= e^x(x^2 - 2x + 2) + M,$$

where M is an arbitrary constant.

7.7 Exercises

Find each of the following antiderivatives, using Table 6.

1. $\int \ln 4x \, dx$
2. $\int \ln \frac{3}{5} x \, dx$
3. $\int \frac{-4}{\sqrt{x^2 + 36}} \, dx$
4. $\int \frac{9}{\sqrt{x^2 + 9}} \, dx$
5. $\int \frac{6}{x^2 - 9} \, dx$
6. $\int \frac{-12}{x^2 - 16} \, dx$
7. $\int \frac{-4}{x\sqrt{9 - x^2}} \, dx$
8. $\int \frac{3}{x\sqrt{121 - x^2}} \, dx$
9. $\int \frac{-2x}{3x + 1} \, dx$
10. $\int \frac{6x}{4x - 5} \, dx$
11. $\int \frac{2}{3x(3x - 5)} \, dx$
12. $\int \frac{-4}{3x(2x + 7)} \, dx$
13. $\int \frac{4}{4x^2 - 1} \, dx$
14. $\int \frac{-6}{9x^2 - 1} \, dx$
15. $\int \frac{3}{x\sqrt{1 - 9x^2}} \, dx$
16. $\int \frac{-2}{x\sqrt{1 - 16x^2}} \, dx$
17. $\int x^4 \ln x \, dx$
18. $\int 4x^2 \ln x \, dx$
19. $\int \frac{\ln x}{x^2} \, dx$
20. $\int \frac{-2 \ln x}{x^3} \, dx$
21. $\int xe^{-2x} \, dx$
22. $\int xe^{3x} \, dx$
23. $\int 2x^2 e^{-2x} \, dx$
24. $\int -3x^2 e^{-4x} \, dx$
25. $\int x^3 e^x \, dx$
26. $\int x^3 e^{2x} \, dx$

7.8* Integration by Parts

In this section we introduce a technique of integration which often makes it possible to reduce a complicated integral to a simpler integral. If u and v are both differentiable functions, then uv is also differentiable and, by the product rule for derivatives,

$$\frac{d(uv)}{dx} = u \frac{dv}{dx} + v \frac{du}{dx}.$$

We can rewrite this expression, using differentials, as

$$d(uv) = u \, dv + v \, du.$$

Now, integrating both sides of this last equation, we have

$$\int d(uv) = \int u \, dv + \int v \, du,$$

or

$$uv = \int u \, dv + \int v \, du.$$

---

* Material in this section will not be used in later chapters.

7.8 Integration by Parts

By rearranging terms, we obtain the formula:

$$\int u\, dv = uv - \int v\, du.$$

The technique of using this formula to find integrals, called **integrating by parts**, is illustrated in the following examples.

Example 27 Find $\int x\sqrt{1-x}\, dx$.

To use integration by parts here, first write the expression $x\sqrt{1-x}$ as a product of functions u and dv in such a way that $\int dv$ can be found. Let us select $u = x$ and $dv = \sqrt{1-x}\, dx$. Then $du = dx$. We find v by integrating dv. Therefore,

$$v = \int \sqrt{1-x}\, dx = -\frac{2}{3}(1-x)^{3/2} + C.$$

(Verify this.) For simplicity, we shall ignore the constant C, and just add it at the end. Thus, $v = -\frac{2}{3}(1-x)^{3/2}$. Now substitute into the formula for integration by parts as follows.

$$\int u\, dv = uv - \int v\, du,$$

$$\int x\sqrt{1-x}\, dx = x\left[-\frac{2}{3}(1-x)^{3/2}\right] - \int -\frac{2}{3}(1-x)^{3/2}\, dx$$

$$= -\frac{2}{3}x(1-x)^{3/2} - \frac{2}{3}\int -(1-x)^{3/2}\, dx$$

$$= -\frac{2}{3}x(1-x)^{3/2} - \frac{2}{3}\left[\frac{2}{5}(1-x)^{5/2}\right]$$

$$\int x\sqrt{1-x}\, dx = -\frac{2}{3}x(1-x)^{3/2} - \frac{4}{15}(1-x)^{5/2} + C.$$

We added a constant C in the last step.

Example 28 Find $\int x^3\sqrt{x^2+4}\, dx$.

To be able to find v, we must be able to integrate dv. This can be done here by the substitution technique of Section 7.6. If we choose $dv = x\sqrt{x^2+4}\, dx$, then

$$v = \int x\sqrt{x^2+4}\, dx = \int x(x^2+4)^{1/2}\, dx$$

$$= \frac{1}{3}(x^2+4)^{3/2}.$$

Then $u = x^2$ and $du = 2x\,dx$. Now using the formula for integration by parts, we have

$$\int u\,dv = uv - \int v\,du$$

$$\int x^3\sqrt{x^2+4}\,dx = x^2\left[\frac{1}{3}(x^2+4)^{3/2}\right] - \int \frac{1}{3}(x^2+4)^{3/2}(2x\,dx). \quad (1)$$

Now find the integral on the right side in equation (1).

$$\int \frac{1}{3}(x^2+4)^{3/2}(2x\,dx) = \frac{1}{3}\int (x^2+4)^{3/2}(2x\,dx)$$

$$= \frac{2}{5}\cdot\frac{1}{3}(x^2+4)^{5/2}$$

$$= \frac{2}{15}(x^2+4)^{5/2}$$

Substitute this result back into equation (1) and add a constant C. This gives

$$\int x^3\sqrt{x^2+4}\,dx = \frac{1}{3}x^2(x^2+4)^{3/2} - \frac{2}{15}(x^2+4)^{5/2} + C.$$

Sometimes it is necessary to use the technique of integrating by parts more than once as in the following example.

Example 29 Find $\int 2x^2 e^x\,dx$.

Since $\int e^x\,dx$ can be found, let us choose $dv = e^x\,dx$, so that $v = e^x$. Then $u = 2x^2$ and $du = 4x\,dx$. Now we substitute into the formula for integrating by parts.

$$\int u\,dv = uv - \int v\,du$$

$$\int 2x^2 e^x\,dx = 2x^2 e^x - \int e^x(4x\,dx). \quad (2)$$

We must find $\int e^x(4x\,dx)$ by parts. Again, we choose $dv = e^x\,dx$, which gives $v = e^x$, $u = 4x$, and $du = 4\,dx$. Substituting again into the formula for integrating by parts, we have the following.

$$\int e^x(4x\,dx) = 4xe^x - \int e^x(4\,dx)$$

$$= 4xe^x - 4e^x.$$

Now we must substitute back into our first result, equation (2), to get the final answer.

$$\int 2x^2 e^x\,dx = 2x^2 e^x - (4xe^x - 4e^x)$$

$$= 2x^2 e^x - 4xe^x + 4e^x + C,$$

where a constant C was added at the last step.

7.8 Integration by Parts

The method of integration by parts requires choosing the factor dv so that $\int dv$ can be found. If this is not possible, or if the remaining factor, which becomes u, does not have a differential du such that $v\,du$ can be integrated, the technique cannot be used. For example, to integrate

$$\int \frac{1}{4-x^2}\,dx$$

we might choose $dv = dx$ and $u = (4-x^2)^{-1}$. Then $v = x$ and $du = 2x/(4-x^2)^2$. Then we have

$$\int \frac{1}{4-x^2}\,dx = \frac{x}{4-x^2} - \int \frac{2x^2\,dx}{(4-x^2)^2},$$

where the integral on the right is more complicated than the original integral was. A second use of integration by parts on the new integral would make matters even worse. Since we cannot choose $dv = (4-x^2)^{-1}$ because we cannot integrate it, integration by parts is not suitable for this problem. In fact, there are many functions whose integrals cannot be found by any of the methods we have described. Many of these can be found by more advanced methods and are available in tables of integrals, discussed in Section 7.7. However, there are functions which cannot be integrated at all, for example, $f(x) = e^{-x^2}$, a function which is important in statistical theory, as mentioned in Section 7.7.

7.8 Exercises

Use integration by parts to find the following integrals.

1. $\int x(x+1)^5\,dx$
2. $\int x\sqrt{5x-1}\,dx$
3. $\int x^3(4-x^2)^{3/2}\,dx$
4. $\int x^5\sqrt{2x^3+1}\,dx$
5. $\int x^3(x^2-1)^8\,dx$
6. $\int x^3(1+x^2)^{1/4}\,dx$
7. $\int 2x^5(1-3x^3)^{1/2}\,dx$
8. $\int \frac{x}{\sqrt{x-1}}\,dx$
9. $\int \frac{2x}{(x+5)^6}\,dx$
10. $\int \frac{x^3\,dx}{\sqrt{3-x^2}}$
11. $\int \frac{x\,dx}{\sqrt{1-x}}$
12. $\int \frac{2x^3\,dx}{(x^2-5)^5}$
13. $\int \frac{x^7\,dx}{(x^4+3)^{2/3}}$
14. $\int 2x^2(x+1)^{2/3}\,dx$
15. $\int 3x^2(x+2)^{1/2}\,dx$
16. $\int xe^x\,dx$

248 *Integration*

17. $\int (x + 1)e^x \, dx$

18. $\int (x^2 + 1)e^x \, dx$

19. $\int 3x^2 e^{-x} \, dx$

20. $\int (1 - x^2)e^{2x} \, dx$

21. Find the area between the curves $y = e^x$ and $y = x + 4$ from $x = 0$ to $x = 2$.

22. Find the area between $y = xe^x$ and the x-axis from $x = 0$ to $x = 1$.

7.9* Numerical Integration

As mentioned in Section 7.8, some integrals cannot be evaluated by any technique or found in any table. Since $\int_a^b f(x) \, dx$ represents an area, any approximation of that area approximates the definite integral. Many methods of approximating definite integrals by areas are in use today, made feasible by the availability of pocket calculators and the high-speed computer. These methods are referred to as **numerical integration**. We shall discuss two methods of numerical integration, the *trapezoidal rule* and *Simpson's rule*.

Figure 7.14

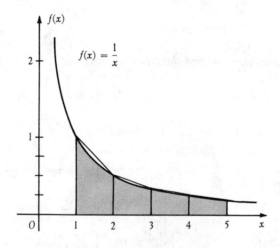

As an example, let us consider

$$\int_1^5 \frac{1}{x} \, dx.$$

* Material in this section will not be used in later chapters.

7.9 Numerical Integration

Figure 7.14 shows the area under the graph of $f(x) = 1/x$, above the x-axis, and between the lines $x = 1$ and $x = 5$. Since $\int (1/x) \, dx = \ln x + C$, we have

$$\int_1^5 \frac{1}{x} \, dx = \ln x \Big]_1^5$$
$$= \ln 5 - \ln 1$$
$$= \ln 5 - 0$$
$$= \ln 5,$$

a number which can be found in Table 4. Using that table, we find that $\ln 5 \approx 1.6094$. How was this number in the table obtained? One way to find this number is to approximate the area under the curve in Figure 7.14. This can be done by the trapezoidal rule. To get a first approximation to $\ln 5$ by the trapezoidal rule, we might first find the sum of the areas of the four trapezoids indicated in Figure 7.14. The area of a trapezoid is given by one-half the product of the sum of the bases and the altitude. Each of the trapezoids in Figure 7.14 has height 1. Thus, we have

$$\ln 5 = \int_1^5 \frac{1}{x} \, dx \approx \frac{1}{2}\left(\frac{1}{1} + \frac{1}{2}\right)(1) + \frac{1}{2}\left(\frac{1}{2} + \frac{1}{3}\right)(1) + \frac{1}{2}\left(\frac{1}{3} + \frac{1}{4}\right)(1) + \frac{1}{2}\left(\frac{1}{4} + \frac{1}{5}\right)(1)$$

$$\approx \frac{1}{2}\left(\frac{3}{2} + \frac{5}{6} + \frac{7}{12} + \frac{9}{20}\right)$$

$$\approx 1.68.$$

To get a better approximation, we would divide the interval $1 \le x \le 5$ into more subintervals. The larger the number of subintervals, the better the approximation will be. A general statement of the trapezoidal rule follows.

Trapezoidal Rule Let f be a continuous function on $a \le x \le b$. Let $a \le x \le b$ be divided into n equal subintervals by the points $a = x_0, x_1, x_2, \ldots, x_n = b$. Then

$$\int_a^b f(x) \, dx$$

$$\approx \frac{b-a}{n}\left[\frac{1}{2}f(x_0) + f(x_1) + \cdots + f(x_{n-1}) + \frac{1}{2}f(x_n)\right].$$

Another numeration method, called **Simpson's rule**, approximates consecutive portions of the curve with portions of parabolas, rather than with line segments as in the trapezoidal rule. As shown in Figure 7.15, a parabola is fitted through points A, B, and C, another through C, D, and E, and so on. (It is necessary to have an even number of intervals for this process to come out right.) Then the sum of the areas under these parabolas will approximate the area under the graph of the function.

Figure 7.15

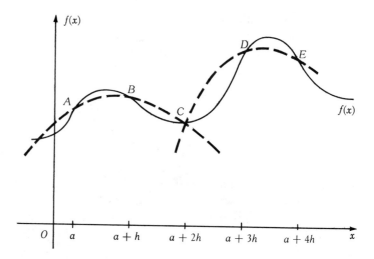

Simpson's Rule Let f be a continuous function on $a \leq x \leq b$. Let $a \leq x \leq b$ be divided into an even number n of equal subintervals by the points $a = x_0, x_1, x_2, \ldots, x_n = b$. Then,

$$\int_a^b f(x)\, dx \approx \frac{b-a}{3n} [f(x_0) + 4f(x_1) + 2f(x_2) + 4f(x_3) + \cdots$$
$$+ 2f(x_{n-2}) + 4f(x_{n-1}) + f(x_n)].$$

Note that it is necessary for n to be an *even* number. For example, let us use the same function, $\frac{1}{x}$, and divide the interval $1 \leq x \leq 5$ into 4 subintervals, in the same way as in the example of the trapezoidal rule. By Simpson's rule,

$$\int_1^5 \frac{1}{x}\, dx \approx \frac{4}{12}\left[\frac{1}{1} + 4\left(\frac{1}{2}\right) + 2\left(\frac{1}{3}\right) + 4\left(\frac{1}{4}\right) + \frac{1}{5}\right]$$
$$\approx \frac{1}{3}\left(1 + 2 + \frac{2}{3} + 1 + \frac{1}{5}\right)$$
$$\approx 1.62.$$

Simpson's rule gives a better approximation than the trapezoidal rule. However, as n is increased, they would differ by less and less.

7.9 Exercises

In Exercises 1–10, using $n = 4$, approximate the value of each of the given integrals (a) by the trapezoidal rule; (b) by Simpson's rule.

1. $\int_0^2 x^2\, dx$

2. $\int_0^2 (2x + 1)\, dx$

3. $\int_{0}^{4} \sqrt{x+1}\, dx$

4. $\int_{1}^{5} \frac{1}{x+1}\, dx$

5. $\int_{-2}^{2} (2x^2 + 1)\, dx$

6. $\int_{0}^{3} (2x^2 + 1)\, dx$

7. $\int_{1}^{5} \frac{1}{x^2}\, dx$

8. $\int_{-1}^{3} \frac{1}{4-x}\, dx$

9. $\int_{-4}^{-2} \frac{1}{x^3}\, dx$

10. $\int_{0}^{4} x\sqrt{2x-1}\, dx$

11. The table below shows the results from a chemical experiment.

Concentration of chemical A	1	2	3	4	5	6	7
Rate of formation of chemical B	12	16	18	21	24	27	32

(a) Plot these points. Connect the points with line segments.
(b) Use the trapezoidal rule to find the area bounded by the broken line of part (a), the x-axis, the line $x = 1$ and the line $x = 7$.
(c) Find the area with Simpson's rule.

12. The results from a research study in psychology were as follows.

Number of hours of study	1	2	3	4	5	6	7
Number of extra points earned on a test	4	7	11	9	15	16	23

Repeat steps (a)–(c) of Exercise 11 for this data.

● **Case 10 Estimating Depletion Dates for Minerals**

It is becoming more and more obvious that the earth contains only a finite quantity of minerals. The "easy and cheap" sources of minerals are being used up, forcing an ever more expensive search for new sources. For example, oil from the North Slope of Alaska would never have been used in the United States during the 1930's since there was so much Texas and California oil readily available.

We said in Chapter 6 that population tends to follow an exponential growth curve. Mineral usage also follows such a curve. Thus, if q represents the rate of consumption of a certain mineral at time t, while q_0 represents consumption when $t = 0$, then

$$q = q_0 e^{kt},$$

where k is the annual rate of increase in usage of the mineral. For example, the world consumption of petroleum in 1973 was about 19,600 million barrels, with the annual rate of increase running about 6%. If we let $t = 0$ correspond to 1973, then $q_0 = 19,600$, $k = 0.06$, and

$$q = 19,600 e^{0.06t}$$

is the rate of consumption at time t, assuming that all present trends continue.

Based on estimates of the National Academy of Science, we shall use 2,000,000 as the number of millions of barrels of oil that are now in provable reserves or that are likely to be discovered in the future. At the present rates of consumption, how many years would be necessary to deplete these estimated reserves? We can use the integral calculus of this chapter to find out.

To begin, we need to know the total quantity of petroleum that would be used between time $t = 0$ and some future time $t = t_1$. Figure 1 shows a typical graph of the function $q = q_0 e^{kt}$.

Figure 1

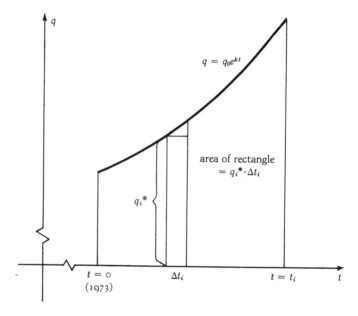

Following the work we did in Section 7.2, divide the time interval from $t = 0$ to $t = t_1$ into n subintervals. Let the ith subinterval have width Δt_i. Let the rate of consumption for the ith subinterval be approximated by q_i^*. Thus, the approximate total consumption for the subinterval is given by

$$q_i^* \cdot \Delta t_i,$$

and the total consumption over the interval from time $t = 0$ to $t = t_1$ is approximated by

$$\sum_{i=1}^{n} q_i^* \cdot \Delta t_i.$$

Case 10 Estimating Depletion Dates for Minerals   253

The limit of this sum as each of the Δt_i's approach 0 gives the total consumption from time $t = 0$ to $t = t_1$. That is,

$$\text{total consumption} = \lim_{\Delta t_i \to 0} \sum q_i^* \cdot \Delta t_i.$$

However, we have seen that this limit is the definite integral of the function $q = q_0 e^{kt}$ from $t = 0$ to $t = t_1$, or

$$\text{total consumption} = \int_0^{t_1} q_0 e^{kt}\, dt.$$

We can now evaluate this definite integral.

$$\int_0^{t_1} q_0 e^{kt}\, dt = q_0 \int_0^{t_1} e^{kt}\, dt$$

$$= q_0 \left[\frac{1}{k} e^{kt}\right]\Big|_0^{t_1}$$

$$= \frac{q_0}{k} e^{kt}\Big|_0^{t_1}$$

$$= \frac{q_0}{k} e^{kt_1} - \frac{q_0}{k} e^0$$

$$= \frac{q_0}{k} e^{kt_1} - \frac{q_0}{k} \quad (1)$$

$$= \frac{q_0}{k} (e^{kt_1} - 1). \qquad (1)$$

Now let us return to the numbers we gave for petroleum. We said that $q_0 = 19{,}600$ million barrels where q_0 represents consumption in 1973. We have $k = 0.06$, with total petroleum reserves estimated as 2,000,000 million barrels. Thus, using equation (1) we have

$$2{,}000{,}000 = \frac{19{,}600}{0.06}(e^{0.06 t_1} - 1).$$

Multiply both sides of the equation by 0.06:

$$120{,}000 = 19{,}600(e^{0.06 t_1} - 1).$$

Divide both sides of the equation by 19,600.

$$6.1 = e^{0.06 t_1} - 1.$$

Add 1 to both sides.

$$7.1 = e^{0.06 t_1}.$$

Take natural logarithms of both sides:

$$\ln 7.1 = \ln e^{0.06 t_1}$$
$$\ln 7.1 = 0.06 t_1 \ln e$$
$$\ln 7.1 = 0.06 t_1 \qquad \text{(since } \ln e = 1\text{)}.$$

Finally,

$$t_1 = \frac{\ln 7.1}{0.06}.$$

From Table 5, estimate $\ln 7.1$ as about 1.96. Thus,

$$t_1 = \frac{1.96}{0.06}$$
$$= 33.$$

By this result, petroleum reserves will last the world until about $1973 + 33 = 2006$.

The results of mathematical analyses such as this must be used with great caution. By the analysis above, the world would use all the petroleum that it wants in the year 2005, but there would be none at all in the year 2007. This is not at all realistic. As petroleum reserves decline, the price will increase, causing demand to decline, and supplies to increase.

Case 10   Exercises

1. Find the number of years that the estimated petroleum reserves would last if used at the 1973 rate.
2. How long would the estimated petroleum reserves last if the rate of growth of use was only 2% instead of 6%?

Estimate the length of time until depletion for each of the following minerals.

3. Bauxite (the ore from which aluminum is obtained) estimated reserves in 1973, 15,000,000 thousand tons, 1973 rate of consumption 63,000 thousand tons, annual rate of increase of consumption 6%.
4. Bituminous coal, estimated world reserves 2,000,000 million tons, rate of consumption in 1973, 2200 million tons, annual rate of increase of consumption 4%.

● Case 11   How Much Does a Warranty Cost?*

This case uses some of the ideas of probability. The probability of an event is a number p, where $0 \le p \le 1$, such that p is the ratio of the number of ways that

---

* Based on "Determination of Warranty Reserves," by Warren W. Menks, *Management Science*, June 1969, pp. B542–549.

Case 11 How Much Does a Warranty Cost? 255

the event can happen divided by the total number of possible outcomes. For example, the probability of drawing a red card from a deck of 52 cards (of which 26 are red) is given by

$$P \text{ (red card)} = \frac{26}{52} = \frac{1}{2}.$$

In the same way, the probability of drawing a black queen from a deck of 52 cards is

$$P \text{ (black queen)} = \frac{2}{52} = \frac{1}{26}.$$

In this case we find the cost of a warranty program to a manufacturer. This cost depends on the quality of the products made, as we might expect. This case uses the following variables.

$c =$ constant product price, per unit, including cost of warranty (We assume the price charged per unit is constant, since this price is likely to be fixed by competition.)

$m =$ expected lifetime of the product

$w =$ length of the warranty period

$N =$ size of a production lot, such as a year's production

$r =$ warranty cost per unit

$C(t) =$ pro rata customer rebate at time t

$P(t) =$ probability of product failure at any time t

$F(t) =$ number of failures occurring at time t.

We shall assume that the warranty is of the pro rata customer rebate type, in which the customer is paid for the proportion of the warranty left. Hence, if the product has a warranty period of w, and fails at time t, then the product worked for the fraction t/w of the warranty period. Hence, the customer is reimbursed for the unused portion of the warranty, or the fraction

$$1 - \frac{t}{w}.$$

If we assume the product cost c originally, and if we use $C(t)$ to represent the customer rebate at time t, we have

$$C(t) = c\left(1 - \frac{t}{w}\right).$$

For many different types of products, it has been shown by experience that

$$P(t) = 1 - e^{-t/m}$$

provides a good estimate of the probability of product failure at time t. The total number of failures at time t is given by the product of $P(t)$ and N, the total number of items per batch. If we use $F(t)$ to represent this total, we have

$$F(t) = N \cdot P(t) = N(1 - e^{-t/m}).$$

The total number of failures in some "tiny time interval" of width dt can be shown to be the derivative of $F(t)$,

$$F'(t) = \left(\frac{N}{m}\right)e^{-t/m},$$

while the cost for the failures in this "tiny time interval" is

$$C(t) \cdot F'(t) = c\left(1 - \frac{t}{m}\right)\left(\frac{N}{m}\right)e^{-t/m}.$$

The total cost for all failures during the warranty period is thus given by the definite integral

$$\int_0^w c\left(1 - \frac{t}{w}\right)\left(\frac{N}{m}\right)e^{-t/m}\,dt.$$

Using integration by parts, this definite integral can be shown to equal

$$Nc\left[-e^{-t/m} + \frac{t}{w}\cdot e^{-t/m} + \frac{m}{w}(e^{-t/m})\right]\bigg|_0^w$$

or

$$Nc\left(1 - \frac{m}{w} + \frac{m}{w}e^{-w/m}\right).$$

This last quantity is the total warranty cost for all the units manufactured. Since there are N units per batch, the warranty cost per item is

$$r = \frac{1}{N}\left[Nc\left(1 - \frac{m}{w} + \frac{m}{w}e^{-w/m}\right)\right] = c\left(1 - \frac{m}{w} + \frac{m}{w}e^{-w/m}\right).$$

For example, suppose a product which costs \$100 has an expected life of 24 months, with a 12-month warranty. Then we have $c = \$100$, $m = 24$, $w = 12$, with r, the warranty cost per unit, given by

$$\begin{aligned}r &= 100(1 - 2 + 2e^{-0.5}) \\ &= 100[-1 + 2(0.6065)] \\ &= 100(0.2130) \\ &= 21.30,\end{aligned}$$

where we found $e^{-0.5}$ from Table 4.

Case 11 Exercises

Find r for each of the following.

1. $c = \$50$, $m = 48$ months, $w = 24$ months
2. $c = \$1000$, $m = 60$ months, $w = 24$ months
3. $c = \$1200$, $m = 30$ months, $w = 30$ months

Chapter 8 Pretest

1. Given $x = 2$ and $y = 1$. Find m and n if $y = 4x - 1 + m$ and $y = 2x^2 - x + 3m + n$. **[1.2, 1.3]**

Find the following indefinite integrals.

2. $\int 5y\, dy$ **[7.3]**

3. $\int (x^2 + 3x + 2)\, dx$ **[7.3]**

4. $\int (-x^3 + 3)\, dx$ **[7.3]**

5. $\int e^y\, dy$ **[7.6]**

6. $\int e^{4x}\, dx$ **[7.6]**

7. $\int \dfrac{1}{y}\, dy$ **[7.6]**

8. Write in exponential form.
   (a) $\ln x = 2x + 3$ (b) $\ln y = x^2 - 5x$
9. Solve $3 = e^x$.
10. If $y = 5e^x$, find y when $x = 0;\ 1;\ -1$. **[8.1]**

8 Differential Equations

Many practical problems involve the rate of change of one variable with respect to another. For example, marginal profit is the rate of change of profit with respect to production. In Chapter 4, we expressed the rate of change of a variable y, with respect to another variable, x, as a derivative. The rate of change of many natural growth processes with respect to time can be expressed as a function of time of the form

$$\frac{dy}{dx} = kx \quad \text{or} \quad \frac{dy}{dx} = ky,$$

where x represents time and k is some constant. Such equations containing a derivative or differential are called **differential equations**. In most differential equations, it is convenient to represent the derivative as dy/dx, rather than y' or $f'(x)$. For a second derivative, the notation d^2y/dx^2 is used. Some examples of differential equations include

$$\frac{dy}{dx} = x^{1/2}, \quad \frac{dy}{dx} = 2y^2, \quad \text{and} \quad \frac{d^2y}{dx^2} = 3x^2 + 4.$$

8.1 General and Particular Solutions

The time-dating of dairy products depends on the solution of a differential equation. The rate of growth of bacteria in such products increases with time. If y is the number of bacteria present at a time t in days, then the rate of growth of bacteria can be expressed as dy/dt and we can write

$$\frac{dy}{dt} = kt,$$

where k is an appropriate constant. For simplicity, let us assume that $k = 10$ for a particular product, so that

$$\frac{dy}{dt} = 10t. \tag{1}$$

8.1 General and Particular Solutions

In Chapter 7, we defined the differential dy as

$$dy = \frac{dy}{dx} dx.$$

Using this idea, we have $dy = (dy/dt)\, dt$, and we can rewrite equation (1) as

$$dy = 10t\, dt.$$

If we now integrate both sides of this equation, we get

$$\int dy = \int 10t\, dt.$$

The differential dy on the left indicates that we should integrate there with respect to y, while the dt on the right indicates that we should integrate with respect to t on that side. Performing the two integrations gives

$$y + C_1 = 5t^2 + C_2.$$
$$y = 5t^2 + C_2 - C_1.$$

We can replace $C_2 - C_1$ with the single constant C to get

$$y = 5t^2 + C. \tag{2}$$

Equation (2) is called the **general solution** of differential equation (1). (This process of replacing two or more arbitrary constants with one can be skipped from now on—we shall just include one arbitrary constant and let it go at that.)

Suppose there is a known number of bacteria present at time $t = 0$, say $y = 50$ in thousands. Such a condition called an **initial condition,** or **boundary condition,** can be used in the general solution (2) to find C as follows:

$$y = 5t^2 + C$$
$$50 = 5(0)^2 + C$$
$$C = 50.$$

With this value of C, equation (2) becomes

$$y = 5t^2 + 50. \tag{3}$$

Since this solution to the differential equation depends on the particular values of t and y that were given, it is called a **particular solution** of differential equation (1). By determining the maximum acceptable value of y, then the number of days, t, the product is usable can be determined. For example, if the maximum value of y is to be 550, from equation (3) we have

$$y = 5t^2 + 50$$
$$550 = 5t^2 + 50$$
$$t^2 = 100$$
$$t = 10.$$

Thus, the product should be dated for 10 days from the date when $t = 0$.

Example 1 Find the particular solution to

$$\frac{dy}{dx} = 3x^2 + 4x + 5,$$

given that $y = -1$ when $x = 0$.

We first find the general solution.

$$\frac{dy}{dx} = 3x^2 + 4x + 5$$

$$dy = (3x^2 + 4x + 5)\,dx$$

$$\int dy = \int (3x^2 + 4x + 5)\,dx$$

$$y = x^3 + 2x^2 + 5x + C.$$

Then, substituting the given values for x and y, we find C.

$$-1 = 0^3 + 2(0)^2 + 5(0) + C$$
$$C = -1.$$

Thus, this particular solution is

$$y = x^3 + 2x^2 + 5x - 1,$$

as can be verified by differentiating.

Example 2 Find a particular solution to

$$\frac{d^2y}{dx^2} = 2x + 4, \tag{4}$$

given that $y = 2$ when $x = 0$, and $y = -4$ when $x = 3$.

Integrating both sides with respect to x, we have

$$\int \frac{d^2y}{dx^2}\,dx = \int (2x + 4)\,dx.$$

The integral of the second derivative becomes

$$\int \frac{d^2y}{dx^2}\,dx = \frac{dy}{dx} + C.$$

Thus, the differential equation can be written as

$$\frac{dy}{dx} = x^2 + 4x + C_1,$$

for some constant C_1. Taking antiderivatives again, we have

$$y = \frac{x^3}{3} + 2x^2 + C_1 x + C_2 \tag{5}$$

8.1 General and Particular Solutions

for some constant C_2. Equation (5) is a general solution of equation (4). To find values of C_1 and C_2, we use the given pairs of values. If $x = 0$, then $y = 2$, so that

$$2 = \frac{0^3}{3} + 2 \cdot 0^2 + C_1 \cdot 0 + C_2$$

$$C_2 = 2.$$

Using $C_2 = 2$, $x = 3$, and $y = -4$, we have

$$-4 = \frac{3^3}{3} + 2 \cdot 3^2 + C_1 \cdot 3 + 2$$

$$-4 = 29 + 3C_1$$

$$C_1 = \frac{-33}{3} = -11.$$

The particular solution is thus

$$y = \frac{x^3}{3} + 2x^2 - 11x + 2.$$

8.1 Exercises

Find general solutions for each of the following differential equations.

1. $\dfrac{dy}{dx} = x^2$
2. $\dfrac{dy}{dx} = -x + 2$
3. $\dfrac{dy}{dx} = -2x + 3x^2$
4. $\dfrac{dy}{dx} = 6x^2 - 4x$
5. $\dfrac{dy}{dx} = -4 + 3x^3$
6. $\dfrac{dy}{dx} = -4x^3 + 3$
7. $\dfrac{dy}{dx} = e^x$
8. $\dfrac{dy}{dx} = e^{3x}$
9. $\dfrac{dy}{dx} = 2e^{-x}$
10. $\dfrac{dy}{dx} = xe^{x^2}$
11. $3\dfrac{dy}{dx} = -4x^2$
12. $3x^3 - 2\dfrac{dy}{dx} = 0$
13. $4\dfrac{dy}{dx} - x = 0$
14. $3x^2 - 3\dfrac{dy}{dx} = 2$
15. $4x - 4 + \dfrac{dy}{dx} = 6x^2$
16. $-8x^3 - 3x^2 + \dfrac{1}{2} \cdot \dfrac{dy}{dx} = 3$
17. $\dfrac{d^2y}{dx^2} = -8x$
18. $\dfrac{d^2y}{dx^2} = 4x$
19. $\dfrac{d^2y}{dx^2} = 5 - 4x$
20. $\dfrac{d^2y}{dx^2} = 2 + 8x$
21. $\dfrac{d^2y}{dx^2} = 4x^2 - 2x$
22. $x^2 - \dfrac{d^2y}{dx^2} = 3x$

23. $5x - 3 + \dfrac{d^2y}{dx^2} = 0$

24. $x\dfrac{d^2y}{dx^2} = 6$

25. $\dfrac{d^2y}{dx^2} = e^x$

26. $\dfrac{d^2y}{dx^2} = e^{2x} + 3$

Find particular solutions for each of the following differential equations.

27. $\dfrac{dy}{dx} + 2x = 3x^2;\quad y = 2$ when $x = 0$

28. $\dfrac{dy}{dx} = 4x + 3;\quad y = -4$ when $x = 0$

29. $\dfrac{dy}{dx} = 5x + 2;\quad y = -3$ when $x = 0$

30. $\dfrac{dy}{dx} = 6 - 5x;\quad y = 6$ when $x = 0$

31. $\dfrac{dy}{dx} = 3x^2 - 4x + 2;\quad y = 3$ when $x = -1$

32. $\dfrac{dy}{dx} = 4x^3 - 3x^2 + x;\quad y = 0$ when $x = 1$

33. $\dfrac{d^2y}{dx^2} = 2x + 1;\quad y = 2$ when $x = 0;\quad y = 3$ when $x = -2$

34. $\dfrac{d^2y}{dx^2} = -3x + 2;\quad y = -3$ when $x = 0;\quad y = 4$ when $x = 3$

35. $\dfrac{d^2y}{dx^2} = e^x + 1;\quad y = 2$ when $x = 0;\quad y = 3/2$ when $x = 1$

36. $\dfrac{d^2y}{dx^2} = -e^x - 2;\quad y = -2$ when $x = 0;\quad y = 3/2$ when $x = 1$

37. $3x^2 - \dfrac{d^2y}{dx^2} = 2;\quad y = 4$ when $x = 0;\quad y = 6$ when $x = 2$

38. $x\dfrac{d^2y}{dx^2} = 1;\quad y = -2$ when $x = 1;\quad y = -1$ when $x = e$

39. In the time-dating of dairy products example discussed in the text, suppose the number of bacteria present at time $t = 0$ was 250.
    (a) Find a particular solution of the differential equation.
    (b) Find the number of days the product can be sold if the maximum value of y is 970.

40. Suppose the rate at which a rumor spreads—that is, the number of people who have heard the rumor over a period of time—increases with the number of days. If y is the number of people who have heard the rumor, then

$$\dfrac{dy}{dt} = kt,$$

where t is the time in days.
(a) If y is 0 when $t = 0$, and y is 100 when $t = 2$, find k.
(b) Using the value of k from part (a), find y when $t = 3; 5; 10$.

8.2 Separation of Variables

The differential equations we have discussed up to this point have been of the form

$$\frac{dy}{dx} = f(x) \quad \text{or} \quad \frac{d^2y}{dx^2} = f(x).$$

One of the most common types of differential equations is of the more general form

$$h(y)\, dy = f(x)\, dx,$$

which can also be solved by taking antiderivatives of both sides. Such a differential equation is said to have **variables separable**. To solve this kind of differential equation it is necessary to separate all terms involving y (including dy) on one side of the equation, and all terms involving x (and dx) on the other side. For example,

$$y\frac{dy}{dx} = x^2$$

is a variables separable equation since we can write it as

$$y\, dy = x^2\, dx.$$

The general solution can be found by taking integrals on both sides.

$$\int y\, dy = \int x^2\, dx$$

$$\frac{y^2}{2} = \frac{x^3}{3} + C.$$

Sometimes the differential equation must be rearranged to separate the variables as in Example 3 below.

Example 3 Find the general solution of $\frac{dy}{dx} = ky$, where k is a constant.

This is a variables separable differential equation, since it can be rewritten as

$$\frac{1}{y}\, dy = k \cdot dx.$$

To solve this equation, we take antiderivatives of both sides.

$$\int \frac{1}{y}\, dy = \int k \cdot dx$$

Evaluating these antiderivatives, we have

$$\ln y = kx + C.$$

This general solution can be rewritten, using the definition of natural logarithms, as

$$y = e^{kx+C}$$
$$= e^{kx}e^{C}$$
$$= Me^{kx},$$

where we replaced the constant e^C with the constant M.

In general, in the equation

$$y = Me^{kx},$$

the constant k gives the rate of change of a population of size M at time $x = 0$. The equation represents growth when k is positive and decay when k is negative. The function $y = Me^{kx}$ is extremely important because of its many applications in management and the social and behavioral sciences. (See the next section.)

Example 4 Find a general solution of

$$x\frac{d^2y}{dx^2} - \frac{dy}{dx} = 0.$$

If we let $\frac{dy}{dx} = u$, then $\frac{d^2y}{dx^2} = \frac{du}{dx}$. Then the given equation becomes,

$$x \cdot \frac{du}{dx} - u = 0$$

$$x \cdot \frac{du}{dx} = u$$

$$\frac{1}{u} du = \frac{1}{x} dx.$$

Taking antiderivatives, we have

$$\int \frac{1}{u} du = \int \frac{1}{x} dx$$

$$\ln u = \ln x + C_1.$$

If we let $C_1 = \ln M$, where M is some constant, then

$$\ln u = \ln x + \ln M$$
$$\ln u = \ln Mx$$
$$u = Mx$$
$$\frac{dy}{dx} = Mx.$$

Now, taking antiderivatives again gives the general solution

$$dy = Mx \, dx$$
$$y = \frac{M}{2} x^2 + K$$

for some constant K.

In Sections 8.1 and 8.2, we have discussed only the simplest techniques for solving differential equations. In actual practice other methods are often required.

8.2 Exercises

Find general solutions for each of the following differential equations.

1. $y\dfrac{dy}{dx} = x$
2. $y^2 \cdot \dfrac{dy}{dx} = x^2 - 1$
3. $y\dfrac{dy}{dx} = x^2 - 1$
4. $2x + x^2 - y\dfrac{dy}{dx} = 0$
5. $\dfrac{dy}{dx} = 2xy$
6. $\dfrac{dy}{dx} = x^2 y$
7. $\dfrac{dy}{dx} = 2x^2 y - xy$
8. $(y^2 - y)\dfrac{dy}{dx} = x$
9. $\dfrac{dy'}{dx} = 4x$
10. $\dfrac{dy'}{dx} = 2x - 3x^2$
11. $2\dfrac{dy'}{dx} - 3x^2 = 0$
12. $2\dfrac{dy'}{dx} = x^2$

13. A radioactive substance decays at a rate given by
$$\dfrac{dy}{dx} = -0.05y$$
where y represents the amount, in grams, present at time x, in months.
   (a) Find the general solution.
   (b) Find a particular solution if $y = 90$ when $x = 0$.
   (c) Find the amount left when $x = 10$.

14. Extensive experiments have shown that under relatively constant market conditions, sales of a product, in the absence of promotional activities such as advertising, decrease at a constant yearly rate. This rate of sales decline varies considerably from product to product, but it seems to be relatively constant for a particular product. Suppose the yearly rate of sales decrease for a certain company is given by
$$\dfrac{dy}{dt} = -0.25y,$$
where t is the time in years.
   (a) Find a particular solution if $y = 80,000$ when $x = 0$.
   (b) Find y when $t = 1; 3$.
   (c) Suppose $y = 50,000$ when $t = 1$ and $y = 60,000$ when $t = 0$. Find k given
$$\dfrac{dy}{dt} = -ky.$$

8.3 Applications of Differential Equations

Differential equations have many practical applications, especially in biology and economics. For example, marginal productivity is the rate at which production changes (increases or decreases) for a unit change in capitalization. Thus, marginal productivity can be expressed as the first derivative of the function which gives production in terms of capitalization. Suppose the marginal productivity of an operation is given by

$$P'(x) = 3x^2 - 10, \qquad (6)$$

where x is the amount of capitalization in hundred-thousands. If the operation produces 100 units per month with its present capitalization of \$300,000 ($x = 3$), how much would production increase if the capitalization is increased to \$500,000?

To obtain an equation for production we can take antiderivatives on both sides of equation (6), to get

$$P(x) = x^3 - 10x + C.$$

Now, using the given initial values, $P(x) = 100$ when $x = 3$, we can find C.

$$100 = 3^3 - (10)(3) + C = 27 - 30 + C$$
$$C = 103.$$

Thus, the production is given by

$$P(x) = x^3 - 10x + 103,$$

and if capitalization is increased to \$500,000, production becomes

$$P(5) = 5^3 - 10(5) + 103 = 178.$$

An increase to \$500,000 in capitalization will increase production from 100 units to 178 units.

Example 5 It is common for a population (of people, bacteria, or insects, for example) to increase at a rate proportional to the number of individuals present, at least until food supply and the accumulation of waste products affects the rate. That is, when only a few individuals are present, the rate of growth of the population is small. When the number of individuals is large, the rate of growth of the population is high. If y is the population at a time x, then such a situation can be expressed as

$$\frac{dy}{dx} = ky,$$

for some constant k. As we have seen, the general solution of this differential equation is

$$y = Me^{kx}.$$

8.3 Applications of Differential Equations

To obtain a particular solution it is necessary to know some boundary conditions for y and x. For example, suppose 1000 individuals are present when $x = 0$. (In other words, $y = 1000$ when $x = 0$.) Then

$$1000 = Me^{k \cdot 0}$$
$$= M \cdot 1$$
$$= M,$$

so that

$$y = 1000e^{kx}.$$

Now suppose $y = 1500$ when $x = 10$. Then

$$1500 = 1000e^{10k},$$
$$1.5 = e^{10k}.$$

Taking natural logarithms of both sides, we have

$$\ln 1.5 = \ln e^{10k}$$
$$= 10k \ln e$$
$$= 10k,$$

from which

$$k = \frac{\ln 1.5}{10}.$$

Using Table 5, we have

$$k \approx \frac{0.41}{10} = 0.041.$$

Hence, $y \approx 1000e^{0.041x}$.

Example 6 Suppose a population is limited (perhaps by a fixed food supply or competition with other species) and cannot exceed some fixed value, such as M. It is plausible to assume that the rate of change of the population is proportional both to the number of individuals present, y, and to the difference $M - y$. That is,

$$\frac{dy}{dx} = ky(M - y) \tag{7}$$

for some constant k. We can rewrite equation (7) as

$$\frac{1}{y(M - y)} dy = k \, dx$$

from which

$$\int \frac{1}{y(M - y)} dy = \int k \, dx. \tag{8}$$

Now we need a little algebra: verify that

$$\frac{1}{y(M - y)} = \frac{1}{M}\left(\frac{1}{y} + \frac{1}{M - y}\right).$$

Thus,

$$\int \frac{1}{y(M-y)} \, dy = \int \frac{1}{M} \left(\frac{1}{y} + \frac{1}{M-y} \right) dy$$

$$= \frac{1}{M} \left(\int \frac{1}{y} \, dy + \int \frac{1}{M-y} \, dy \right)$$

$$= \frac{1}{M} [\ln y - \ln (M-y)].$$

Using this result, and the fact that $\int k \, dx = kx + C$, equation (8) becomes

$$\frac{1}{M} [\ln y - \ln (M-y)] = kx + C.$$

If we multiply both sides by M and use properties of logarithms, we have

$$\ln \frac{y}{M-y} = Mkx + C_1,$$

where $C_1 = MC$. By the definition of logarithms, this yields

$$\frac{y}{M-y} = C_2 e^{Mkx},$$

where C_2 is e^{C_1}. If we let y_0 represent the population at time $x = 0$, we get

$$\frac{y_0}{M-y_0} = C_2 e^{Mk(0)} = C_2(1) = C_2.$$

Hence,

$$\frac{y}{M-y} = \frac{y_0}{M-y_0} e^{Mkx}.$$

After considerable algebraic manipulation, this last result can be solved for y. The result is

$$y = \frac{My_0}{y_0 + (M-y_0)e^{-Mkx}}. \tag{9}$$

As time increases without bound, that is, as $x \to \infty$, the expression e^{-Mkx} approaches 0. Thus,

$$\lim_{x \to \infty} \frac{My_0}{y_0 + (M-y_0)e^{-Mkx}} = \frac{My_0}{y_0 + (M-y_0)(0)} = \frac{My_0}{y_0} = M,$$

so that the population does tend to approach M, the limiting value, as required by the statement of the problem.

According to the result obtained above, population will tend to increase right up to the absolute limit that the land can support, even if that support is only at a sustenance level. When another variable, the quality of life, is introduced, the situation may change. The individuals within a society may seek to

8.3 Applications of Differential Equations

limit the growth of that society so that the quality of life is enhanced. For example, the number M, the upper limit on population, will soon be reached in some underdeveloped countries. On the other hand, people in many countries now seem to believe that a population substantially less than M provides a better overall quality of life. The results of a society permitting population to try to reach the upper limit M are explored in *Limits to Growth*, by Donella H. Meadow, et al., Signet Paperback Y5250, 1972. A good discussion of the flaws in *Limits to Growth* is given in "The Computer That Printed Out W*O*L*F*" by Carl Kaysen, *Foreign Affairs*, July 1972, page 660.

8.3 Exercises

1. The rate of change of demand for a certain product is given by

$$\frac{dy}{dx} = -4x + 40,$$

   where x represents the price of the item. Find the demand at the following price levels if $y = 6$ when $x = 0$.
   - (a) $x = 5$
   - (b) $x = 8$
   - (c) $x = 10$
   - (d) $x = 20$

2. The marginal profit of a certain company is given by

$$\frac{dy}{dx} = 32 - 4x,$$

   where x represents the amount in thousands of dollars spent on advertising. Find the profit for each of the following advertising expenditures if the profit is 1000 when nothing is spent for advertising.
   - (a) $x = 3$
   - (b) $x = 5$
   - (c) $x = 7$
   - (d) $x = 10$

3. The rate at which the bacteria in a culture are changing after the introduction of a bactericide is given by

$$\frac{dy}{dx} = 50 - 10x,$$

   where y is the number of bacteria present at time x. Find the number of bacteria present at each of the following times if there were 1000 bacteria present at time $x = 0$.
   - (a) $x = 2$
   - (b) $x = 5$
   - (c) $x = 10$
   - (d) $x = 15$

4. Suppose the marginal cost of producing x copies of a book is given by

$$\frac{dy}{dx} = \frac{50}{\sqrt{x}}.$$

If $y = 25{,}000$ when $x = 0$, find the cost of producing the following number of books.
(a) 100 books (b) 400 books
(c) 625 books

5. Suppose a radioactive sample decays according to the relationship

$$\frac{dy}{dx} = -0.08y,$$

where x is measured in years. Find the amount left after 5 years if $y = 20$ grams when $x = 0$.

6. A company has found that the rate at which a person new to the assembly line produces items is

$$\frac{dy}{dx} = 7.5e^{-0.3x},$$

where x is the number of days he has worked on the line. How many items can a new worker be expected to produce on the 8th day if he produces none when $x = 0$?

7. Assume the rate of change of the population of a certain city is given by

$$\frac{dy}{dt} = 6000e^{0.06t},$$

where y is the population at time t, measured in years. If the population was 100,000 in 1960 (assume $t = 0$ in 1960), predict the population in 1980.

8. A colony of bacteria grows at a rate proportional to the number present. Initially, 4000 bacteria are present, while 40,000 are present at time $x = 10$.
(a) Find an equation showing the number of bacteria present at time x.
(b) How many bacteria will be present at time $x = 20$?

9. The phosphate compounds found in many detergents are highly water soluble and are excellent fertilizers for algae. Assume that the rate of growth of algae, in the presence of sufficient phosphates, is proportional to the number present. Assume that there are 3000 algae present at time $x = 0$ and 20,000 at time $x = 10$.
(a) Find a function showing the number of algae present at time x.
(b) How many algae will be present at time $x = 20$?

10. Suppose the growth of a population of insects is given by

$$y = \frac{My_0}{y_0 + (M - y_0)e^{-0.001Mx}}$$

where x represents time in weeks, y_0 represents the initial number of insects, and M represents the "bound" or upper limit of the population. Find the number present at each of the following times if $M = 1000$ and $y_0 = 100$.
(a) $x = 1$ (b) $x = 3$ (c) $x = 10$

11. An isolated fish population is bounded by the amount of food available. If $M = 5000$, $y_0 = 150$, $k = 0.0001$, and x is in years, use the equation of Example 8 to find the number present at the end of each of the following times.
    (a) $x = 1$ (b) $x = 2$ (c) $x = 5$
12. If long chain polymers in a cell, such as a protein or nucleic acid, are hit by a beam of ionized particles, some of the polymers may be damaged. Let n_0 be the number of undamaged polymers of a specific protein present in a cell. Let D be the number of ionizing particles penetrating the cell. Let n be the number of chains undamaged after exposure to the radiation. Since n and D are usually very large numbers, we may safely treat them as continuous functions. The higher the value of D, the more damage to the chains of the protein, so that dn/dD must be negative. If dn/dD is proportional to n, we have

$$\frac{dn}{dD} = -kn,$$

for some positive constant k. Solve this equation for n.

● Case 12 Differential Equations in Ecology

Consider a closed area containing only two animal or plant species. Assume that there is effective interaction between the two species. For example, the first species may be a source of food for the second species. A species of trees might reduce the light falling on another plant species. One species might provide shelter for another. A species of insects might pollinate a species of plant. One species may poison the soil for another species. There are many applications of such interaction.

Let $N_1 = N_1(t)$ and $N_2 = N_2(t)$ represent the number of individuals of the two species at time t. Let ΔN_1 represent the change in the population of species 1 during a given time period, while ΔN_2 is the change in the population of species 2 during the same time period. During a given time interval of length Δt, we may describe the populations of the species as follows:

Population of species 1:

$$\Delta N_1 = \begin{pmatrix} \text{any change in the} \\ \text{absence of interaction} \end{pmatrix} + \begin{pmatrix} \text{any change due to} \\ \text{interaction with species 2} \end{pmatrix}.$$

Population of species 2:

$$\Delta N_2 = \begin{pmatrix} \text{any change in the} \\ \text{absence of interaction} \end{pmatrix} + \begin{pmatrix} \text{any change due to} \\ \text{interaction with species 1} \end{pmatrix}.$$

Divide both sides of each of these equations by Δt, and assume that each term approaches a limit as $\Delta t \to 0$. Then, by the definition of derivative, we have

$$\frac{dN_1}{dt} = \begin{pmatrix} \text{rate of change in the} \\ \text{absence of interaction} \end{pmatrix} + \begin{pmatrix} \text{rate of change due to} \\ \text{interaction with species 2} \end{pmatrix}$$

$$\frac{dN_2}{dt} = \begin{pmatrix} \text{rate of change in the} \\ \text{absence of interaction} \end{pmatrix} + \begin{pmatrix} \text{rate of change due to} \\ \text{interaction with species 1} \end{pmatrix}.$$

Let us assume, for simplicity, that the birth and death rates in both populations are constant, so that the rate of change of a population is proportional to the size of the population. We can also assume that the rate of change due to interaction is proportional to the size of the interacting population. These assumptions lead to differential equations of the form

$$\frac{dN_1}{dt} = a \cdot N_1 + b \cdot N_2, \qquad \frac{dN_2}{dt} = c \cdot N_2 + d \cdot N_1,$$

for appropriate constants a, b, c, and d.

The common solution of these two differential equations requires a knowledge of the values of a, b, c, and d. In more advanced courses, it is shown that the solutions will be of the form

$$N_1 = p_1 \cdot e^{qt} + k \cdot p_2 \cdot e^{rt}, \qquad N_2 = k \cdot p_1 \cdot e^{qt} + p_2 \cdot e^{rt},$$

where q and r are solutions of the quadratic equation

$$m^2 - (a+c)m + (ac+bd) = 0, \tag{1}$$

and where p_1, p_2, and k are constants depending on a, b, c, and d.

If either q or r (or both) are positive, then both populations will grow with time, and increase without bound. If both q and r are negative, the populations will both tend to decrease in size. If there are no real-number solutions of the quadratic equation (1) above, then the populations will oscillate, from very large to very small, and so on.

If specific values of a, b, c and d are known, the changes in the populations of the two species can be found from the solutions of the differential equations given above. For example, if the two species are in a predator-prey relationship, then the solutions of the quadratic equation (1) above will not be real numbers, and the populations will oscillate. To see how this works in practice, consider populations of hares and lynx in some limited area. If the hare population starts to increase, the number of lynx will also increase, because of the additional food supply. The lynx will compete for this additional food, causing the number of hares to decline; this causes the number of lynx to decline. As the number of lynx goes down, the hares have more chance to increase their numbers, starting the whole cycle over again. An idealized graph of this situation is shown in Figure 1.

Figure 1

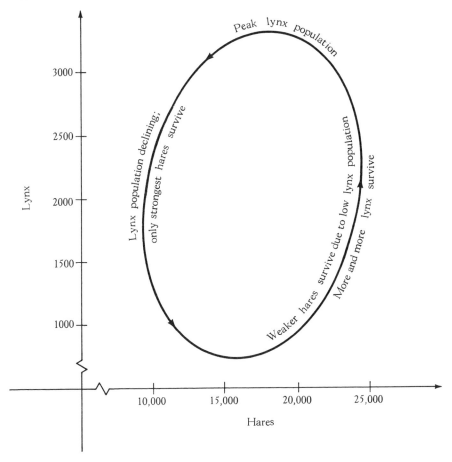

To test the accuracy of this theory, we shall use the records on lynx and hare trappings kept by the Hudson's Bay Company in Canada. For many years, starting in 1847, the company caught as many lynx and hares annually as possible, and kept records of this catch. We shall assume that the company catches are roughly proportional to the total population of each species, although there are some outside influences on the size of the catch (for example, notice the fall-off in the catches during the years 1861–65, the years of the American Civil War). The company data has been graphed in Figure 2, with a smooth curve for each animal drawn through the data points. From the graph we see that if the hare population begins to increase, then the lynx population will also increase, about a year later. As the hare population declines, lynx population also starts to decline, again with a time lag of one year. Thus, the populations oscillate on about a ten-year cycle. Using the data from Figure 2, a graph very similar to the one of Figure 1 could be constructed.

Figure 2*

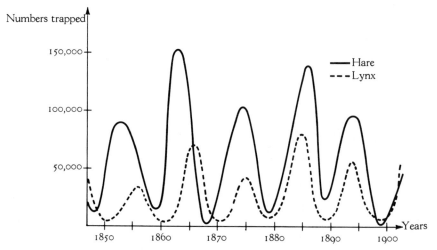

Idealized curves showing abundance of lynx and hare in Canada, 1847–1903.

* The actual data from which Figure 2 was constructed is given in "Some Mathematical Problems in Biology," v. 1 of the series *Lectures on Mathematics in the Life Sciences*, published by the American Mathematical Society, Providence, Rhode Island. This series contains six volumes on mathematical biology, all unfortunately written at a highly advanced level.

Chapter 9 Pretest

Let $f(x) = -4x^2 + 2x - 3$. Find each of the following. [2.1]

1. $f(0)$
2. $f(-2)$
3. $f(3)$

Graph each of the following.

4. $4y + 3x = 12$ [2.2]
5. $y = -2x + 5$ [2.2]
6. $y = 4$ [2.2]
7. $y = -2x^2$ [2.3]
8. $x^2 + y^2 = 16$ [2.3]
9. $y = \sqrt{16 - x^2}$ [2.3]

Find the derivative of each of the following functions.

10. $f(x) = 6\pi x^2 - 4\pi + 12$ [4.2]
11. $-\ln(\pi x + 2\pi)$ [6.4]

Find the second derivative for each of the following functions. [5.2]

12. $f(x) = 8x^2 + 12x + 4x^{-1/2}$
13. $f(x) = kx^3 + 5kx^2 + 6kx + 3$, where k is a constant

Find all relative maximum and relative minimum values of each of the following. [5.1]

14. $f(x) = 2x^2 - 8x + 6$
15. $f(x) = 2x^3 - 3x^2 - 36x + 4$
16. Let $f(t) = 5t^2 - 8t + \ln(t + 1) + 50$, where $f(t)$ represents the population of a certain fish at time t. Find the rate of change of the population with respect to time when
    (a) $t = 0$
    (b) $t = 4$ [4.1]

9
Multivariate Functions

So far in this text, all the functions we have used have involved just one independent variable. For example, we have seen examples that describe sales as a function of time or price, and functions that give populations in terms of time. In many cases, however, the value of a variable may depend on more than one independent variable. For example, the sale of air conditioners depends on their price, the weather, and the availability and price of electricity, among other possible variables. The population of foxes in an area depends on the population of rabbits, squirrels, gophers, or other food resources.

In this chapter, we discuss the mathematics of functions of more than one independent variable. We shall see that much of the material developed for functions of one independent variable applies to functions of more than one independent variable. In particular, the idea of a derivative, a fundamental idea of calculus, generalizes in a very natural way to functions of more than one independent variable.

We shall first discuss the definition of and notation used for these functions. Then we shall develop a means for graphing them. Finally, we shall look at how the ideas of calculus generalize to them. The chapter ends with an important application—the formula for the straight line of "best fit" for a given collection of data values.

9.1 Functions of Several Variables

Suppose that the value of the variable z is related to the values of x and y by the relationship

$$z = 4x^2 + 2xy + 3y.$$

For every pair of values of x and y that we might choose, we can find exactly one value for z. For example, if $x = 1$ and $y = -4$, we have

$$\begin{aligned} z &= 4x^2 + 2xy + 3y \\ &= 4(1)^2 + 2(1)(-4) + 3(-4) \\ &= 4 - 8 - 12 \\ &= -16. \end{aligned}$$

Therefore, if $x = 1$ and $y = -4$, we have $z = -16$. This can be expressed by saying that if $z = f(x, y) = 4x^2 + 2xy + 3y$, then $f(1, -4) = -16$. To calculate $f(3, -2)$, let $x = 3$ and $y = -2$:

$$\begin{aligned} z = f(x, y) &= 4x^2 + 2xy + 3y \\ &= 4(3)^2 + 2(3)(-2) + 3(-2) \\ &= 36 - 12 - 6 \\ &= 18. \end{aligned}$$

Thus, $f(3, -2) = 18$. In the same way, $f(0, 0) = 0$, and $f(2, 5) = 51$.

In general, $z = f(x, y)$ is a **function of two variables** whenever each pair of values for x and y leads to no more than one value of z. The variables x and y are called **independent variables**, with z called the **dependent variable**.

Example 1 Let $f(x, y) = 4x\sqrt{x^2 + y^2}$. Find each of the following: $f(0, 0)$, $f(3, -4)$, and $f(-5, -7)$.

(a) $f(0, 0) = 4(0)\sqrt{0^2 + 0^2} = 0$.
(b) $f(3, -4) = 4(3)\sqrt{3^2 + (-4)^2} = 12\sqrt{9 + 16} = 60$.
(c) $f(-5, -7) = 4(-5)\sqrt{(-5)^2 + (-7)^2} = -20\sqrt{74}$.

Example 2 The amount of money, $M(x, y)$, in dollars, that a twelve-year-old girl will spend in a week on rock-and-roll records depends on her weekly allowance, x, in dollars, and the amount, y, in dollars, that she spends weekly on cosmetics. In a suburb of North Carmichael, it was found that $M(x, y)$ can be closely approximated by

$$M(x, y) = 4x - 2y - 9.$$

(a) Find $M(4, 2.5)$.
We have

$$\begin{aligned} M(4, 2.5) &= 4(4) - 2(2.5) - 9 \\ &= 16 - 5 - 9 \\ &= 2. \end{aligned}$$

Thus, the girl spends $2 on records. This sum, together with the $2.50 spent on cosmetics, exceeds her allowance by 50¢. The girl is therefore getting early training in deficit financing, and should be well qualified to go to Congress.

(b) How much will be spent on records by a girl whose allowance is $3 per week and who spends $1.25 per week on cosmetics?

Here, $x = 3$ and $y = 1.25$. We have

$$M(3, 1.25) = 4(3) - 2(1.25) - 9$$
$$= 12 - 2.50 - 9$$
$$= 0.50.$$

She will spend about 50¢ per week on rock-and-roll records.

While we have defined only a function of two independent variables, similar definitions could be given for functions of three, four, or more independent variables. For example, if

$$f(x, y, z) = 4xz - 3yx^2 + 2z^2,$$

then $f(2, -3, 1)$ is given by

$$f(2, -3, 1) = 4(2)(1) - 3(-3)(2)^2 + 2(1)^2$$
$$= 8 + 36 + 2$$
$$= 46.$$

Also,

$$f(-4, 3, -2) = 4(-4)(-2) - 3(3)(-4)^2 + 2(-2)^2$$
$$= 32 - 144 + 8$$
$$= -104.$$

Example 3 Let $g(x, y, z, w) = 5x^2 - 4xz + 2yw - 4w$. Find $g(-1, 2, -3, 0)$ and $g(4, -2, 3, 1)$.

(a) $g(-1, 2, -3, 0) = 5(-1)^2 - 4(-1)(-3) + 2(2)(0) - 4(0)$
$= 5 - 12 + 0 - 0$
$= -7.$

(b) $g(4, -2, 3, 1) = 5(4)^2 - 4(4)(3) + 2(-2)(1) - 4(1)$
$= 80 - 48 - 4 - 4$
$= 24.$

9.1 Exercises

Let $f(x, y) = 4x + 5y + 3$. Find each of the following.

1. $f(2, -1)$
2. $f(3, -2)$
3. $f(-4, 1)$
4. $f(-2, -3)$
5. $f(0, 0)$
6. $f(-1, 0)$

Let $g(x, y) = -x^2 - 4xy + y^3$. Find each of the following.

7. $g(-2, 4)$
8. $g(3, -2)$
9. $g(-1, -2)$
10. $g(2, -3)$
11. $g(-2, 3)$
12. $g(0, 0)$

Let $h(x, y) = \sqrt{x^2 + 2y^2}$. Find each of the following.

13. $h(5, 3)$
14. $h(2, 4)$
15. $h(-1, -3)$
16. $h(-3, -1)$

The population of cats on a certain farm, $C(x, y)$, is approximated by

$$C(x, y) = x^2 + 200y - 1200,$$

where x is the population, in hundreds, of small mice, and y is the population, in tens, of large rats. Find each of the following.

17. $C(50, 0)$
18. $C(30, 4)$
19. How many cats will be present if there are 1400 small mice and 150 large rats?
20. If the farm has 3000 small mice and 200 large rats, how many cats will be present?

The labor charge, $L(x, y)$, for assembling a precision camera is given by

$$L(x, y) = 12x + 6y + 2xy + 40,$$

where x is the number of work hours required by a skilled crafts person, and y is the number of hours required of a semi-skilled person. Find each of the following.

21. $L(3, 5)$
22. $L(5, 2)$
23. If a skilled crafts person requires 7 hours and a semi-skilled person needs 9 hours, find the total labor charge.
24. Find the total labor charge if a skilled worker needs 12 hours and a semi-skilled worker requires 4 hours.

9.2 Graphing in Three Dimensions

To graph functions of one independent variable, such as $f(x) = 4x^2 + 2x$, we use **ordered pairs** of numbers (x, y), where $y = f(x)$. For example, if $x = -2$, then $y = f(-2) = 4(-2)^2 + 2(-2) = 12$. This result gives the ordered pair $(-2, 12)$, which can be used to help sketch the graph of the function $f(x) = 4x^2 + 2x$.

The situation is similar for functions of two independent variables. Thus, if $z = f(x, y) = 4x + 2y$, we obtain one value of z for each pair of values of x and y that we might choose. For example, if $x = 2$ and $y = -6$, we get

$$z = f(2, -6) = 4(2) + 2(-6)$$
$$= -4.$$

Thus, if $x = 2$ and $y = -6$, we have $z = -4$. This result is written as the **ordered triple** $(2, -6, -4)$. Other ordered triples for the function $f(x, y) = 4x + 2y$ include $(-3, 5, -2)$, $(6, -4, 16)$, $(0, 0, 0)$, and so on. In the ordered triple (x, y, z), we have $z = f(x, y)$.

280 Multivariate Functions

To graph ordered pairs, we used two axes, an x-axis and a y-axis. To graph ordered triples, we need three coordinate axes, an x-axis, y-axis, and z-axis, each perpendicular to the other two. Figure 9.1 shows one possible way to draw the graph of these three axes. In the figure, the plane containing the y-axis and z-axis is the plane of the page, with the x-axis perpendicular to the plane of the page.

Figure 9.1

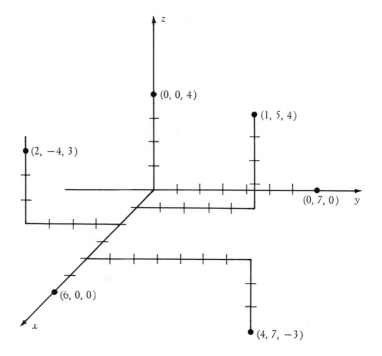

To locate the point corresponding to the ordered triple $(2, -4, 3)$, start at the origin and go 2 units along the positive x-axis. Then go 4 units parallel to the y-axis in the negative direction. Finally, go 3 units parallel to the z-axis. The point representing $(2,-4, 3)$, and other sample points, are shown in Figure 9.1. The region in three-dimensional space in which the x-, y-, and z-coordinates are all positive is called the *first octant*.

In the rest of this section, we shall look at various types of functions and the graphs that they produce in three-dimensional space. We shall not prove the statements we make about these graphs; for more information (and proofs) on three-dimensional graphs, see the books listed in the bibliography.

Planes The graph of $ax + by = c$ in two-dimensional space is a line (assuming not both a and b are zero). In three-dimensional space, the graph of $ax + by + cz = d$ is a plane. That is, the equation

$$ax + by + cz = d,$$

where a, b, c, and d are real numbers, with not all of a, b, and c equal to 0, has a graph which is a **plane**.

Example 4 Graph $z = -2x - y + 6$.

This equation can be written as $2x + y + z = 6$. Therefore, by the remark above, this function has a graph which is a plane. To graph this plane, find some representative ordered triples for the function. Some ordered triples here include $(0, 0, 6)$, $(3, 0, 0)$, and $(0, 6, 0)$. If we plot these ordered triples, we get the graph shown in Figure 9.2. We shall normally graph only the first octant when graphing planes such as this.

Figure 9.2

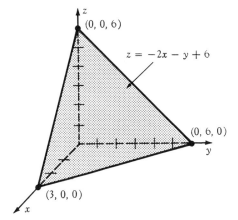

Example 5 Graph $x + 4y + 2z = 12$.

This graph is again a plane. To obtain this graph, we can complete some ordered triples. For example, $(0, 0, 6)$, $(0, 3, 0)$, and $(12, 0, 0)$ satisfy the given function. Using these points we can draw the graph shown in Figure 9.3.

Figure 9.3

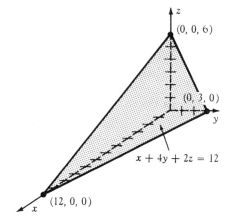

Example 6 Graph $x + z = 6$.

The graph again is a plane. Verify that the plane goes through $(6, 0, 0)$, $(0, 0, 6)$, and $(2, 0, 4)$, for example, and is parallel to the y-axis. The graph of this plane is shown in Figure 9.4.

Figure 9.4

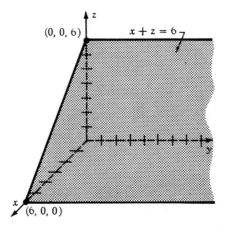

Example 7 Graph $x = 3$.

This function has a graph which is parallel to both the y-axis and the z-axis, and goes through $(3, 0, 0)$, as shown in Figure 9.5.

Figure 9.5

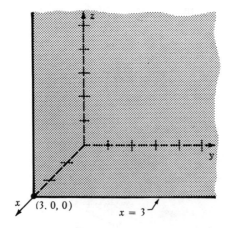

Spheres In two-dimensional space, the equation $x^2 + y^2 = 25$ has a graph which is a circle. The radius of the circle is 5, with the center at $(0, 0)$. In general, the graph of $(x - h)^2 + (y - k)^2 = r^2$ is a circle of radius r, centered at the point (h, k). In three-dimensional space, the graph of $x^2 + y^2 + z^2 = 25$ is the sphere with radius 5 and center at $(0, 0, 0)$. In general, the **sphere** with center at (h, k, j) and radius r is given by

$$(x - h)^2 + (y - k)^2 + (z - j)^2 = r^2.$$

Example 8 Graph $(x - 2)^2 + (y + 3)^2 + (z - 4)^2 = 4$.

This equation represents a sphere of radius 2 with center at $(2, -3, 4)$, as shown in Figure 9.6. This sphere is not the graph of a function—if $x = 2$ and $y = -3$, then z can equal either 2 or 6. The sphere could be divided horizontally into two half-spheres, each of which would be the graph of a function, but the entire sphere does not represent a function.

Figure 9.6

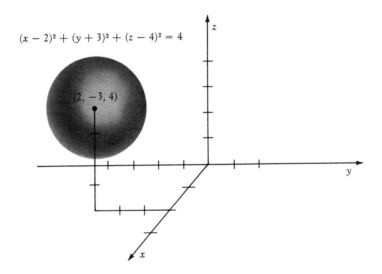

$(x - 2)^2 + (y + 3)^2 + (z - 4)^2 = 4$

Cylinders* The equation of a **right circular cylinder** of radius r and whose axis coincides with the z-, x-, or y-axis, is given by

$$x^2 + y^2 = r^2, \qquad y^2 + z^2 = r^2, \qquad \text{or} \qquad x^2 + z^2 = r^2,$$

respectively.

Example 9 Graph $x^2 + y^2 = 4$.

The equation $x^2 + y^2 = 4$ leads to a cylinder whose axis is the z-axis. The cylinder contains the point $(0, 2, 0)$ and $(2, 0, 0)$, as shown in Figure 9.7. The radius of the cylinder is 2.

* Optional section.

Figure 9.7

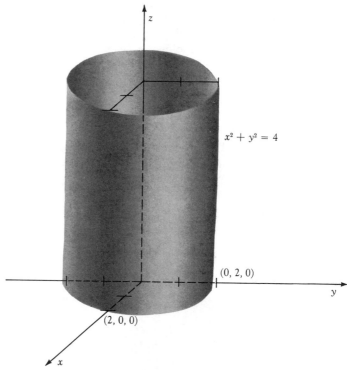

Example 10 Graph $y^2 + z^2 = 25$.

The graph of this equation is a cylinder whose axis is the x-axis, with radius 5. See Figure 9.8.

Figure 9.8

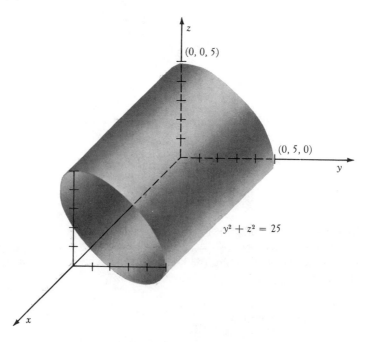

*Paraboloids** The graph of

$$z = x^2 + y^2, \quad y = x^2 + z^2, \quad \text{or} \quad x = y^2 + z^2$$

is called a **paraboloid**. Cross sections formed by planes perpendicular to the axis of the paraboloid are circles, and cross sections formed by planes containing the axis are parabolas.

Example 11 Graph $z = x^2 + y^2$.

The graph of this paraboloid is shown in Figure 9.9. A plane perpendicular to the z-axis, and above the origin, forms circular cross sections as it intersects the graph. Cross sections formed by planes containing the z-axis are parabolas. Finding such cross sections is often a good way to find the graphs of functions of two variables.

Figure 9.9

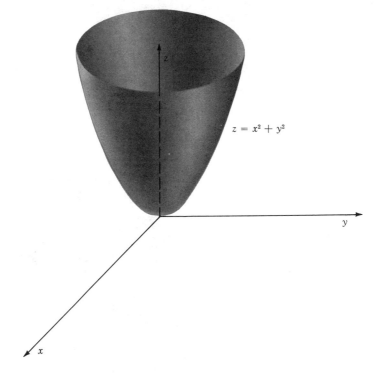

Example 12 Graph $x = y^2 + z^2$.

Here planes perpendicular to the positive x-axis form circular cross sections. Planes containing the x-axis form parabola cross sections. The graph of $x = y^2 + z^2$ is shown in Figure 9.10.

* Optional section.

Figure 9.10

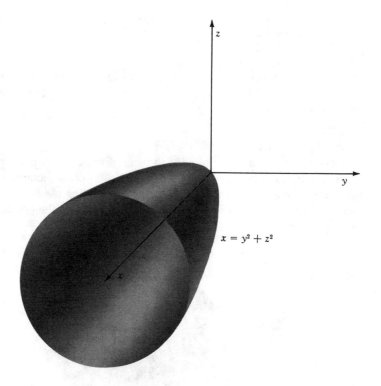

$x = y^2 + z^2$

Figure 9.11

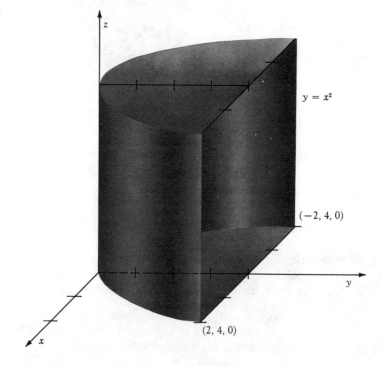

$y = x^2$

$(-2, 4, 0)$

$(2, 4, 0)$

Example 13 Graph $y = x^2$.

In the plane containing the x-axis and y-axis, the cross-section of this graph is a parabola. In fact, the cross-section formed by any plane perpendicular to the z-axis is a parabola, as shown in Figure 9.11.

9.2 Exercises

Graph each of the following planes or spheres. Identify the graph.

1. $x + y + z = 6$
2. $x + y + z = 12$
3. $2x + 3y + 4z = 12$
4. $4x + 2y + 3z = 24$
5. $3x - 2y + z = 18$
6. $4x - 3y - z = 12$
7. $-5x + y - 2z = 10$
8. $-3x - 7y + z = 21$
9. $4x - 3y + 5z = 10$
10. $-3x + 5y - 6z = 20$
11. $x + y = 6$
12. $y + z = 4$
13. $x + z = 3$
14. $-x + y = 5$
15. $x = 2$
16. $z = -3$
17. $y = -4$
18. $y = 2$
19. $x^2 + y^2 + z^2 = 49$
20. $x^2 + y^2 + z^2 = 16$
21. $(x - 5)^2 + (y - 3)^2 + (z + 4)^2 = 9$
22. $(x + 3)^2 + (y + 2)^2 + (z - 1)^2 = 16$
23. $(x - 2)^2 + (y - 5)^2 + (z + 3)^2 = 25$
24. $(x + 2)^2 + (y - 3)^2 + (z - 2)^2 = 1$

Graph each of the following cylinders or paraboloids. Identify the graph.

25. $x^2 + z^2 = 16$
26. $x^2 + y^2 = 25$
27. $z^2 + y^2 = 49$
28. $x^2 + z^2 = 36$
29. $x^2 + y^2 = 121$
30. $y^2 + z^2 = 100$
31. $2z = x^2 + y^2$
32. $5y = x^2 + z^2$
33. $2y = x^2 + z^2$
34. $3y = x^2 + z^2$
35. $4x = y^2 + z^2$
36. $2x = y^2 + z^2$
37. $-3z = x^2 + y^2$
38. $-2y = x^2 + z^2$
39. $z = x^2$
40. $x = y^2$
41. $y = (x - 2)^2$
42. $x = (z - 3)^2$

Graph each of the following. Identify each graph. (Hint: Square both sides.)

43. $z = \sqrt{16 - x^2 - y^2}$
44. $y = \sqrt{25 - x^2 - z^2}$
45. $-x = \sqrt{1 - y^2 - z^2}$
46. $-z = \sqrt{100 - x^2 - y^2}$
47. $z = \sqrt{x - y^2}$
48. $x = \sqrt{z - y^2}$

9.3 Partial Derivatives

If $C(x)$ represents the cost of producing x items, then $C'(x)$ gives the rate of change of the cost with respect to a change in the number of items produced. Derivatives are also useful for functions of two variables, as we shall see in this section. To define the derivative for a function of one variable, we needed the idea of limit. For functions of more than one independent variable, limits can be defined in much the same way that we defined them in Chapter 3. With such a definition, we could then write theorems on limits very similar to those given in Chapter 3. Using such a definition of limit, we can give the following definition of a derivative for functions of two independent variables.

Given a function $z = f(x, y)$, we can take a derivative either with respect to the independent variable x (with y held constant), or with respect to the independent variable y (with x held constant). Each of these two possible derivatives is called a partial derivative, and defined as follows:

If $z = f(x, y)$ is a function of two variables, the **partial derivatives** of z with respect to x, and with respect to y, are defined by

$$\frac{\partial z}{\partial x} = f_x = \lim_{\Delta x \to 0} \frac{f(x + \Delta x, y) - f(x, y)}{\Delta x}$$

$$\frac{\partial z}{\partial y} = f_y = \lim_{\Delta x \to 0} \frac{f(x, y + \Delta y) - f(x, y)}{\Delta y}$$

if these limits exist.

To calculate a partial derivative with respect to x, written $\partial z/\partial x$, or f_x, treat y as a constant, and find the derivative using the rules we developed for one variable. For example, if

$$z = f(x, y) = 4x^2 - 9xy + 6y^3,$$

then

$$\frac{\partial z}{\partial x} = f_x = 8x - 9y.$$

Since y is treated as a constant, the partial derivative of the term $-9xy$ with respect to x is $-9y$, and the partial derivative of $6y^3$ with respect to x is 0. The partial derivative of z with respect to y is given by

$$\frac{\partial z}{\partial y} = f_y = -9x + 18y^2.$$

Example 14 Calculate f_x and f_y for each of the following functions.
(a) $f(x, y) = 9x^3 + 2x^2y - 5y^3x^2$.
Treating y as a constant, and differentiating with respect to x, we have

$$f_x = 27x^2 + 4xy - 10y^3x.$$

If x is held constant, we have

$$f_y = 2x^2 - 15y^2x^2.$$

(b) $f(x, y) = \ln(x^2 + y)$.
Recall that if $g(x) = \ln x$, then $g'(x) = 1/x$. Using this fact here we have

$$f_x = \frac{2x}{x^2 + y} \quad \text{and} \quad f_y = \frac{1}{x^2 + y}.$$

The notation $f_x(a, b)$, or $(\partial z/\partial x)_{(a,b)}$, represents the value of a partial derivative at the point (a, b). For example, if

$$f(x, y) = 2x^2 + 3xy^3 + 2y + 5,$$

then $f_x(-1, 2)$ is found by first calculating f_x:

$$f_x = 4x + 3y^3.$$

Now substitute -1 for x and 2 for y:

$$\begin{aligned} f_x(-1, 2) &= 4(-1) + 3(2)^3 \\ &= -4 + 24 \\ &= 20. \end{aligned}$$

In the same way, $f_y(-4, -3)$ is found by first calculating f_y:

$$f_y = 9xy^2 + 2.$$

Upon substituting -4 for x and -3 for y, we find that $f_y(-4, -3) = -322$.

Example 15 Let $z = f(x, y) = 2x^2 - 4xy + 3y^2$.
(a) Find $\partial z/\partial x$.
Here we have

$$\frac{\partial z}{\partial x} = 4x - 4y.$$

This partial derivative gives the rate of change of z with respect to x, given that y is held fixed.
(b) Find $\partial z/\partial y$.
We get

$$\frac{\partial z}{\partial y} = -4x + 6y,$$

which gives the rate of change of z with respect to y, when x is held fixed.

We can give a geometric interpretation of partial derivatives. The graph of a function $z = f(x, y)$ is a surface in three-dimensional space, as we saw in Section 9.2. If we hold x constant, say $x = a$, then we get the intersection of the surface $z = f(x, y)$ and the plane $x = a$. (See Figure 9.12.) This intersection is a curve in space. The partial derivative $\partial z/\partial y$, evaluated at the point (a, y), gives the slope of the line in the plane $x = a$ which is tangent to the curve at the point $(a, y, f(a, y))$.

Figure 9.12

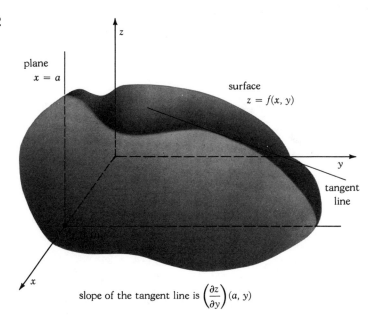

Example 16 Suppose that the temperature of the water at the point on a river where a nuclear power plant discharges its hot waste water is approximated by $T(x, y)$, where

$$T(x, y) = 2x + 5y + xy - 40.$$

Here x represents the temperature of the river water in degrees Centigrade before it reaches the power plant, and y is the number of megawatts of electricity, in hundreds, that are being produced. For example, if $x = 8°C$, and $y = 3$ (representing a production of 300 megawatts of electricity), we have

$$T(8, 3) = 2(8) + 5(3) + 8(3) - 40$$
$$= 15.$$

Thus, the water at the outlet from the plant is at a temperature of 15°C.

Here $\partial T/\partial x = 2 + y$. If we are given a specific value of x and y, such as $x = 9°C$, $y = 3$, then $(\partial T/\partial x)_{(9,3)}$ represents the change in temperature resulting from a change in x from 9° to 10°. Also, $(\partial T/\partial y)_{(9,3)}$ represents the change in temperature resulting from an increase in production of electricity from $y = 3$ to $y = 4$ hundred megawatts.

When working with functions of one variable, we found second derivatives by taking the derivative of the first derivative. In much the same way, we can define **second partial derivatives**:

$$\frac{\partial}{\partial x}\left(\frac{\partial z}{\partial x}\right) = \frac{\partial^2 z}{\partial x^2} = f_{xx} \qquad \frac{\partial}{\partial y}\left(\frac{\partial z}{\partial y}\right) = \frac{\partial^2 z}{\partial y^2} = f_{yy}$$

$$\frac{\partial}{\partial y}\left(\frac{\partial z}{\partial x}\right) = \frac{\partial^2 z}{\partial y\, \partial x} = f_{xy} \qquad \frac{\partial}{\partial x}\left(\frac{\partial z}{\partial y}\right) = \frac{\partial^2 z}{\partial x\, \partial y} = f_{yx}.$$

Note the difference between these last symbols. If we first find the partial of z with respect to x, and then with respect to y, we have found $(f_x)_y$, or f_{xy}. We have also found

$$\frac{\partial}{\partial y}\left(\frac{\partial z}{\partial x}\right) = \frac{\partial^2 z}{\partial y\, \partial x}.$$

Thus,

$$\frac{\partial^2 z}{\partial y\, \partial x} = f_{xy},$$

where the order of the symbols x and y is reversed.

Example 17 Find f_{xx}, f_{yy}, f_{xy}, and f_{yx} for each of the following functions.

(a) $f(x, y) = -4x^3 - 3x^2 y^3 + 2y^2$.

We have

$$f_x = -12x^2 - 6xy^3 \quad \text{and} \quad f_y = -9x^2 y^2 + 4y.$$

To find f_{xx}, take the partial derivative of f_x with respect to x:

$$f_{xx} = -24x - 6y^3.$$

To find f_{yy}, take the partial derivative of f_y with respect to y:

$$f_{yy} = -18x^2 y + 4.$$

The partial f_{xy} is found by taking the partial derivative of f_x with respect to y:

$$f_{xy} = -18xy^2.$$

The partial f_{yx} is found by taking the partial derivative of f_y with respect to x:

$$f_{yx} = -18xy^2.$$

(b) $f(x, y) = 2e^x - 8x^3 y^2 + 6x^2$.

Here $f_x = 2e^x - 24y^2 x^2 + 12x$ and $f_y = -16yx^3$. Now we can find the second partial derivatives, as described above.

$$f_{xx} = 2e^x - 48y^2 x + 12 \qquad f_{yy} = -16x^3$$
$$f_{xy} = -48yx^2 \qquad f_{yx} = -48yx^2$$

In both of these examples, we found that $f_{xy} = f_{yx}$. This happens with most common functions, as shown by the next theorem, whose proof is given in most of the calculus books listed in the bibliography.

Theorem 9.1 If f, f_x, f_y, f_{xy}, and f_{yx} are all continuous at a point (a, b), then

$$f_{xy}(a, b) = f_{yx}(a, b)$$

for the function $z = f(x, y)$.

9.3 Exercises

Find f_x, f_y, f_{xx}, f_{yy}, and f_{xy} for each of the following.

1. $f(x, y) = 8x + 9y^2$
2. $f(x, y) = 12x^2 - 11y$
3. $f(x, y) = 4x^2y - 12xy^2$
4. $f(x, y) = 18xy^3 - 9x^2y$
5. $f(x, y) = 15x^2 + 12y^3 + 18xy^3 + 12$
6. $f(x, y) = 9x^3 - 9y^2 - 12x^2y - 4$
7. $f(x, y) = 4x^2 - 5xy^3 + 12y^2x^2$
8. $f(x, y) = -8 + 10x + 30y + 5x^2y + 12xy^2$
9. $f(x, y) = 6x^3y^2 - 4xy^4 - 16xy^3$
10. $f(x, y) = 12xy^4 - 3x^2y^2 + 6xy^2$
11. $f(x, y) = e^x \cdot e^y + xe^{2y} + 2y^3e^x$
12. $f(x, y) = 3e^{4x} \cdot e^{2y} + 6x^3e^{2y}$
13. $f(x, y) = \sqrt{x^2 + y^2}$
14. $f(x, y) = \sqrt{4x^2 + 9y^2}$
15. $f(x, y) = \dfrac{x + y}{x - y}$
16. $f(x, y) = \dfrac{2x + 3y}{4x - 3y}$
17. $f(x, y) = \ln(x^2 + y^2)$
18. $f(x, y) = \ln(2x^3 - 3y^2)$
19. $f(x, y) = \ln x^2 y^3$
20. $f(x, y) = \ln x^5 y^2$
21. $f(x, y) = x \cdot e^y$
22. $f(x, y) = y \cdot e^x$

In each of the exercises below, find $f_x(-1, 2)$, $f_y(2, 0)$, and $f_{xy}(3, -2)$.

23. $f(x, y) = 5x^3 - 2y^2$
24. $f(x, y) = 8x^2 - 11y^3$
25. $f(x, y) = 9x^2y + 8xy^3 + 12$
26. $f(x, y) = 4xy^3 - 9x^2y - 4$
27. $f(x, y) = \sqrt{x^2 + y^2}$
28. $f(x, y) = \sqrt{4x^2 + 9y^2}$
29. $f(x, y) = \ln x^2 y^3$
30. $f(x, y) = \ln x^5 y^2$
31. $f(x, y) = x \cdot e^y$
32. $f(x, y) = y \cdot e^x$

33. Suppose that the manufacturing cost, $M(x, y)$, of a precision electronic calculator is approximated by

$$M(x, y) = 40x^2 + 30y^2 - 10xy + 30,$$

where x is the cost of the necessary electronic chips, and y is the cost of labor. Find and interpret each of the following:

(a) $(\partial M / \partial y)_{(4,2)}$
(b) $(\partial M / \partial x)_{(3,6)}$

34. The total number of matings, $M(x, y)$, per day, between individuals of a certain species of Arkansas grasshoppers is approximated by

$$M(x, y) = 2xy + 10xy^2 + 30y^2 + 20,$$

where x represents the temperature in degrees Centigrade, and y represents the number of days until the next full moon. Find and interpret each of the following:

(a) $(\partial M / \partial x)_{(20,4)}$
(b) $(\partial M / \partial y)_{(24,10)}$

9.4 Maxima and Minima

One of the most important applications of calculus for functions of one variable is in finding maxima and minima for the functions, as we saw in Chapter 5. Such maxima and minima also provide important applications for functions of two variables. As we shall see, partial derivatives play an important part in finding maxima and minima.

Just as we defined extrema for a function of one variable, we define extrema for functions of two independent variables by using the idea of a neighborhood. Figure 9.13(a) shows a neighborhood of a point a on a number line. Figure 9.13(b) shows a neighborhood of a point (a, b) in a plane. A **neighborhood** of a point (a, b) is a disc having (a, b) as center. The edge of the disc is not included in the neighborhood.

Figure 9.13

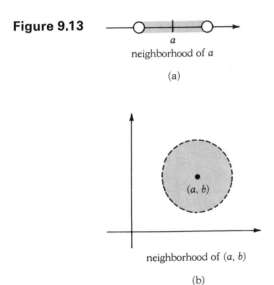

neighborhood of a

(a)

neighborhood of (a, b)

(b)

To define relative maxima and minima, let $z = f(x, y)$ be a function defined for each point in some region in the plane containing the x-axis and the y-axis. The function $z = f(x, y)$ has a **relative maximum** at the point (a, b) if there exists a neighborhood of (a, b) such that

$$f(x, y) \leq f(a, b)$$

for every point (x, y) of that neighborhood. See Figure 9.14(a). A relative minimum is defined in a similar way. An example of a relative minimum is shown in Figure 9.14(b).

Figure 9.14

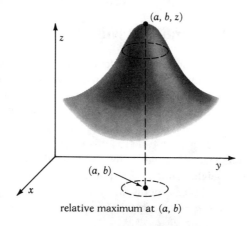

relative maximum at (a, b)

(a)

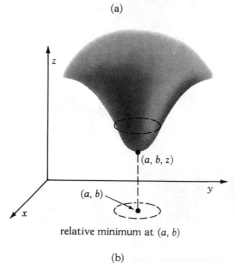

relative minimum at (a, b)

(b)

A function of one variable, such as $y = f(x)$, can have a relative maximum or minimum at a point a such that $f'(a) = 0$. (But the function does not necessarily have a relative extremum at a just because $f'(a) = 0$.) A similar situation holds with functions of two variables, as shown by the next theorem.

Theorem 9.2 Let a function $z = f(x, y)$ have a relative maximum or minimum at the point (a, b). Let $f_x(a, b)$ and $f_y(a, b)$ both exist. Then

$$f_x(a, b) = 0 \quad \text{and} \quad f_y(a, b) = 0.$$

To use this theorem, it is necessary to consider all points (a, b) at which both $f_x(a, b)$ and $f_y(a, b)$ are equal to 0. However, just because $f_x(a, b) = 0$ and $f_y(a, b) = 0$, we have no assurance at all that $f(a, b)$ is a relative maximum or a relative minimum. For example, Figure 9.15(a) shows a graph of the function

$z = f(x, y) = x^2 - y^2$. Both $f_x(0, 0) = 0$ and $f_y(0, 0) = 0$, and yet $f(0, 0)$ is neither a relative maximum nor a relative minimum for the function. Here the point $(0, 0, 0)$ of the graph is called a **saddle point**. Figure 9.15(b) shows the graph of $f(x, y) = x^3$. Here $f_x(0, 0) = 0$, and $f_y(0, 0) = 0$. However, $(0, 0)$ leads to neither a maximum, a minimum, nor a saddle point.

Figure 9.15

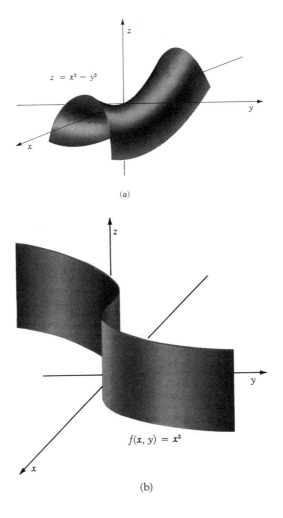

Example 18 At what points might the function

$$f(x, y) = 6x^2 + 6y^2 + 6xy + 36x - 5$$

have any maxima or minima?

We must first find all points (a, b) at which $f_x(a, b) = 0$ and $f_y(a, b) = 0$. Here we have

$$f_x(x, y) = 12x + 6y + 36 \quad \text{and} \quad f_y(x, y) = 12y + 6x.$$

Putting these two partial derivatives equal to 0 gives

$$12x + 6y + 36 = 0 \quad \text{and} \quad 12y + 6x = 0.$$

This is a system of linear equations. We can solve this system as follows. From the equation $12y + 6x = 0$, we get $-6x = 12y$, or $x = -2y$. Substituting $-2y$ for x in the first equation gives

$$12x + 6y + 36 = 0$$
$$12(-2y) + 6y + 36 = 0$$
$$-24y + 6y + 36 = 0$$
$$-18y + 36 = 0$$
$$y = 2.$$

Since $x = -2y$, we have $x = -2(2)$, or $x = -4$. Thus, the solution of the system is $(-4, 2)$. Based on this, if the function $f(x, y) = 6x^2 + 6y^2 + 6xy + 36x - 5$ has any relative maxima or relative minima, they occur at the point $(-4, 2)$. To decide whether the point $(-4, 2)$ leads to a relative maximum, relative minimum, or neither, we need to use the results of the next theorem, whose proof can be found in the calculus books in the bibliography.

Theorem 9.3 Let all the first and second partial derivatives of the function $z = f(x, y)$ exist at all points in some neighborhood of the point (a, b). Suppose also that

$$f_x(a, b) = 0 \quad \text{and} \quad f_y(a, b) = 0.$$

Let

$$M = f_{xx}(a, b) \cdot f_{yy}(a, b) - [f_{xy}(a, b)]^2.$$

Then

(i) $f(a, b)$ is a relative maximum if $M > 0$ and $f_{xx}(a, b) < 0$.
(ii) $f(a, b)$ is a relative minimum if $M > 0$ and $f_{xx}(a, b) > 0$.
(iii) $f(a, b)$ is a saddle point (neither a maximum nor a minimum) if $M < 0$.
(iv) if $M = 0$, this test gives no information.

Example 19 In the example above, we found that if the function

$$z = f(x, y) = 6x^2 + 6y^2 + 6xy + 36x - 5$$

has any relative maxima or relative minima, it has them at the point $(-4, 2)$. Does $(-4, 2)$ lead to a relative maximum, a relative minimum, or neither?

We can find out by using Theorem 9.3. We already know that

$$f_x(-4, 2) = 0 \quad \text{and} \quad f_y(-4, 2) = 0.$$

We now need to find the various second partial derivatives required to find M. From $f_x = 12x + 6y + 36$ and $f_y = 12y + 6x$, we get

$$f_{xx} = 12, \quad f_{yy} = 12, \quad \text{and} \quad f_{xy} = 6.$$

(If these second partial derivatives had not all been equal to constants, it would have been necessary to evaluate each of them at the point $(-4, 2)$.) Here we have

$$M = f_{xx}(-4, 2) \cdot f_{yy}(-4, 2) - [f_{xy}(-4, 2)]^2$$
$$= 12 \cdot 12 - 6^2$$
$$= 108.$$

Since $M > 0$ and $f_{xx}(-4, 2) = 12 > 0$, part (ii) of Theorem 9.3 applies. Thus, $z = f(x, y) = 6x^2 + 6y^2 + 6xy + 36x - 5$ has a relative minimum at the point $(-4, 2)$. This relative minimum is given by $f(-4, 2) = -77$.

Example 20 Find any relative maxima or relative minima for the function

$$f(x, y) = x^2 - 4x + y^2 + 6y - 6.$$

Here we have

$$f_x = 2x - 4 \quad \text{and} \quad f_y = 2y + 6.$$

Both these partial derivatives are equal to 0 only when $x = 2$ and $y = -3$. (Place each partial derivative equal to 0 to see this.) Thus, any possible maximum or minimum will occur at the point $(2, -3)$. To find out whether $(2, -3)$ leads to a relative maximum, a relative minimum, or neither, we use Theorem 9.3. We have

$$f_{xx} = 2, \quad f_{yy} = 2, \quad f_{xy} = 0.$$

Thus,

$$M = 2 \cdot 2 - 0^2 = 4,$$

which is positive. Here $M > 0$ and $f_{xx} > 0$. Thus, by part (ii) of Theorem 9.3, $f(2, -3)$ is a relative minimum.

Example 21 Find any relative maxima or relative minima for the function

$$f(x, y) = 9xy - x^3 - y^3 - 6.$$

Here we have

$$f_x = 9y - 3x^2 \quad \text{and} \quad f_y = 9x - 3y^2.$$

If we place each partial derivative equal to 0, we get

$$9y - 3x^2 = 0 \quad \text{and} \quad 9x - 3y^2 = 0$$
$$3y = x^2 \qquad\qquad 3x = y^2.$$

From the left-hand equation, $y = x^2/3$. Substituting this into the right-hand equation, we have

$$3x = y^2$$
$$3x = \left(\frac{x^2}{3}\right)^2$$
$$3x = \frac{x^4}{9}.$$

Multiplying both sides by 9 gives

$$27x = x^4 \quad \text{or} \quad x^4 - 27x = 0.$$

Factoring, we have $x(x^3 - 27) = 0$, from which

$$x = 0 \quad \text{or} \quad x^3 - 27 = 0.$$

The only real number solution to the equation $x^3 - 27 = 0$ is $x = 3$. If $x = 0$, then $y = 0$. If $x = 3$, then $y = 3$. Thus, the only possible relative maxima or relative minima for the given function occur at $(0, 0)$ or at $(3, 3)$. To find out which, we use Theorem 9.3. We have

$$f_{xx} = -6x, \quad f_{yy} = -6y, \quad \text{and} \quad f_{xy} = 9.$$

Thus,

for $(0, 0)$

$f_{xx}(0, 0) = 0$
$f_{yy}(0, 0) = 0,$

so that

$M = 0 \cdot 0 - 9^2 = -81.$

Since $M < 0$, by part (iii) of Theorem 9.3, we have a saddle point at $(0, 0)$.

for $(3, 3)$

$f_{xx}(3, 3) = -6(3) = -18$
$f_{yy}(3, 3) = -6(3) = -18,$

so that

$M = -18(-18) - 9^2 = 243.$

Here $M > 0$ and $f_{xx}(3, 3) = -18 < 0$. Thus, by part (i) of Theorem 9.4, $f(3, 3)$ is a relative maximum.

Example 22 Find any relative maxima or relative minima for the function

$$f(x, y) = 50 + 4x - 5y + x^2 + y^2 + xy.$$

Here we have

$$f_x = 4 + 2x + y \quad \text{and} \quad f_y = -5 + 2y + x.$$

If we place the partial derivatives equal to 0, and rearrange terms somewhat, we get

$$2x + y = -4$$
$$x + 2y = 5.$$

One way to solve this system of equations is to multiply both sides of the top equation by -2. We then add the two equations. This gives

$$-4x - 2y = 8$$
$$\underline{x + 2y = 5}$$
$$-3x \quad\quad = 13$$

From this last result, we have $x = -13/3$. Substitute $-13/3$ for x in either equation and verify that $y = 14/3$. Thus, the only possible point leading to a relative maximum or minimum is $(-13/3, 14/3)$.

For the function given above, we have

$$f_{xx} = 2, \quad f_{yy} = 2, \quad \text{and} \quad f_{xy} = 1.$$

Thus,

$$M = 2 \cdot 2 - 1^2 = 3,$$

which is positive. Since $M > 0$ and $f_{xx}(-13/3, 14/3) = 2 > 0$, we see that $f_{xx}(-13/3, 14/3)$ is a relative minimum.

9.4 Exercises

Find any relative maxima or relative minima for each of the following functions.

1. $f(x, y) = x^2 - 2xy + 2y^2 + x - 5$
2. $f(x, y) = x^2 + xy + y^2 - 6x - 3$
3. $f(x, y) = x^2 - xy + y^2 + 2x + 2y + 6$
4. $f(x, y) = x^2 + xy + y^2 + 3x - 3y$
5. $f(x, y) = x^2 + 3xy + 3y^2 - 6x + 3y$
6. $f(x, y) = 5xy - 7x^2 - y^2 + 3x - 6y - 4$
7. $f(x, y) = 4xy - 10x^2 - 4y^2 + 8x + 8y + 9$
8. $f(x, y) = x^2 + xy + 3x + 2y - 6$
9. $f(x, y) = y^2 + xy - 2x - 2y + 2$
10. $f(x, y) = x^2 + xy + y^2 - 3x - 5$
11. $f(x, y) = xy + x - y$
12. $f(x, y) = x^2 - y^2 - 2x + 4y - 5$
13. $f(x, y) = 4x + 2y - x^2 + xy - y^2 + 3$
14. Suppose that the profit, $P(x, y)$, of a certain firm is approximated by

    $$P(x, y) = 1000 + 24x - x^2 + 80y - y^2,$$

    where x is the cost of a unit of labor and y is the cost of a unit of goods. Find the value of x and y that will maximize profit. Find the maximum profit.
15. The labor cost, $L(x, y)$, for manufacturing a precision camera can be approximated by

    $$L(x, y) = \tfrac{3}{2}x^2 + y^2 - 2x - 2y - 2xy + 68,$$

    where x is the number of days required by a skilled crafts person, and y is the number of days required by a semi-skilled crafts person. Find the value of x and y that will minimize the labor charge. Find the minimum labor charge.

16. The number of roosters, $R(x, y)$, that can be fed from x pounds of Super-Hen chicken feed and y pounds of Super-Rooster feed is given by

$$R(x, y) = 800 - 2x^3 + 12xy - y^2.$$

Find the number of pounds of each kind of feed that will produce the maximum number of roosters.

9.5 Lagrange Multipliers

In the previous section we saw how to find any relative maxima or relative minima for functions of two variables. It is common, however, for such functions to be given with a secondary condition, or **constraint**. For example, suppose we need to find any relative maximum or minimum values of the function

$$f(x, y) = 5x^2 + 6y^2 - xy,$$

subject to the constraint or condition that $x + 2y = 24$. It would be possible, but very difficult, to solve this problem using the methods of the last section. However, it is easier to work the problem using the method of **Lagrange multipliers**, discussed in this section. This method looks fairly complicated when written down as a theorem, so we shall first look at examples of the method, and then at the end of the section state the method as a theorem.

One difficulty with the method of Lagrange multipliers is that it is not at all easy to tell whether the answer is a relative maximum or a relative minimum.* Usually, it is necessary to try to decide from the context of the problem—a cost function would probably lead to a minimum, while a problem involving the volume of a box would probably lead to a maximum.

Let us find the minimum value of $f(x, y) = 5x^2 + 6y^2 - xy$, subject to the constraint $x + 2y = 24$.

Step 1. Write the constraint so that it equals 0. Let $g(x, y)$ equal the constraint.

In this example, the constraint, $x + 2y = 24$, can be rewritten as $x + 2y - 24 = 0$, so that

$$g(x, y) = x + 2y - 24.$$

Step 2. Form the **Lagrange function** $F(x, y, \lambda)$. The symbol λ is the Greek letter lambda. The Lagrange function is the sum of $f(x, y)$ and λ times $g(x, y)$.

In our example we have

$$\begin{aligned} F(x, y, \lambda) &= f(x, y) + \lambda \cdot g(x, y) \\ &= 5x^2 + 6y^2 - xy + \lambda(x + 2y - 24). \end{aligned}$$

---

* Details on deciding whether a relative maximum or a relative minimum has been obtained are given in Jean E. Draper and Jane S. Klingman, *Mathematical Analysis*, second edition, New York: Harper and Row, 1972, pages 376 and 602.

9.5 Lagrange Multipliers

Step 3. Form the system of equations $F_x = 0$, $F_y = 0$, and $F_\lambda = 0$.
In our example, we have

$$F_x = 10x - y + \lambda = 0 \qquad (1)$$
$$F_y = 12y - x + 2\lambda = 0 \qquad (2)$$
$$F_\lambda = x + 2y - 24 = 0 \qquad (3)$$

Step 4. Solve the system of equations obtained in Step 3 for x and y.

To solve the system of Step 3, we can multiply both sides of equation (1) by 12, and add the result to equation (2). Doing this, and rearranging terms, we have

$$\begin{array}{r} 120x - 12y + 12\lambda = 0 \\ -x + 12y + 2\lambda = 0 \\ \hline 119x \qquad\quad + 14\lambda = 0 \end{array} \qquad (4)$$

To eliminate y from equations (2) and (3), multiply both sides of equation (3) by -6 and add the result to equation (2). After rearranging terms, we have

$$\begin{array}{r} -x + 12y + 2\lambda = 0 \\ -6x - 12y \qquad\quad = -144 \\ \hline -7x \qquad\quad + 2\lambda = -144 \end{array} \qquad (5)$$

We must now solve equations (4) and (5) together to find x. If we multiply both sides of equation (5) by -7, we get

$$\begin{array}{r} 119x + 14\lambda = 0 \\ 49x - 14\lambda = 1008 \\ \hline 168x \qquad\quad = 1008 \end{array}$$

from which $x = 6$. Substituting 6 for x in equation (3), we have $y = 9$. Thus, the minimum value for $f(x, y) = 5x^2 + 6y^2 - xy$, subject to the constraint $x + 2y = 24$, is at the point $(6, 9)$.

Example 23 Find two numbers whose sum is 50 and whose product is a maximum.

If we let x and y represent the two numbers, then we wish to maximize the product

$$f(x, y) = xy,$$

subject to the constraint $x + y = 50$. We go through the four steps presented above.

Step 1. $g(x, y) = x + y - 50$.
Step 2. $F(x, y, \lambda) = xy + \lambda(x + y - 50)$.
Step 3. $F_x = y + \lambda = 0$ \hfill (6)
$F_y = x + \lambda = 0$ \hfill (7)
$F_\lambda = x + y - 50 = 0.$ \hfill (8)

Multivariate Functions

Step 4. From equation (6), we have $\lambda = -y$. Substituting $-y$ for λ in equation (7), we have

$$x + \lambda = 0$$
$$x + (-y) = 0$$
$$x = y.$$

Substituting y for x in equation (8), we have

$$y + y - 50 = 0$$
$$y = 25.$$

Since $x = y$, we have $x = 25$. Thus, 25 and 25 are the two numbers whose sum is 50 and whose product is a maximum. The maximum product is given by $25 \cdot 25 = 625$.

Example 24 Find the dimensions of the rectangular box of maximum volume that can be produced from 6 square feet of material.

Figure 9.16

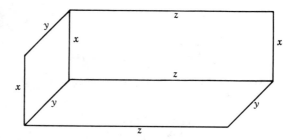

Let x, y, and z represent the dimensions of the box, as shown in Figure 9.16. The volume of the box is given by

$$f(x, y, z) = xyz.$$

As shown in Figure 9.16, the total amount of material required for the two ends of the box is $2xy$, the total needed for the sides is $2xz$, and the total needed for the top and bottom is $2yz$. Since 6 square feet of material is available, we have

$$2xy + 2xz + 2yz = 6 \quad \text{or} \quad xy + xz + yz = 3.$$

In summary, we must maximize $f(x, y, z) = xyz$, subject to the constraint $xy + xz + yz = 3$. This problem involves functions of *three* variables instead of two, but the method of Lagrange multipliers can be used in much the same way as before.

Step 1. $g(x, y, z) = xy + xz + yz - 3$.
Step 2. $F(x, y, z, \lambda) = xyz + \lambda(xy + xz + yz - 3)$.
Step 3. $F_x = yz + \lambda y + \lambda z = 0$
$F_y = xz + \lambda x + \lambda z = 0$
$F_z = xy + \lambda x + \lambda y = 0$
$F_\lambda = xy + xz + yz - 3 = 0$.

Step 4. Verify that the solution of the system of equations of Step 3 is $x = 1$, $y = 1$, $z = 1$. In other words, the desired box is a cube, 1 foot on a side.

Now we can state the method of Lagrange multipliers as a theorem. As we said above, the method works with two independent variables, or three, or four, or more. For simplicity, we shall state the theorem only for two independent variables.

Theorem 9.4 Lagrange multipliers. Any relative minimum or relative maximum values of the function $z = f(x, y)$, subject to a constraint $g(x, y) = 0$, will be found among those points (x, y) for which there exists a value of λ such that

$$F_x(x, y, \lambda) = 0$$
$$F_y(x, y, \lambda) = 0$$
$$F_\lambda(x, y, \lambda) = 0,$$

where

$$F(x, y, \lambda) = f(x, y) + \lambda g(x, y).$$

We assume that all indicated partial derivatives exist.

A proof of this theorem can be found in most advanced calculus texts.*

9.5 Exercises

Find each of the following relative maxima or relative minima.

1. Maximum of $f(x, y) = 2xy$, subject to $x + y = 12$
2. Maximum of $f(x, y) = 4xy + 2$, subject to $x + y = 24$
3. Maximum of $f(x, y) = x^2 y$, subject to $2x + y = 4$
4. Maximum of $f(x, y) = 4xy^2$, subject to $3x - 2y = 5$
5. Minimum of $f(x, y) = x^2 + 2y^2 - xy$, subject to $x + y = 8$
6. Minimum of $f(x, y) = 3x^2 + 4y^2 - xy - 2$, subject to $2x + y = 21$
7. Maximum of $f(x, y) = x^2 - 10y^2$, subject to $x - y = 18$
8. Maximum of $f(x, y) = 12xy - x^2 - 3y^2$, subject to $x + y = 16$
9. A farmer has 200 feet of fencing. Find the dimensions of the rectangular field of largest area that can be enclosed by the fencing material.
10. A rectangular box with no top is to be constructed from 500 square feet of material. Find the dimensions of such a box that will enclose the maximum volume.
11. Find three numbers whose sum is 90 such that their product is a maximum.
12. Find three numbers whose sum is 240 such that their product is a maximum.

* One proof is given in T. M. Apostol, *Mathematical Analysis*, Second Edition, Reading, Massachusetts, Addison-Wesley, 1974.

304 *Multivariate Functions*

13. The total cost to produce x units of Delightful Doggie needlepoint kits and y units of Petunia Pussy kits is $C(x, y)$, where

$$C(x, y) = 2x^2 + 6y^2 + 4xy + 10.$$

If a total of 20 units of kits must be made, how should production be allocated so that cost is minimized?

9.6 An Application—The Least Squares Line

In trying to predict future sales of a product, or the total number of matings between two species of insects, it is common to gather as much past data as possible, and then draw a graph showing the data. This graph, called a **scatter diagram**, can then be inspected to see if a reasonably simple mathematical curve can be found which fits fairly well through all the given data points. Figure 9.17 shows a collection of data values, and the scatter diagram that results when these points are graphed. As shown in the figure, a straight line fits reasonably well through the points of the scatter diagram.

Figure 9.17

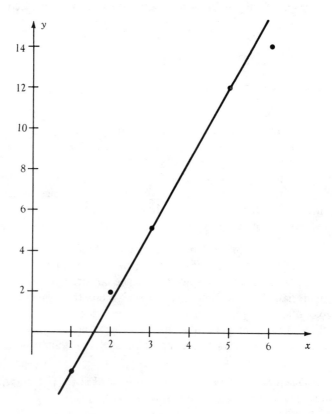

9.6 An Application—The Least Squares Line

If all the data points lie on the same straight line (which rarely happens in practice) then the methods of Section 2.2 could be used to find the equation of the line through the data points. In practice, however, it is rare to have a collection of data that fits perfectly on a straight line. Then we are faced with the problem of deciding on the "best" straight line through the data points. In Figure 9.17 above, it would be easy enough to say that some straight line other than the one shown provided a little better fit to the data points.

How do we decide on the "best" line? Figure 9.18 below shows four data values, (x_1, y_1), (x_2, y_2), (x_3, y_3), and (x_4, y_4). A straight line with equation $y = mx + b$ has been drawn which fits fairly well through the given data points. Vertical line segments have been drawn, connecting the data points and the straight line. The vertical segments cut the straight line at the points $(x_1, \hat{y}_1)$, $(x_2, \hat{y}_2)$, $(x_3, \hat{y}_3)$, and $(x_4, \hat{y}_4)$, where $\hat{y}_1$ (read "y-sub-one hat") is a common statistical symbol used to represent the "predicted" values of y.

Figure 9.18

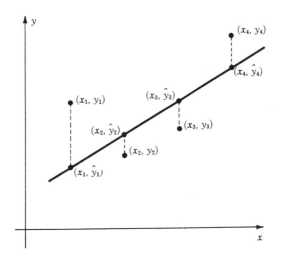

We might agree that the "best" straight line through the data points is the one in which the sum of the differences between the $\hat{y}$ values on the straight line and the y values of the data points themselves is as small as possible. In other words, we might choose to minimize

$$(y_1 - \hat{y}_1) + (y_2 - \hat{y}_2) + \cdots + (y_n - \hat{y}_n)$$

if there are n data points. This sum can be written in the shorthand **sigma notation** as

$$\sum_{i=1}^{n} (y_i - \hat{y}_i),$$

or simply as $\sum (y - \hat{y})$. ($\sum$ is the Greek letter sigma.) However, the sum $\sum (y - \hat{y})$ can be made to equal 0, or even any desired negative number, by choosing a straight line which fits the data points very poorly. It has been found through much practical experience that the "best" line through the data points is found by minimizing the sum

$$\Sigma (y - \hat{y})^2, \quad \text{or} \quad \sum_{i=1}^{n} (y_i - \hat{y}_i)^2.$$

The **least squares regression line** is defined as that straight line that makes the sum $\Sigma (y - \hat{y})^2$ a minimum. The equation of the least squares regression line can be found using partial derivatives and the theory we have developed for finding maxima and minima.

Let the equation of the least squares regression line be $y = mx + b$. Here the letters m and b are the unknowns; if we can find m and b we will have the equation we want. If (x_1, y_1) is one of the given data values, then the corresponding point on the least squares regression line can be found by substituting the x-value, x_1, for x in the equation $y = mx + b$. This gives $y = mx_1 + b$. We have used the notation $\hat{y}_1$ to denote the corresponding y-value on the least squares regression line, so that we now have

$$\hat{y}_1 = mx_1 + b,$$

and for any value of i, where $1 \leq i \leq n$, we have

$$\hat{y}_i = mx_i + b.$$

We want to minimize the sum

$$\sum_{i=1}^{n} (y_i - \hat{y}_i)^2$$

(using the more complete sigma notation). If we substitute $mx_i + b$ for $\hat{y}_i$, we get the sum

$$\sum_{i=1}^{n} [y_i - (mx_i + b)]^2 = \sum_{i=1}^{n} (y_i - mx_i - b)^2.$$

Since the x_i and y_i values represent known data points, the unknowns in this sum are the numbers m and b. To emphasize this fact, we can write the sum in function notation:

$$f(m, b) = \sum_{i=1}^{n} (y_i - mx_i - b)^2$$
$$= (y_1 - mx_1 - b)^2 + (y_2 - mx_2 - b)^2 + \cdots + (y_n - mx_n - b)^2.$$

To find the minimum value of this function, find the partial derivatives with respect to m and to b, and place each partial derivative equal to 0:

$$\frac{\partial f}{\partial m} = -2x_1(y_1 - mx_1 - b) - 2x_2(y_2 - mx_2 - b) + \cdots + -2x_n(y_n - mx_n - b) = 0$$

$$\frac{\partial f}{\partial b} = -2(y_1 - mx_1 - b) - 2(y_2 - mx_2 - b) + \cdots + -2(y_n - mx_n - b) = 0.$$

9.6 An Application—The Least Squares Line

Using some algebra, and rearranging terms, these two equations become

$$(x_1^2 + x_2^2 + \cdots + x_n^2)m + (x_1 + x_2 + \cdots + x_n)b$$
$$= x_1y_1 + x_2y_2 + \cdots + x_ny_n$$
$$(x_1 + x_2 + \cdots + x_n)m + nb = y_1 + y_2 + \cdots + y_n.$$

We can rewrite these last two equations using sigma notation as follows:

$$(\Sigma x^2)m + (\Sigma x)b = \Sigma xy \qquad (1)$$

$$(\Sigma x)m + nb = \Sigma y. \qquad (2)$$

To solve this system of equations, multiply the first on both sides by $-n$ and the second by Σx. This gives

$$-n(\Sigma x^2)m - n(\Sigma x)b = -n(\Sigma xy)$$
$$(\Sigma x)(\Sigma x)m + (\Sigma x)nb = (\Sigma x)(\Sigma y)$$
$$\overline{(\Sigma x)(\Sigma x)m - n(\Sigma x^2)m = (\Sigma x)(\Sigma y) - n(\Sigma xy)}$$

We can write the product $(\Sigma x)(\Sigma x)$ as $(\Sigma x)^2$. Doing this, the last equation becomes (upon solving for m)

$$m = \frac{(\Sigma x)(\Sigma y) - n(\Sigma xy)}{(\Sigma x)^2 - n(\Sigma x^2)}. \qquad (3)$$

Normally, we would substitute this value of m into either equation (1) or equation (2) and find a solution for b. However, it is easier to solve equation (2) for b, obtaining

$$b = \frac{\Sigma y - m(\Sigma x)}{n}. \qquad (4)$$

This way we find b after finding m.

Example 25 Find the least squares regression line for the data of Figure 9.17 above.

From Figure 9.17 we find the data values shown in the following chart. The chart also shows the numerical values needed for finding m and b.

x	y	x^2	xy
1	−2	1	−2
2	2	4	4
3	5	9	15
5	12	25	60
6	14	36	84
Total 17	31	75	161

Hence, we have $\sum x = 17$, $\sum y = 31$, $\sum x^2 = 75$, $\sum xy = 161$, and $n = 5$. Using equation (3), we have

$$m = \frac{(17)(31) - 5(161)}{17^2 - 5(75)}$$

$$= \frac{527 - 805}{289 - 375}$$

$$= \frac{-278}{-86}$$

$$m = 3.2.$$

We can find b using equation (4):

$$b = \frac{31 - (3.2)(17)}{5}$$

$$= -4.7.$$

The least squares regression equation is thus $y = mx + b$, or

$$y = 3.2x - 4.7.$$

This equation can be used to predict values of y for given values of x. For example, if $x = 4$, we have

$$y = 3.2x - 4.7$$
$$= 3.2(4) - 4.7$$
$$= 12.8 - 4.7$$
$$= 8.1.$$

If $x = 12$, we have $y = 3.2(12) - 4.7 = 38.4 - 4.7 = 33.7$.

9.6 Exercises

In each of the following exercises, draw a scatter diagram for the given set of data points and find the least squares regression line.

1.
x	1	2	3	5	9
y	9	13	18	25	41

2.
x	2	3	5	6	8
y	8	13	23	28	38

3.
x	4	5	8	12	14
y	3	7	17	28	35

4.
x	3	4	5	6	8
y	8	12	16	18	28

5. The ACT test scores of eight students were compared to their GPA's after one year in college. The results are shown below.

ACT score (x)	19	20	22	24	25	26	27	29
GPA (y)	2.2	2.4	2.7	2.6	3.0	3.5	3.4	3.8

(a) Plot the 8 points on a scatter diagram.

(b) Find the least squares regression equation and graph it on the scatter diagram of part (a).
(c) Using the results of (b), predict the GPA of a student with an ACT score of 28.

6. Records show the annual sales for the Sweet Palms Life Insurance Company in 5-year periods for the last 20 years were as follows.

Year (x)	Sales (in millions) (y)
1955	1.0
1960	1.3
1965	1.7
1970	1.9
1975	2.1

The company wishes to estimate sales from these records for the next few years. Code the years so that $1955 = 0$, $1956 = 1$, $1957 = 2$, and so on.
(a) Plot the 5 points on a scatter diagram.
(b) Find the least squares regression line and graph it on the scatter diagram of part (a).
(c) Predict the company's sales for 1976 and 1977.

7. McDonald's[*] wishes to find the relationship between annual store sales and pre-tax profit in order to estimate increases in profit due to increased sales volume. The data shown below was obtained from a sample of McDonald's stores throughout the country.

Annual store sales in $1000 ($x$)	250	300	375	425	450	475	500	575	600	650
Pre-tax percent profit	9.3	10.8	14.8	14.0	14.2	15.3	15.9	19.1	19.2	21.0

(a) Plot the ten pairs of values on a scatter diagram.
(b) Find the equation of the least squares regression line and draw it on the scatter diagram of part (a).
(c) Using the equation of part (b), predict the pre-tax percent profit for annual sales of $700,000; for annual sales of $750,000.

• Case 13 Lagrange Multipliers for a Predator[**]

A predator is an animal that feeds upon another animal. Foxes, coyotes, wolves, weasels, lions, and tigers are well-known predators, but many other animals

---

[*] This example supplied by Harvey Lubelchek, Operations Research Manager, McDonald's System, Inc.
[**] Based on Gerald G. Martens, An Optimization Equation for Predation, *Ecology*, Winter 1973, pp. 92–101.

also fall into this category. In this case, we set up a mathematical model for predation, and then use Lagrange multipliers to minimize the difference between the desired and the actual levels of food consumption. There are several research studies which show that animals *do* control their activities so as to maximize or minimize variables—lobsters orient their bodies by minimizing the discharge rate from certain organs, for example.

In this case, we assume that the predator has a diet consisting of only two foods, food 1 and food 2. We also assume that the predator will hunt only in two locations, location 1 and location 2. To make this mathematical model meaningful, we would have to gather data on the predators and the prey that we wished to study. For example, suppose that we have gathered the following data:

$u_{11} = 0.4$ = rate of feeding on food 1 in location 1
$u_{12} = 0.1$ = rate of feeding on food 1 in location 2
$u_{21} = 0.3$ = rate of feeding on food 2 in location 1
$u_{22} = 0.3$ = rate of feeding on food 2 in location 2.

Let x_1 = proportion of time spent feeding in location 1
x_2 = proportion of time spent feeding in location 2
x_3 = proportion of time spent on nonfeeding activities.

Using these variables, the total quantity of food 1 consumed is given by Y_1, where

$$Y_1 = u_{11}x_1 + u_{12}x_2 \\ Y_1 = 0.4x_1 + 0.1x_2. \tag{1}$$

The total quantity of food 2 consumed is given by Y_2, where

$$Y_2 = u_{21}x_1 + u_{22}x_2 \\ Y_2 = 0.3x_1 + 0.3x_2. \tag{2}$$

The total amount of food consumed is thus given by

$$Y_1 + Y_2 = 0.4x_1 + 0.1x_2 + 0.3x_1 + 0.3x_2 \\ Y_1 + Y_2 = 0.7x_1 + 0.4x_2. \tag{3}$$

Again, by gathering experimental data, suppose that

$z_t = 0.4$ = desired level of total food consumption
$z_1 = 0.15$ = desired level of consumption of food 1
$z_2 = 0.25$ = desired level of consumption of food 2.

The predator wishes to find values of x_1 and x_2 such that the difference between the desired level of consumption (the z's) and the actual level of consumption (the Y's) is minimized. However, much experience has shown that the difference between desired food consumption and actual consumption is not perceived by the animal as linear, but rather perhaps as a square. That is, a 5%

Case 13 Lagrange Multipliers for a Predator

shortfall in the actual consumption of food, as compared to the desired consumption, may be perceived by the predator as a 20–25% shortfall. (This phenomenon is known as *Weber's Law*—many psychology books discuss it.) Thus, to have our model approximate reality, we must find values of x_1 and x_2 that will minimize

$$G = [z_t - (Y_1 + Y_2)]^2 + w_1[z_1 - Y_1]^2 + w_2[z_2 - Y_2]^2 + w_3(1 - x_3)^2. \quad (4)$$

The term $w_3(1 - x_3)^2$ represents the fact that the predator does not wish to spend the total available time in searching for food—recall that x_3 represents the proportion of available time that is spent on nonfeeding activities. The variables w_1, w_2, and w_3 represent the relative importance, or weights, assigned by the animal to food 1, food 2, and nonfood activities, respectively. Again, it is necessary to gather experimental data: reasonable values of w_1, w_2, and w_3 are as follows:

$$w_1 = 0.05 \qquad w_2 = 0.02 \qquad w_3 = 0.03.$$

If we substitute 0.4 for z_t, 0.15 for z_1, 0.25 for z_2, 0.05 for w_1, 0.02 for w_2, and 0.03 for w_3, the results of equation (1) for Y_1, (2) for Y_2, and (3) for $Y_1 + Y_2$ into equation (4), we have

$$G = [0.4 - 0.7x_1 - 0.4x_2]^2 + 0.05[0.15 - 0.4x_1 - 0.1x_2]^2 \\ + 0.02[0.25 - 0.3x_1 - 0.3x_2]^2 + 0.03(1 - x_3)^2.$$

We want to minimize G, subject to the constraint $x_1 + x_2 + x_3 = 1$, or $x_1 + x_2 + x_3 - 1 = 0$. This can be done with Lagrange multipliers. First, form the function F:

$$F = [0.4 - 0.7x_1 - 0.4x_2]^2 + 0.05[0.15 - 0.4x_1 - 0.1x_2]^2 \\ + 0.02[0.25 - 0.3x_1 - 0.3x_2]^2 + 0.03(1 - x_3)^2 + \lambda(x_1 + x_2 + x_3 - 1).$$

Now we must find the partial derivatives of F with respect to x_1, x_2, x_3, and λ. Doing this we have

$$\frac{\partial F}{\partial x_1} = 2(-0.7)(0.4 - 0.7x_1 - 0.4x_2) + 2(0.05)(-0.4)(0.15 - 0.4x_1 - 0.1x_2) \\ + 2(0.02)(-0.3)(0.25 - 0.3x_1 - 0.3x_2) + \lambda$$

$$\frac{\partial F}{\partial x_2} = 2(-0.4)(0.4 - 0.7x_1 - 0.4x_2) + 2(0.05)(-0.1)(0.15 - 0.4x_1 - 0.1x_2) \\ + 2(0.02)(-0.3)(0.25 - 0.3x_1 - 0.3x_2) + \lambda$$

$$\frac{\partial F}{\partial x_3} = -2(0.03)(1 - x_3) + \lambda$$

$$\frac{\partial F}{\partial \lambda} = x_1 + x_2 + x_3 - 1.$$

Both $\partial F/\partial x_1$ and $\partial F/\partial x_2$ can be simplified, using some rather tedious algebra.

Multivariate Functions

Doing this, and placing each partial derivative equal to 0, we get the following system of equations:

$$\frac{\partial F}{\partial x_1} = -0.569 + 0.9996x_1 + 0.5676x_2 + \lambda = 0$$

$$\frac{\partial F}{\partial x_2} = -0.338 + 0.5646x_1 + 0.3246x_2 + \lambda = 0$$

$$\frac{\partial F}{\partial x_3} = -0.06 + 0.06x_3 + \lambda = 0$$

$$\frac{\partial F}{\partial \lambda} = x_1 + x_2 + x_3 - 1 = 0.$$

Although we shall not go through the details of the solution here, it can be found from the system above that the values of the x's which minimize G are given by

$$x_1 = 0.3 \quad x_2 = 0.4 \quad x_3 = 0.3.$$

(These values have been rounded to the nearest tenth.) Thus, the predator should spend about 0.3 of the available time searching in location 1, and about 0.4 of the time searching in location 2. This will leave about 0.3 of the time free for nonfeeding activities.

Appendix

Table 1 Selected Powers of Numbers

n	n^2	n^3	n^4	n^5	n^6
2	4	8	16	32	64
3	9	27	81	243	729
4	16	64	256	1024	4096
5	25	125	625	3125	
6	36	216	1296		
7	49	343	2401		
8	64	512	4096		
9	81	729	6561		
10	100	1000	10,000		

Table 2 Squares and Square Roots

n	n^2	$\sqrt{n}$	$\sqrt{10n}$	n	n^2	$\sqrt{n}$	$\sqrt{10n}$
1	1	1.000	3.162	51	2601	7.141	22.583
2	4	1.414	4.472	52	2704	7.211	22.804
3	9	1.732	5.477	53	2809	7.280	23.022
4	16	2.000	6.325	54	2916	7.348	23.238
5	25	2.236	7.071	55	3025	7.416	23.452
6	36	2.449	7.746	56	3136	7.483	23.664
7	49	2.646	8.367	57	3249	7.550	23.875
8	64	2.828	8.944	58	3364	7.616	24.083
9	81	3.000	9.487	59	3481	7.681	24.290
10	100	3.162	10.000	60	3600	7.746	24.495
11	121	3.317	10.488	61	3721	7.810	24.698
12	144	3.464	10.954	62	3844	7.874	24.900
13	169	3.606	11.402	63	3969	7.937	25.100
14	196	3.742	11.832	64	4096	8.000	25.298
15	225	3.873	12.247	65	4225	8.062	25.495
16	256	4.000	12.649	66	4356	8.124	25.690
17	289	4.123	13.038	67	4489	8.185	25.884
18	324	4.243	13.416	68	4624	8.246	26.077
19	361	4.359	13.784	69	4761	8.307	26.268
20	400	4.472	14.142	70	4900	8.367	26.458
21	441	4.583	14.491	71	5041	8.426	26.646
22	484	4.690	14.832	72	5184	8.485	26.833
23	529	4.796	15.166	73	5329	8.544	27.019
24	576	4.899	15.492	74	5476	8.602	27.203
25	625	5.000	15.811	75	5625	8.660	27.386
26	676	5.099	16.125	76	5776	8.718	27.568
27	729	5.196	16.432	77	5929	8.775	27.749
28	784	5.292	16.733	78	6084	8.832	27.928
29	841	5.385	17.029	79	6241	8.888	28.107
30	900	5.477	17.321	80	6400	8.944	28.284
31	961	5.568	17.607	81	6561	9.000	28.460
32	1024	5.657	17.889	82	6724	9.055	28.636
33	1089	5.745	18.166	83	6889	9.110	28.810
34	1156	5.831	18.439	84	7056	9.165	28.983
35	1225	5.916	18.708	85	7225	9.220	29.155
36	1296	6.000	18.974	86	7396	9.274	29.326
37	1369	6.083	19.235	87	7569	9.327	29.496
38	1444	6.164	19.494	88	7744	9.381	29.665
39	1521	6.245	19.748	89	7921	9.434	29.833
40	1600	6.325	20.000	90	8100	9.487	30.000
41	1681	6.403	20.248	91	8281	9.539	30.166
42	1764	6.481	20.494	92	8464	9.592	30.332
43	1849	6.557	20.736	93	8649	9.644	30.496
44	1936	6.633	20.976	94	8836	9.695	30.659
45	2025	6.708	21.213	95	9025	9.747	30.822
46	2116	6.782	21.448	96	9216	9.798	30.984
47	2209	6.856	21.679	97	9409	9.849	31.145
48	2304	6.928	21.909	98	9604	9.899	31.305
49	2401	7.000	22.136	99	9801	9.950	31.464
50	2500	7.071	22.361	100	10000	10.000	31.623

Table 3 Common Logarithms

n	0	1	2	3	4	5	6	7	8	9
1.0	.0000	.0043	.0086	.0128	.0170	.0212	.0253	.0294	.0334	.0374
1.1	.0414	.0453	.0492	.0531	.0569	.0607	.0645	.0682	.0719	.0755
1.2	.0792	.0828	.0864	.0899	.0934	.0969	.1004	.1038	.1072	.1106
1.3	.1139	.1173	.1206	.1239	.1271	.1303	.1335	.1367	.1399	.1430
1.4	.1461	.1492	.1523	.1553	.1584	.1614	.1644	.1673	.1703	.1732
1.5	.1761	.1790	.1818	.1847	.1875	.1903	.1931	.1959	.1987	.2014
1.6	.2041	.2068	.2095	.2122	.2148	.2175	.2201	.2227	.2253	.2279
1.7	.2304	.2330	.2355	.2380	.2405	.2430	.2455	.2480	.2504	.2529
1.8	.2553	.2577	.2601	.2625	.2648	.2672	.2695	.2718	.2742	.2765
1.9	.2788	.2810	.2833	.2856	.2878	.2900	.2923	.2945	.2967	.2989
2.0	.3010	.3032	.3054	.3075	.3096	.3118	.3139	.3160	.3181	.3201
2.1	.3222	.3243	.3263	.3284	.3304	.3324	.3345	.3365	.3385	.3404
2.2	.3424	.3444	.3464	.3483	.3502	.3522	.3541	.3560	.3579	.3598
2.3	.3617	.3636	.3655	.3674	.3692	.3711	.3729	.3747	.3766	.3784
2.4	.3802	.3820	.3838	.3856	.3874	.3892	.3909	.3927	.3945	.3962
2.5	.3979	.3997	.4014	.4031	.4048	.4065	.4082	.4099	.4116	.4133
2.6	.4150	.4166	.4183	.4200	.4216	.4232	.4249	.4265	.4281	.4298
2.7	.4314	.4330	.4346	.4362	.4378	.4393	.4409	.4425	.4440	.4456
2.8	.4472	.4487	.4502	.4518	.4533	.4548	.4564	.4579	.4594	.4609
2.9	.4624	.4639	.4654	.4669	.4683	.4698	.4713	.4728	.4742	.4757
3.0	.4771	.4786	.4800	.4814	.4829	.4843	.4857	.4871	.4886	.4900
3.1	.4914	.4928	.4942	.4955	.4969	.4983	.4997	.5011	.5024	.5038
3.2	.5051	.5065	.5079	.5092	.5105	.5119	.5132	.5145	.5159	.5172
3.3	.5185	.5198	.5211	.5224	.5237	.5250	.5263	.5276	.5289	.5302
3.4	.5315	.5328	.5340	.5353	.5366	.5378	.5391	.5403	.5416	.5428
3.5	.5441	.5453	.5465	.5478	.5490	.5502	.5514	.5527	.5539	.5551
3.6	.5563	.5575	.5587	.5599	.5611	.5623	.5635	.5647	.5658	.5670
3.7	.5682	.5694	.5705	.5717	.5729	.5740	.5752	.5763	.5775	.5786
3.8	.5798	.5809	.5821	.5832	.5843	.5855	.5866	.5877	.5888	.5899
3.9	.5911	.5922	.5933	.5944	.5955	.5966	.5977	.5988	.5999	.6010
4.0	.6021	.6031	.6042	.6053	.6064	.6075	.6085	.6096	.6107	.6117
4.1	.6128	.6138	.6149	.6160	.6170	.6180	.6191	.6201	.6212	.6222
4.2	.6232	.6243	.6253	.6263	.6274	.6284	.6294	.6304	.6314	.6325
4.3	.6335	.6345	.6355	.6365	.6375	.6385	.6395	.6405	.6415	.6425
4.4	.6435	.6444	.6454	.6464	.6474	.6484	.6493	.6503	.6513	.6522
4.5	.6532	.6542	.6551	.6561	.6571	.6580	.6590	.6599	.6609	.6618
4.6	.6628	.6637	.6646	.6656	.6665	.6675	.6684	.6693	.6702	.6712
4.7	.6721	.6730	.6739	.6749	.6758	.6767	.6776	.6785	.6794	.6803
4.8	.6812	.6821	.6830	.6839	.6848	.6857	.6866	.6875	.6884	.6893
4.9	.6902	.6911	.6920	.6928	.6937	.6946	.6955	.6964	.6972	.6981
5.0	.6990	.6998	.7007	.7016	.7024	.7033	.7042	.7050	.7059	.7067
5.1	.7076	.7084	.7093	.7101	.7110	.7118	.7126	.7135	.7143	.7152
5.2	.7160	.7168	.7177	.7185	.7193	.7202	.7210	.7218	.7226	.7235
5.3	.7243	.7251	.7259	.7267	.7275	.7284	.7292	.7300	.7308	.7316
5.4	.7324	.7332	.7340	.7348	.7356	.7364	.7372	.7380	.7388	.7396
n	0	1	2	3	4	5	6	7	8	9

Table 3 cont'd

n	0	1	2	3	4	5	6	7	8	9
5.5	.7404	.7412	.7419	.7427	.7435	.7443	.7451	.7459	.7466	.7474
5.6	.7482	.7490	.7497	.7505	.7513	.7520	.7528	.7536	.7543	.7551
5.7	.7559	.7566	.7574	.7582	.7589	.7597	.7604	.7612	.7619	.7627
5.8	.7634	.7642	.7649	.7657	.7664	.7672	.7679	.7686	.7694	.7701
5.9	.7709	.7716	.7723	.7731	.7738	.7745	.7752	.7760	.7767	.7774
6.0	.7782	.7789	.7796	.7803	.7810	.7818	.7825	.7832	.7839	.7846
6.1	.7853	.7860	.7868	.7875	.7882	.7889	.7896	.7903	.7910	.7917
6.2	.7924	.7931	.7938	.7945	.7952	.7959	.7966	.7973	.7980	.7987
6.3	.7993	.8000	.8007	.8014	.8021	.8028	.8035	.8041	.8048	.8055
6.4	.8062	.8069	.8075	.8082	.8089	.8096	.8102	.8109	.8116	.8122
6.5	.8129	.8136	.8142	.8149	.8156	.8162	.8169	.8176	.8182	.8189
6.6	.8195	.8202	.8209	.8215	.8222	.8228	.8235	.8241	.8248	.8254
6.7	.8261	.8267	.8274	.8280	.8287	.8293	.8299	.8306	.8312	.8319
6.8	.8325	.8331	.8338	.8344	.8351	.8357	.8363	.8370	.8376	.8382
6.9	.8388	.8395	.8401	.8407	.8414	.8420	.8426	.8432	.8439	.8445
7.0	.8451	.8457	.8463	.8470	.8476	.8482	.8488	.8494	.8500	.8506
7.1	.8513	.8519	.8525	.8531	.8537	.8543	.8549	.8555	.8561	.8567
7.2	.8573	.8579	.8585	.8591	.8597	.8603	.8609	.8615	.8621	.8627
7.3	.8633	.8639	.8645	.8651	.8657	.8663	.8669	.8675	.8681	.8686
7.4	.8692	.8698	.8704	.8710	.8716	.8722	.8727	.8733	.8739	.8745
7.5	.8751	.8756	.8762	.8768	.8774	.8779	.8785	.8791	.8797	.8802
7.6	.8808	.8814	.8820	.8825	.8831	.8837	.8842	.8848	.8854	.8859
7.7	.8865	.8871	.8876	.8882	.8887	.8893	.8899	.8904	.8910	.8915
7.8	.8921	.8927	.8932	.8938	.8943	.8949	.8954	.8960	.8965	.8971
7.9	.8976	.8982	.8987	.8993	.8998	.9004	.9009	.9015	.9020	.9025
8.0	.9031	.9036	.9042	.9047	.9053	.9058	.9063	.9069	.9074	.9079
8.1	.9085	.9090	.9096	.9101	.9106	.9112	.9117	.9122	.9128	.9133
8.2	.9138	.9143	.9149	.9154	.9159	.9165	.9170	.9175	.9180	.9186
8.3	.9191	.9196	.9201	.9206	.9212	.9217	.9222	.9227	.9232	.9238
8.4	.9243	.9248	.9253	.9258	.9263	.9269	.9274	.9279	.9284	.9289
8.5	.9294	.9299	.9304	.9309	.9315	.9320	.9325	.9330	.9335	.9340
8.6	.9345	.9350	.9355	.9360	.9365	.9370	.9375	.9380	.9385	.9390
8.7	.9395	.9400	.9405	.9410	.9415	.9420	.9425	.9430	.9435	.9440
8.8	.9445	.9450	.9455	.9460	.9465	.9469	.9474	.9479	.9484	.9489
8.9	.9494	.9499	.9504	.9509	.9513	.9518	.9523	.9528	.9533	.9538
9.0	.9542	.9547	.9552	.9557	.9562	.9566	.9571	.9576	.9581	.9586
9.1	.9590	.9595	.9600	.9605	.9609	.9614	.9619	.9624	.9628	.9633
9.2	.9638	.9643	.9647	.9652	.9657	.9661	.9666	.9671	.9675	.9680
9.3	.9685	.9689	.9694	.9699	.9703	.9708	.9713	.9717	.9722	.9727
9.4	.9731	.9736	.9741	.9745	.9750	.9754	.9759	.9763	.9768	.9773
9.5	.9777	.9782	.9786	.9791	.9795	.9800	.9805	.9809	.9814	.9818
9.6	.9823	.9827	.9832	.9836	.9841	.9845	.9850	.9854	.9859	.9863
9.7	.9868	.9872	.9877	.9881	.9886	.9890	.9894	.9899	.9903	.9908
9.8	.9912	.9917	.9921	.9926	.9930	.9934	.9939	.9943	.9948	.9952
9.9	.9956	.9961	.9965	.9969	.9974	.9978	.9983	.9987	.9991	.9996
n	0	1	2	3	4	5	6	7	8	9

Table 4 Powers of e

x	e^x	e^{-x}	x	e^x	e^{-x}
0.00	1.00000	1.00000	1.60	4.95302	0.20189
0.01	1.01005	0.99004	1.70	5.47394	0.18268
0.02	1.02020	0.98019	1.80	6.04964	0.16529
0.03	1.03045	0.97044	1.90	6.68589	0.14956
0.04	1.04081	0.96078	2.00	7.38905	0.13533
0.05	1.05127	0.95122			
0.06	1.06183	0.94176			
0.07	1.07250	0.93239	2.10	8.16616	0.12245
0.08	1.08328	0.92311	2.20	9.02500	0.11080
0.09	1.09417	0.91393	2.30	9.97417	0.10025
0.10	1.10517	0.90483	2.40	11.02316	0.09071
			2.50	12.18248	0.08208
0.11	1.11628	0.89583	2.60	13.46372	0.07427
0.12	1.12750	0.88692	2.70	14.87971	0.06720
0.13	1.13883	0.87810	2.80	16.44463	0.06081
0.14	1.15027	0.86936	2.90	18.17412	0.05502
0.15	1.16183	0.86071	3.00	20.08551	0.04978
0.16	1.17351	0.85214			
0.17	1.18530	0.84366	3.50	33.11545	0.03020
0.18	1.19722	0.83527	4.00	54.95815	0.01832
0.19	1.20925	0.82696	4.50	90.01713	0.01111
0.20	1.22140	0.81873	5.00	148.41316	0.00674
0.30	1.34985	0.74081	5.50	224.69193	0.00409
0.40	1.49182	0.67032			
0.50	1.64872	0.60653	6.00	403.42879	0.00248
0.60	1.82211	0.54881	6.50	665.14163	0.00150
0.70	2.01375	0.49658			
0.80	2.22554	0.44932	7.00	1096.63316	0.00091
0.90	2.45960	0.40656	7.50	1808.04241	0.00055
1.00	2.71828	0.36787			
			8.00	2980.95799	0.00034
			8.50	4914.76884	0.00020
1.10	3.00416	0.33287			
1.20	3.32011	0.30119	9.00	8130.08393	0.00012
1.30	3.66929	0.27253	9.50	13359.72683	0.00007
1.40	4.05519	0.24659			
1.50	4.48168	0.22313	10.00	22026.46579	0.00005

Table 5 Natural Logarithms

x	$\ln x$	x	$\ln x$	x	$\ln x$
		4.5	1.5041	9.0	2.1972
0.1	7.6974 − 10	4.6	1.5261	9.1	2.2083
0.2	8.3906 − 10	4.7	1.5476	9.2	2.2192
0.3	8.7960 − 10	4.8	1.5686	9.3	2.2300
0.4	9.0837 − 10	4.9	1.5892	9.4	2.2407
0.5	9.3069 − 10	5.0	1.6094	9.5	2.2513
0.6	9.4892 − 10	5.1	1.6292	9.6	2.2618
0.7	9.6433 − 10	5.2	1.6487	9.7	2.2721
0.8	9.7769 − 10	5.3	1.6677	9.8	2.2824
0.9	9.8946 − 10	5.4	1.6864	9.9	2.2925
1.0	0.0000	5.5	1.7047	10	2.3026
1.1	0.0953	5.6	1.7228	11	2.3979
1.2	0.1823	5.7	1.7405	12	2.4849
1.3	0.2624	5.8	1.7579	13	2.5649
1.4	0.3365	5.9	1.7750	14	2.6391
1.5	0.4055	6.0	1.7918	15	2.7081
1.6	0.4700	6.1	1.8083	16	2.7726
1.7	0.5306	6.2	1.8245	17	2.8332
1.8	0.5878	6.3	1.8405	18	2.8904
1.9	0.6419	6.4	1.8563	19	2.9444
2.0	0.6931	6.5	1.8718	20	2.9957
2.1	0.7419	6.6	1.8871		
2.2	0.7885	6.7	1.9021	25	3.2189
2.3	0.8329	6.8	1.9169	30	3.4012
2.4	0.8755	6.9	1.9315	35	3.5553
				40	3.6889
2.5	0.9163	7.0	1.9459		
2.6	0.9555	7.1	1.9601	45	3.8067
2.7	0.9933	7.2	1.9741	50	3.9120
2.8	1.0296	7.3	1.9879		
2.9	1.0647	7.4	2.0015	55	4.0073
				60	4.0943
3.0	1.0986	7.5	2.0149	65	4.1744
3.1	1.1314	7.6	2.0281		
3.2	1.1632	7.7	2.0412	70	4.2485
3.3	1.1939	7.8	2.0541	75	4.3175
3.4	1.2238	7.9	2.0669	80	4.3820
				85	4.4427
3.5	1.2528	8.0	2.0794	90	4.4998
3.6	1.2809	8.1	2.0919		
3.7	1.3083	8.2	2.1041	95	4.5539
3.8	1.3350	8.3	2.1163	100	4.6052
3.9	1.3610	8.4	2.1281		
4.0	1.3863	8.5	2.1401		
4.1	1.4110	8.6	2.1518		
4.2	1.4351	8.7	2.1633		
4.3	1.4586	8.8	2.1748		
4.4	1.4816	8.9	2.1861		

Table 6 Integrals

(C is an arbitrary constant; in all expressions involving ln x, it is assumed that $x > 0$.)

1. $\int x^n \, dx = \dfrac{1}{n+1} x^{n+1} + C \qquad \text{(if } n \neq -1\text{)}$

2. $\int e^{kx} \, dx = \dfrac{1}{k} e^{kx} + C$

3. $\int \dfrac{1}{x} \, dx = \ln x + C$

4. $\int \ln x \, dx = x(\ln x - 1) + C$

5. $\int \dfrac{1}{\sqrt{x^2 + a^2}} \, dx = \ln\left(\dfrac{x + \sqrt{x^2 + a^2}}{a}\right) + C$

6. $\int \dfrac{1}{\sqrt{x^2 - a^2}} \, dx = \ln\left(\dfrac{x + \sqrt{x^2 - a^2}}{a}\right) + C$

7. $\int \dfrac{1}{a^2 - x^2} \, dx = \dfrac{1}{2a} \cdot \ln\left(\dfrac{a + x}{a - x}\right) + C \qquad (x^2 < a^2)$

8. $\int \dfrac{1}{x^2 - a^2} \, dx = \dfrac{1}{2a} \cdot \ln\left(\dfrac{x - a}{x + a}\right) + C \qquad (x^2 > a^2)$

9. $\int \dfrac{1}{x\sqrt{a^2 - x^2}} \, dx = -\dfrac{1}{a} \cdot \ln\left(\dfrac{a + \sqrt{a^2 - x^2}}{x}\right) + C \qquad (0 < x < a)$

10. $\int \dfrac{1}{x\sqrt{a^2 + x^2}} \, dx = -\dfrac{1}{a} \cdot \ln\left(\dfrac{a + \sqrt{a^2 + x^2}}{x}\right) + C$

11. $\int \dfrac{x}{ax + b} \, dx = \dfrac{x}{a} - \dfrac{b}{a^2} \cdot \ln(ax + b) + C \qquad (a \neq 0)$

12. $\int \dfrac{x}{(ax + b)^2} \, dx = \dfrac{b}{a^2(ax + b)} + \dfrac{1}{a^2} \cdot \ln(ax + b) + C \qquad (a \neq 0)$

13. $\int \dfrac{1}{x(ax + b)} \, dx = \dfrac{1}{b} \cdot \ln\left(\dfrac{x}{ax + b}\right) + C \qquad (b \neq 0)$

14. $\int \dfrac{1}{x(ax + b)^2} \, dx = \dfrac{1}{b(ax + b)} + \dfrac{1}{b^2} \cdot \ln\left(\dfrac{x}{ax + b}\right) + C \qquad (b \neq 0)$

15. $\int \sqrt{x^2 + a^2} \, dx = \dfrac{x}{2} \sqrt{x^2 + a^2} + \dfrac{a^2}{2} \cdot \ln(x + \sqrt{x^2 + a^2}) + C$

16. $\int x^n \cdot \ln x \, dx = x^{n+1} \left[\dfrac{\ln x}{n+1} - \dfrac{1}{(n+1)^2}\right] + C \qquad (n \neq -1)$

17. $\int x^n e^{ax} \, dx = \dfrac{x^n e^{ax}}{a} - \dfrac{n}{a} \cdot \int x^{n-1} e^{ax} \, dx + C \qquad (a \neq 0)$

Answers

Chapter 1 Pretest *(page 1)*

1. 3
2. 2
3. 10
4. [number line with points at $-2, -1, 0, 1, 2$]
5. [number line with point at 2]
6. $x = 1/3$
7. $x = 4$
8. $a > 4$
9. $m > -9/2$
10. $(3x+1)(2x-3)$
11. $(3y+2)(3y-2)$
12. $m = 5, m = -2$
13. $x = 1/3; x = -1$
14. $m < -1$ or $m > 3$
15. $2/[m(m+1)]$
16. $(3+4p-p^2)/[(p-1)(p+1)]$
17. $(a^2-ab+b)/(a^2-b^2)$
18. $1/25$
19. $3/2$
20. 4
21. 4
22. $7/18$
23. x^3
24. $1/(9m)$
25. 25
26. $y^{7/3}$

Section 1.1 *(page 6)*

1. T
3. T
5. T
7. F
9. [number line, points at $-5,-4,-3,-2,-1,0,1,2,3,4,5$]
11. [number line, points at $-1, 0, 1, 2, 3, 4$]
13. [number line, points at $0, 1, 2, 3, 4$]
15. [number line, open circle at 4]
17. [number line, open circle at 5]
19. [number line, closed circle at 6]
21. [number line, open circles at -5 and -4]
23. [number line, closed circles at -3 and 6]
25. [number line, open circles at 1 and 6]
27. $=$
29. $\leqslant$
31. $=$
33. $=$
35. $=$
37. $=$
39. $=$
41. $\leqslant$
43. $=$
45. $=$
47. yes

Section 1.2 *(page 11)*

1. 3
3. -9
5. 2
7. 3
9. 4
11. $2/3$
13. 7
15. 3
17. -3
19. 5
21. $x \leqslant 4$
23. $p \geqslant -1$
25. $k < 1$
27. $m > -1$
29. $y < 1$
31. $p > 1/5$

Section 1.3 *(pages 15–16)*

1. $4m(3m+2)$
3. $2z(3z-4z^2-6)$
5. $(4p-1)(p+1)$
7. $(2x-5)(x+6)$
9. $(2q-3)(3q+4)$
11. $3(2r-1)(2r+5)$
13. $(3r+2)(6r-5)$
15. $(m+n)(m-n)$
17. $(1-a)(1+a)$
19. $(5x-2y)(5x+2y)$
21. $-3, 5$
23. $-1, 4$
25. $3, -8$
27. $3, -3$
29. $2/3, -3$
31. $0, 1/2$
33. $5/2, 1$
35. $4, 2$
37. $(3 \pm \sqrt{21})/(-2)$ or $(-3 \pm \sqrt{21})/2$
39. $4/3, 1/2$
41. $2, -5$
43. $3, -4$
45. $-3 < x < 3$
47. $0 < y < 10$
49. $r \leqslant -5$ or $r \geqslant 1$
51. $-2 \leqslant x \leqslant 3$
53. $k < 1/2$ or $k > 4$

Section 1.4 *(pages 19–20)*

1. $4/(x-1)$
3. $a + 2$
5. 1
7. $(n-4)/(n+4)$
9. $(x+y)(x-y)/(x^2-xy+y^2)$
11. 1
13. $(2y+1)/[y(y+1)]$
15. $3/[2(a+b)]$

17. $-2/[(a+1)(a-1)]$
19. $2(m^2+1)/[(m+1)(m-1)]$
21. $4/(a-2)$
23. $5/[(a+2)(a-2)(a-3)]$
25. $2x/[(x-3)(x+4)(x-4)]$
27. $(p+5)/[p(p+1)]$
29. $-1/[x(x+h)]$
31. $(x+1)/(x-1)$
33. $(2-b)(1+b)/[b(1-b)]$

Section 1.5 *(pages 23–24)*

1. 81
3. 1/64
5. 1/8
7. 4/9
9. 8
11. 6
13. 36
15. -2
17. -8
19. 1/125
21. 3/4
23. 9
25. 3^6
27. $1/3^3$
29. 1
31. 5
33. 4
35. 3
37. $4^{1/3}$
39. 5; 32

Case 1 *(page 27)*

1. (a) $C_c = 12.5 + 3.75A$
   (b) $1/70$; $180 + .1435T$; $22.1 + .00769T$
   (c) $14.8 + .94A$
   (d) $1/70$; $180 + .1435T$; $17.2 + .00193T$
3. (a) $22.2 + .00861T$, labor, materials, $15.2 + .00185T$, land
   (b) 22,200, 8.61; 15,200, 1.85; 37,400, 10.46

Chapter 2 Pretest *(page 28)*

1. $x = 4, x = -4$
2. (a) $y = 3 - 2x$
   (b) $y = (15-x)/(-6)$ or $y = (x-15)/6$
3. $x = (y+3)/(y-2)$
4. $y = 4$
5. 5/3; 5; $-5/17$; $-5/197$
6. $x = 22/5$
7. (a) $x = 0, x = 3/2$
   (b) $x = (-5 \pm \sqrt{33})/2$
8. 0; 5, 2
9. 2; $\sqrt{14}$; $2\sqrt{7}$
10. $-\sqrt{3}$; -2; -3
11. -11

Section 2.1 *(pages 33–35)*

1.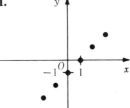

3.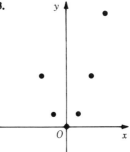

5.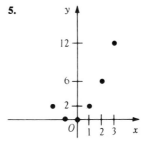

7.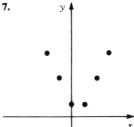

9. function
11. not a function
13. function
15. not a function
17. not a function
19. function
21. (a) -22 (b) 20 (c) 2
    (d) $-6a + 2$ (e) $-6a - 10$

23. (a) 21 (b) −7 (c) 5 (d) 4a +5 (e) 4a + 13
25. −3
27. 2a − 3
29. 2a + 1
31. (a) 0 (b) 0 (c) 2 (d) 20 (e) 56
(f)

33. (a)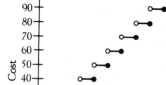

(b) domain: $x > 0$; range, $\{10, 20, 30, 40, \ldots\}$

Section 2.2 *(pages 41–43)*

1.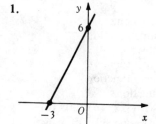

3.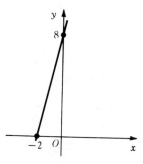

5.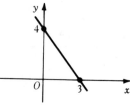

7.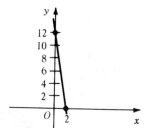

9.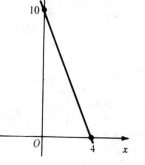

11.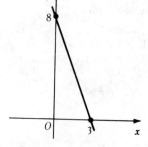

13. $-1/5$
15. $2/3$
17. $-3/2$
19. no slope
21. no slope
23. 0
25. 3
27. -4
29. $-3/4$
31. 0

33.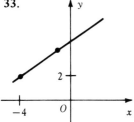

35.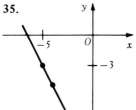

37.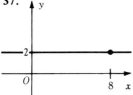

39.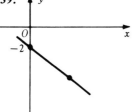

41. $3y = -2x + 12$
43. $2y = -x - 4$
45. $4y = 6x + 5$
47. $y = 2x + 9$
49. $4y = x + 5$
51. $2y = 3x + 12$
53. $3y = 4x + 7$
55. $3y = -2x$
57. $x = -8$

59. (a) 8 (b) 0
    (c)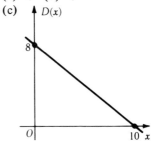

61. (a)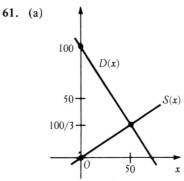

   (b) $x = 50$ (c) $100/3$

Section 2.3 *(pages 51–52)*

1.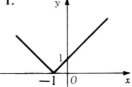

3.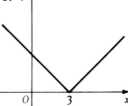

5.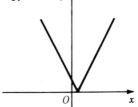

7.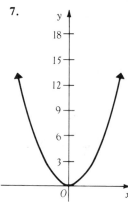

17.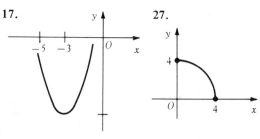

27.

19.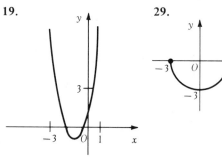

29.

9.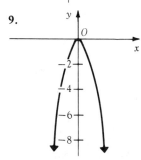

21.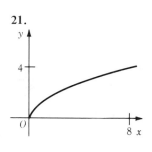

11.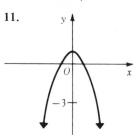

23.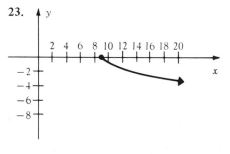

13.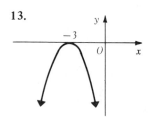

25.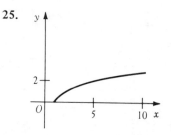

15.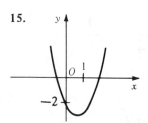

325

31. (a)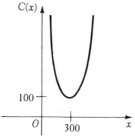

(b) vertex is (300, 100); 300 sandwiches

The answers in 33 and 35 are given in thousands.

33. (a) 2 (b) $\sqrt{3} \approx 1.73$ (c) 0
    (d)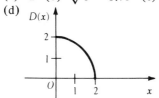

35. (a) 10.9; 6.6; 0
    (b) 7.1; 10; 10.95
    (c)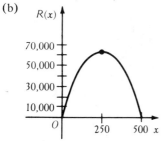

    (d) 8

37. (a) $R(x) = x(500-x) = 500x - x^2$
    (b)

    (c) 250

Section 2.4 (pages 58–59)

1.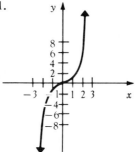

3.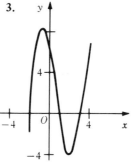

5.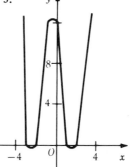

7.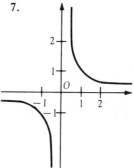

9.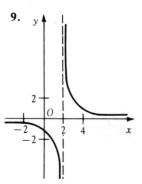

11.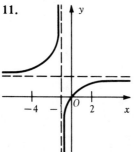

13.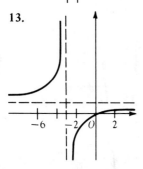

15.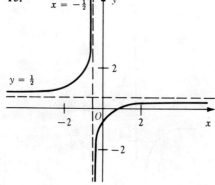

17.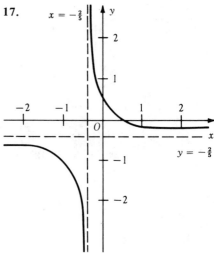

19. (a) 12.50 (b) 10 (c) 6.25 (d) 5 (e)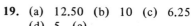

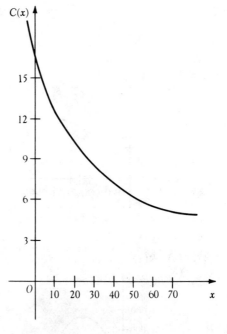

21. (a) 6.70 (b) 15.63 (c) 26.80
(d) 60.30 (e) 77.05
(f) 127.30 (g) 328.30
(h) 663.30 (i) no

(j)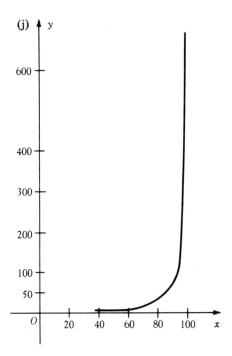

23. 4, −1, −2 **27.** 3
25. 2, 3, −3 **29.** −1, 2, 3, −3

Section **2.5** *(pages 61–62)*

1. $[f+g](x) = \sqrt{x} + x^2 - 1;\ x \geq 0$
3. $[fg](x) = x^2\sqrt{x} - \sqrt{x};\ x \geq 0$
5. $[g/f](x) = (x^2-1)/(\sqrt{x});\ x > 0$
7. $[g \circ f](x) = x - 1;\ x \geq 0$
9. if $g(x) = 1$ and $f(x) = x$, then $[g/f](x) = 1/x$
11. if $f(x) = 3$ and $g(x) = |x|$, then $[fg](x) = 3|x|$
13. if $g(x) = |x|$, and $f(x) = |x|$, then $[fg](x) = |x|^2$
15. if $f(x) = x^3 + 1$ and $g(x) = x$, then $[f/g](x) = (x^3+1)/x$
17. if $f(x) = \sqrt{x}$ and $g(x) = x^2 - 1$, then $[f \circ g](x) = \sqrt{x^2 - 1}$
19. if $f(x) = 1/x$ and $g(x) = x^2 - 1$, then $[f \circ g](x) = 1/(x^2 - 1)$
21. if $f(x) = \sqrt{x} - 1$ and $g(x) = 2x^2 + 1$, then $[f \circ g](x) = \sqrt{2x^2+1} - 1$
23. $18a^2 + 24a + 9$

Case **2** *(page 63)*

1. \$440 **3.** \$232

Chapter 3 Pretest *(page 64)*

1. $(x-2)(x-3)$
2. $x - 2$
3. (a) $1/(\sqrt{x}+4)$ (b) $1/(1+\sqrt{x})$
   (c) $1/(\sqrt{x}-5)$
4. (a) $x - 2/x$ (b) $2/x - 3/x^2 + 2/x^3$
5. (a) $x^{1/2}$ (b) $(x-5)^{1/2}$
   (c) $(x+5)^{3/2}$

6.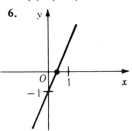

7.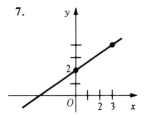

8.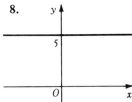

9.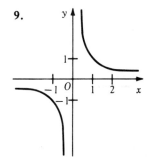

10.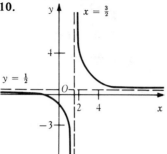

11.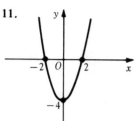

12.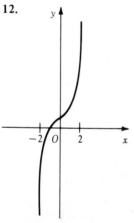

13.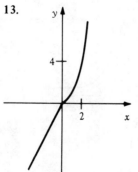

14.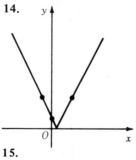

15.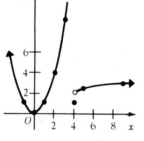

Section 3.1 *(pages 71–75)*

1. 3
3. does not exist
5. 2
7. 0
9. -2

11. -4;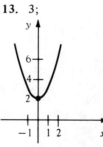

13. 3;

15. 0;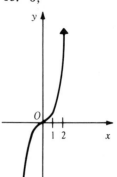

17. does not exist;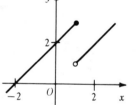

19. 1;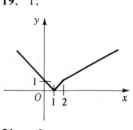

21. -2;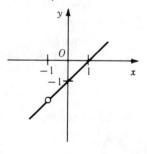

23. 1; 4; 10
25. 0; 1; 4
27. 1/2; 1; 5; does not exist
29. 2; does not exist; −1; 1/2
31. 10
33. 0

Section 3.2 *(pages 79–80)*

1. 12   21. −1
3. 29   23. 4
5. −4   25. 0
7. 56   27. 1/10
9. 9/7   29. 1/8
11. 3/2   31. 4
13. 0   33. −6x
15. 6   35. $8x - 3$
17. −5   37. $3x^2 + 2$
19. 7   39. 0

Section 3.3 *(pages 85–86)*

1. 3   19. 2
3. 0   21. (−2,1); (1,2)
5. 3; −5   23. (−2,2)
7. −1/2; 2/3   25. (−2,2)
9. −2   27. (−2,1); (1,2)
11. none   29. (−2,2)
13. none   31. (0,1); (1,2); (2,3);
15. none   (3,4), etc.
17. 2

Section 3.4 *(page 89)*

1. 3/5   15. does not exist
3. 2   17. 0
5. 4   19. does not exist
7. 1/3   21. $y = 1/2$
9. 1/3   23. $y = 3$
11. 0   25. 0
13. does not exist   27. none

Case 3 *(pages 93–94)*

1. 5, 10, 20, 40, 80   9. 93/4
3. 16, −8, 4, −2, 1   11. 3/2
5. 64, 16, 4, 1, 1/4   13. 2
7. 31   15. 1/5

17. not a geometric sequence, formula not valid
19. 1/3

21. $r = 1.2 > 1$, formula not valid
23. 99
25. 48 feet down, 36 feet up, 84 feet total
27. 4

Chapter 4 Pretest *(page 95)*

1. 7; 31; $3m + 1$
2. (a) $5x^2 + 3x - 2$
   (b) $2x + h$
   (c) $-1/[x(x+h)]$
3. (a) −1   (b) 1   (c) −2
4. 0; 9; 9; 9
5. −2; $2 - a^2$; $4 - a^2$; $2 + a$; 2
6. −1/3
7. $4y + 3x = 6$
8. (a) discontinuous at 0
   (b) $x = 3$
   (c) none
9. $(1/x+1)^2$; $1/(x+1)^2$

Section 4.1 *(pages 102–103)*

1. $f'(x) = 3$ 5. $f'(x) = 0$
3. $f'(x) = 4x$ 7. $f'(x) = 6x - 4$
9. $f'(x) = 3/x^2$
11. $f'(x) = 3x^2 + 2x$
13. $f'(x) = 1/(2\sqrt{x})$
15. (a) 0   (b) 8   (c) −12
17. (a) −4 (b) 8   (c) −22
19. (a) 3   (b) 3   (c) 3
21. (a) does not exist   (b) 3/4
    (c) 1/3
23. (a) 0   (b) 16 (c) 21
25. (a) does not exist   (b) $1/(2\sqrt{2})$
    (c) does not exist
27. 1   29. 0   31. −8
33. (a) 30 (b) 20 (c) 10 (d) 0
    (e) −10 (f) $t = 5$

Section 4.2 *(pages 106–107)*

1. $f'(x) = 10x$
3. $f'(x) = 8$
5. $f'(x) = 6x - 4$
7. $f'(x) = -12x^2 - 2x + 2$
9. $f'(x) = 4x^3 - 3x^2 + 2x - 4$
11. $f'(x) = 2x^{-1/2}$
13. $f'(x) = \dfrac{3}{2}x^{-1/2} + 3x^{1/2}$

15. $f'(x) = 2x^{-1/2} - 3$
17. $f'(x) = x^{-1/2} - \frac{3}{2}x^{1/2}$
19. $f'(x) = 12x^{-3}$
21. $f'(x) = 9x^{-4}$
23. $f'(x) = -2x^{-2} - 6x^{-3}$
25. $f'(x) = -12x^{-5/2}$
27. $f'(x) = -12x^{-5/2} - 2x^{-1/2} - \frac{1}{2}x^{-3/2}$
29. $f'(x) = 15x^{-6}$
31. $f'(x) = 9x^{-4}$
33. (a) 22 (b) 13 (c) −13 (d) −313/5
35. (a) $A = 16\pi t^2$ (b) 128π; 640π; 1920π; 3200π

Section 4.3 (pages 114–116)

1. 3
3. 4
5. 24
7. 36
9. 3/4
11. 1/4
13. 1/8
15. $y + 12x + 18 = 0$
17. $y + 16x = 16$
19. $5x + y + 9 = 0$
21. $y = 1$
23. $y = 32x + 50$
25. (a) $C'(50) = 82$ (b) $C'(11) = 4$
27. (a) $55/3 \approx 18.33$ (b) $151/5 \approx 30.20$
29. (a) $3 (b) $4 (c) −$3 (d) −$2

Section 4.4 (pages 119–120)

1. $f'(x) = 2x - 1$
3. $f'(x) = -4x - 3$
5. $f'(x) = 9x^2 - 10x + 2$
7. $f'(x) = 60x^4 - 40x^3 - 12x + 5$
9. $f'(x) = 4x^3 - 16x$
11. $f'(x) = 3x^{1/2} - 1/(2x^{1/2})$
13. $f'(x) = 8 + x^{-1/2}$
15. $f'(x) = 24x^{1/2} + 8x^{-2}$
17. $f'(x) = -2/(x-1)^2$
19. $f'(x) = 6/(3+x)^2$
21. $f'(x) = -27/(4+3x)^2$
23. $f'(x) = (6x^2 - 12x - 2)/(x-1)^2$
25. $f'(x) = (2x^2 + 10x - 1)/(2x^2 + 1)^2$
27. $f'(x) = 1$
29. $f'(x) = (1-2x)/[2\sqrt{x}(2x+1)^2]$
31. $f'(x) = (2x^2 - 3x + 3)/[2\sqrt{x}(x-1)^2]$
33. (a) $G'(20) < 0$—go faster; (b) $G'(40) > 0$—go slower

Section 4.5 (pages 122–123)

1. $f'(x) = 2(x-5)$
3. $f'(x) = 30x(5x^2 + 6)^2$
5. $f'(x) = (x-1)(x^2 - 2x)^{-1/2}$
7. $f'(x) = 6(4x-1)^{1/2}$
9. $f'(x) = 108x^2 - 96x + 16$
11. $f'(x) = (2x-1)^{-1/2}$
13. $f'(x) = (15x-5)/\sqrt{2x-1}$
15. $f'(x) = (-15x^2 - 4x)/\sqrt{3x+1}$
17. $f'(x) = (-4x-10)/(x-5)^4$
19. $f'(x) = \frac{3}{2}(3x^2 - 3x)^{1/2}(6x-3)$
21. $f'(x) = (3x + 1/x)/[2(x+1/x)^{1/2}]$
23. $f'(x) = -1/[(x+1)^{1/2}(x-1)^{3/2}]$
25. 92; the marginal receipts from the 201st unit
27. −2

Section 4.6 (pages 128–129)

1. $-4x/(3y)$
3. $-y/(x+y)$
5. $-3y^2/(6xy - 4)$
7. $2/y$
9. -1
11. $(-6x - 4y)/(4x+y)$
13. $3x^2/(2y)$
15. $-2y/x$
17. $-y/x$
19. $(-2xy - 2y^2)/(x^2 + 4xy + 3y^2)$
21. $-y^{1/2}/x^{1/2}$
23. $-y/[x + 2(xy)^{1/2}]$
25. $4y = 3x + 25$
27. $y = x + 2$
29. $-7/6$ feet per minute

Case 4 (page 131)

1. 4.8 million units
3. In the interval under discussion (3.1 to 5.7 million units) the marginal cost always exceeds the selling price.

Chapter 5 Pretest (page 132)

1. −1
2. $(7 \pm \sqrt{17})/8$
3. 2/3, −2/3
4. $f'(x) = 0$
5. $f'(x) = 6$
6. $f'(x) = 6x - 2$
7. $f'(x) = 9x^2 - 8x + 2$
8. $f'(x) = x^{-1/2} + \frac{3}{2}x^{1/2}$

9. $f'(x) = -5/(x+1)^2 + 3$
10. $1, -1$
11. derivative always exists
12. 1
13. -1
14. -1
15. $5/6$
16. 2
17. 5
18. -10

Section 5.1 *(pages 141–142)*

1. minimum of 2 at $x = 2$
3. maximum of 18 at $x = 4$
5. minimum of -26 at $x = 2$; maximum of 82 at $x = -4$
7. minimum of -32 at $x = -2$; maximum of -31 at $x = -3$
9. abs. and rel. minimum of -3 at $x = 2$; abs. maximum of 61 at $x = -6$
11. abs. and rel. maximum of 14 at $x = 4$; abs. minimum of -67 at $x = -5$
13. abs. maximum of 14 at $x = 2$; abs. minimum of -46 at $x = -3$
15. abs. and rel. maximum of 59/3 at $x = -4$; abs. and rel. minimum of $-7/6$ at $x = 1$
17. $(4,-12)$
19. $(2,-5)$
21. $(2,7)$
23. $(5/4,-9/8)$
25. 10, 10
27. 10; 180
29. 10 units

Section 5.2 *(pages 144–145)*

1. 8; 20; -10
3. -2; -26; -146
5. 16; 16; 16
7. -12; 0; -30
9. -4; 4; $-1/16$
11. minimum at $x = 6$
13. maximum at $x = -5/3$
15. minimum at $x = 1$
17. maximum at $x = 0$; minimum at $x = 2$
19. maximum at $x = 0$; minimum at $x = 2/3$
21. minimum at 2, maximum at -5
23. maximum at $x = -1$; minimum at $x = 6$
25. minimum at $x = 4/3$; maximum at $x = -2$
27. the only critical point is at $x = 0$, which is neither a maximum or a minimum
29. maximum at $x = 0$; minimum at $x = -2$ and at $x = 2$
31. no maximums or minimums
33. minimum at $x = 2$; maximum at $x = -2$; neither at $x = 0$
35. neither at $x = -1$
37. 80/3, 40/3
39. maximum at $x = 8$; minimum at $x = 4/3$

Section 5.3 *(pages 149–153)*

1. 50, 50
3. 50, 50
5. 300 by 600; area is 180,000 square feet
7. 6 units; $P(6) = 44$
9. $x = 10$
11. rent = \$250; maximum income = \$12,500
13. 90 seats, \$405 profit
15. 20 by 20 by 40; cost is \$7200
17. $3\sqrt{6} + 3$ by $2\sqrt{6} + 2$
19. $\sqrt[3]{18}$ by $2\sqrt[3]{18}$ by $12/\sqrt[3]{18^2}$
21. when $t = 4$; $R(4) = 1,160,000$
23. 125 by 125
25. (a) The length of the piece of wire to be made into a circle is $12\pi/(4+\pi)$, or about 5.28.
    (b) Make no square at all; use the entire piece of wire to make a circle.

Section 5.4 *(page 161)*

1. increasing on $(2,+\infty)$; decreasing on $(-\infty,2)$
3. increasing on $(-\infty,9/8)$; decreasing on $(9/8,+\infty)$
5. increasing on $(-\infty,-1)$ and $(2,+\infty)$; decreasing on $(-1,2)$
7. increasing on $(-\infty,-1)$ and $(5/2,+\infty)$; decreasing on $(-1,5/2)$
9. always increasing
11. always decreasing

In Exercises 13–27, the answers are given in this order: increasing, decreasing, concave upward, concave downward, and inflection points.

13. always; never; $(0,+\infty)$; $(-\infty,0)$; $x = 0$

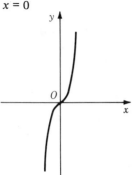

15. $(-2,+\infty)$; $(-\infty,-2)$; always; never; none

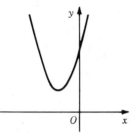

17. $(0,+\infty)$; $(-\infty,0)$; never; always; none

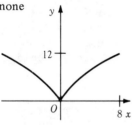

19. $(-\infty,0)$ and $(2,+\infty)$; $(0,2)$; $(1,+\infty)$; $(-\infty,1)$; $x = 1$

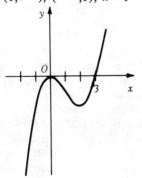

21. $(-\infty,-1)$ and $(6,+\infty)$; $(-1,6)$; $(5/2,+\infty)$; $(-\infty,5/2)$; $x = 5/2$

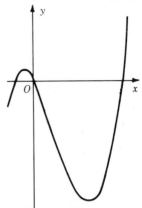

23. always; never; $(1,+\infty)$; $(-\infty,1)$; $x = 1$

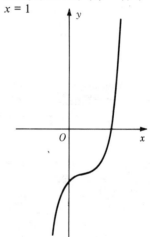

25. never; always except at $x = -1$; $(-1,+\infty)$; $(-\infty,-1)$; none

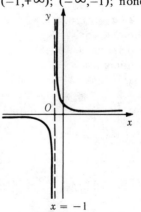

27. $(-\infty,-1)$ and $(1,+\infty)$; $(-1,0)$ and $(0,1)$; $(0,+\infty)$; $(-\infty,0)$; none

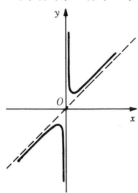

Section 5.5 *(page 165)*

1. .67
3. 2.33
5. −2.20
7. 3.10
9. .67; 3.5; −2
11. −1; 3.67
13. 1.414
15. 3.317
17. 15.811
19. 2.080
21. 4.642

Section 5.6 *(page 168)*

1. 7
3. 4
5. 15
7. 21
9. 60
11. 0
13. −13/3
15. 6
17. 5
19. 3/2
21. 3/2
23. 0
25. no limit
27. −1
29. no limit

Case 5 *(page 170)*

1. 10
3. $T''(x) = 3kMx^{-3}/2$, which is positive when $x = \sqrt{kM/(2f)}$

Case 6 *(page 171)*

1. 1600 square miles

Case 7 *(page 174)*

1. (a) $520 million, or $4.33 million per plane
   (b) $80 million
3. (a) $310 million
   (b) −4 million, or a loss of $4 million

Chapter 6 Pretest *(page 175)*

1. (a) 2 (b) 0 (c) 4
2. (a) 1 (b) 2 (c) 1/8 (d) 1/4
   (e) $\sqrt{2}$
3. (a) 1/81 (b) 2 (c) 1/16
4. (a) 1/10 (b) 1 (c) $1/\sqrt{10}$
   (d) 1000 (e) 1/1000
5. (a) 3 (b) −2 (c) 1/2
6. (a) −.8 (b) $2.8x$ (c) $8x - 2$
   (d) $1/\sqrt{2x-5}$

Section 6.1 *(page 182)*

1. and 7. $y = 3^x$; $y = (\frac{1}{3})^{-x}$

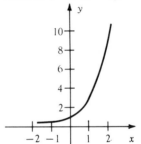

3. $y = 3^{-x}$

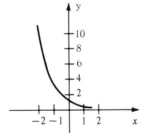

5. $y = (\frac{1}{4})^x$

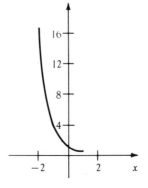

9. $y = 3^{|x|}$

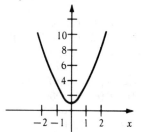

11. $y = 2^{-|x|}$

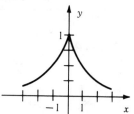

13. $y = 2^{x^2}$

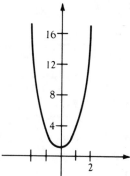

15. $y = 3^{-x^2}$

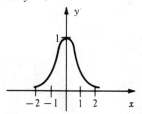

17. $y = 2^{x+1}$

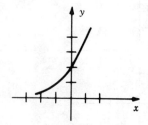

$y = 2^{1-x}$

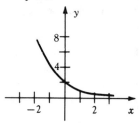

21.

23.

Section **6.2** *(pages 187–189)*

1. $\log_2 8 = 3$
3. $\log_3 81 = 4$
5. $\log_{1/3} 9 = -2$
7. $2^3 = 8$
9. $10^2 = 100$
11. $10^5 = 100{,}000$
13. 4
15. 3
17. -2
19. $\log_4 16 = 2$

21. $\log_7 15/11$
23. $\log_{10} 30 \cdot 2$
25. $\ln e = 1$
27. $\ln(5^{.3}/6^{.4})$
29. 2.9957
31. 4.0943
33. 6.6847
35. 6.2767
37. 6.6439
39. 10.9777

335

41.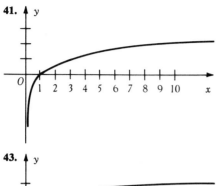

43.

45.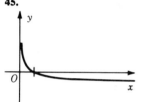

47. (a) 0 (b) 8959 (c) 16,095
    (d) 19,560

Section 6.3 *(pages 193–195)*

1. 1,000,000
3. 1,080,000
5. 500
7. 335
9. 25,000
11. 37,300
13. 50,000
15. 40,900
17. 1000
19. 4460
21. 5000
23. 0
25. 432
27. 500
29. 527
31. 6020
33. 1800 years
35. 18,600 years
37. 1000
39. 3700/404 or about 9
41. .125
43. about 12.5
45. about 20.6

Section 6.4 *(pages 197–199)*

1. $f'(x) = 4e^{4x}$
3. $f'(x) = 12e^{-2x}$
5. $f'(x) = -16e^{2x}$
7. $f'(x) = -16e^{x+1}$
9. $f'(x) = 2xe^{x^2}$
11. $f'(x) = 16xe^{2x^2 - 4}$
13. $f'(x) = xe^x + e^x = e^x(x+1)$
15. $f'(x) = 2(x-3)(x-2)e^{2x}$
17. $f'(x) = (2x-2)/(x^2-2x)$
19. $f'(x) = 2x$
21. $f'(x) = 1/[2(x+5)]$
23. $f'(x) = [x(\ln x)e^x - e^x]/[x(\ln x)^2]$
25. $f'(x) = [x(e^x - e^{-x}) - (e^x + e^{-x})]/x^2$
27. $f'(x) = -20{,}000e^{.4x}/(1+10e^{.4x})^2$
29. $f'(x) = 8000e^{-.2x}/(8+4e^{-.2x})^2$
    $= 500e^{-.2x}/(2+e^{-.2x})^2$
31. $200e^{.4} \approx 298$; $200e^{1.6} \approx 991$
33. (a) $e^{-.02} \approx .98$ (b) $e^{-.2} \approx .82$
    (c) $e^{-2} \approx .14$
    (d) $-.02e^{-2} \approx -.0027$, the rate of change in the proportion wearable at $x = 100$
35. (a) .005 (b) .0007
    (c) .000014 (d) $-.022$
    (e) $-.0029$ (f) $-.000054$

Chapter 6 Appendix *(pages 202–203)*

1. 2.8745
3. 0.9872
5. 0.8531
7. $9.9085 - 10$
9. $7.5416 - 10$
11. 2.24
13. 888
15. 30.8
17. .0862
19. .000850
21. 35.0
23. 273
25. 186
27. 8.83
29. 558
31. 19,400
33. 70,900
35. 395,000

Case 8 *(pages 205–206)*

(The answers here may differ by a few cents due to rounding.)

1. $1123.60
3. $2165.71
5. $2699.70
7. $4555.28
9. $276.28
11. $2214.00
13. about $1960

Case 9 *(page 208)*

1. 722   3. 956

Chapter 7 Pretest *(page 209)*

1. $0, -5/4, -2n - n^2, -2/n - 1/n^2$
3. $-5/2, 3$
5. $(5 \pm \sqrt{13})/6$
7. $8x - 3$
9. $(2 - 3x^2)/(2\sqrt{2x - x^3})$
11. $6e^{3x}$
12. $-4x/(x^2 - 1)^2$
13. $6x/(x^2 + 1)$
14. $1/2$

Section 7.1 *(page 213)*

1. $2x^2 + C$
3. $5x^3/3 + C$
5. $6x + C$
7. $x^2 + 3x + C$
9. $x^3/3 + 3x^2 + C$
11. $x^3/3 - 2x^2 + 5x + C$
13. $2x^{3/2}/3 + C$
15. $2x^{3/2}/3 + 2x^{5/2}/5 + 2x^{7/2}/7 + C$
17. $2x^{7/2}/7 + 6x^{5/2}/5 + 4x^{3/2}/3 + C$
19. $4x^{5/2} - 4x^{7/2} + C$
21. $6x^{3/2} + 4x + C$
23. $-1/x + C$
25. $-1/2x^2 - 2x^{1/2} + C$
27. $2 \ln x + C$
29. $9/x - 2 \ln x + C$
31. $P(x) = -x^2 + 20x - 50$
33. $f(x) = 2x^3 - 2x^2 + 3x + 1$

Section 7.2 *(page 217)*

1. 32, 38
3. 15, 31/2
5. 20, 30
7. 16, 14
9. (a) 10.03 (b) 15.44 (c) 21

Section 7.3 *(pages 220–221)*

1. 35/2
3. 44
5. 24
7. 42
9. 12
11. 124
13. 46
15. 36
17. 768
19. 14/3 = 4.67
21. 161
23. 21
25. 81/4

Section 7.4 *(pages 225–226)*

1. $C(x) = 2x^2 - 5x + 8$
3. $C(x) = .07x^3 + 10$
5. $C(x) = (2x^{3/2})/3 - 8/3$
7. $C(x) = x^3/3 - x^2 + 3x + 6$
9. $280, $11.67
11. $31, $216
13. no, only 7680

Section 7.5 *(pages 233–234)*

1. 4
3. 8/3
5. 40
7. 1735/6
9. 5/3
11. 31.50
13. 54

15. (a)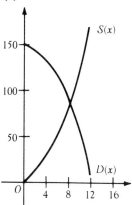

(b) 8 (c) 86
(d) 1024/3 = 341.33
(e) 1288/3 = 429.33
17. (a) 5 (b) 130/3 = 43.33 millions of dollars

Section 7.6 *(pages 240–241)*

1. $5x^7/7 + C$
3. $4x^3/3 - 3x^2 + 2x + C$
5. $-2/x^2 + C$
7. $2x^{3/2}/3 + C$
9. $(x^2+1)^4/4 + C$
11. $(x^2-5)^{3/2}/3 + C$
13. $2(x^2+12x)^{3/2}/3 + C$
15. $2(x^3+1)^{3/2}/9 + C$
17. $e^{2x}/2 + C$
19. $-2e^{2x} + C$
21. $-8 \ln x + C$
23. $5 \ln(x+1) + C$
25. $e^{x^3}/3 + C$
27. $\ln(2x+1)/2 + C$
29. $-1/3(3x^2+2)^3 + C$
31. $-1/2(2x-x^2) + C$
33. $-1/[4(2x^2 - 4x)] + C$
35. $[(1/x)+x]^2/2 + C$
37. $(2x^3-2x)^2/4 + C$
39. 14/3
41. 15/4
43. 1024/3 = 341.33
45. 1
47. $(e^2-1)/2$
49. 1.10

Section 7.7 (page 244)

1. $x \ln 4 + x(\ln x - 1) + C$
3. $-4 \ln[(x+\sqrt{x^2+36})/6] + C$
5. $\ln[(x-3)/(x+3)] + C$ $(x^2 > 9)$
7. $(4/3)\ln[(3+\sqrt{9-x^2})/x] + C$
   $(0 < x < 3)$
9. $-2x/3 + 2\ln(3x+1)/9 + C$
11. $(-2/15)\ln[x/(3x-5)] + C$
13. $\ln[(2x-1)/(2x+1)] + C$
15. $-3\ln[(1+\sqrt{1-9x^2})/3x] + C$
17. $x^5[\ln x/5 - 1/25] + C$
19. $(1/x)[-\ln x - 1] + C$
21. $-\tfrac{1}{2}xe^{-2x} - \tfrac{1}{4}e^{-2x} + C$
23. $-x^2 e^{-2x} - xe^{-2x} - e^{-2x}/2 + C$
25. $x^3 e^x - 3x^2 e^x + 6xe^x - 6e^x + C$

Section 7.8 (pages 247–248)

1. $\dfrac{1}{6}x(x+1)^6 - \dfrac{1}{42}(x+1)^7 + C$

3. $-\dfrac{1}{5}x^2(4-x^2)^{5/2} - \dfrac{2}{35}(4-x^2)^{7/2} + C$

5. $\dfrac{1}{18}x^2(x^2-1)^9 - \dfrac{1}{180}(x^2-1)^{10} + C$

7. $-\dfrac{4}{27}x^3(1-3x^3)^{3/2}$
   $-\dfrac{8}{405}(1-3x^3)^{5/2} + C$

9. $-\dfrac{2}{5}x(x+5)^{-5} - \dfrac{1}{10}(x+5)^{-4} + C$

11. $-2x(1-x)^{1/2} - \dfrac{4}{3}(1-x)^{3/2} + C$

13. $\dfrac{3}{4}x^4(x^4+3)^{1/3} - \dfrac{9}{16}(x^4+3)^{4/3} + C$

15. $2x^2(x+2)^{3/2} - \dfrac{8}{5}x(x+2)^{5/2}$
    $+ \dfrac{16}{35}(x+2)^{7/2} + C$

17. $xe^x + C$
19. $-3x^2 e^{-x} + 6xe^{-x} - 6e^{-x}$
21. $11 - e^2 \approx 3.61$

Section 7.9 (pages 250–251)

1. 2.75; 2.67
3. 6.76; 6.79
5. 16; 14.67
7. .94; .84
9. .19; .10

11. (a)

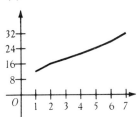

(b) 128 (c) 128

Case 10 (page 254)

1. about 102 years
3. about 45½ years

Case 11 (page 256)

1. 10.65   3. 441.46

Chapter 8 Pretest (page 257)

1. $m = -6, n = 13$
3. $x^3/3 + 3x^2/2 + 2x + C$
5. $e^y + C$
7. $\ln y + C$
9. $x = \ln 3$

Section 8.1 (pages 261–262)

1. $y = x^3/3 + C$
3. $y = -x^2 + x^3 + C$
5. $y = -4x + 3x^4/4$
7. $y = e^x + C$
9. $y = -2e^{-x} + C$
11. $y = -4x^3/9 + C$
13. $y = x^2/8 + C$
15. $y = 2x^3 - 2x^2 + 4x + C$
17. $y = -4x^3/3 + C_1 x + C_2$
19. $y = 5x^2/2 - 2x^3/3 + C_1 x + C_2$
21. $y = x^4/3 - x^3/3 + C_1 x + C_2$
23. $y = 3x^2/2 - 5x^3/6 + C_1 x + C_2$
25. $y = e^x + C_1 x + C_2$
27. $y = x^3 - x^2 + 2$
29. $y = 5x^2/2 + 2x - 3$
31. $y = x^3 - 2x^2 + 2x + 8$
33. $y = x^3/3 + x^2/2 - 5x/6 + 2$
35. $y = e^x + x^2/2 - ex + 1$
37. $y = x^4/4 - x^2 + x + 4$
39. (a) $y = 5t^2 + 250$ (b) 12 days

Section 8.2 *(page 265)*

1. $y^2 = x^2 + C$
3. $y^2 = 2x^3/3 - 2x + K$
5. $y = Ke^{x^2}$
7. $y = Ke^{(2/3)x^3 - (1/2)x^2}$
9. $y = 2x^3/3 + C_1 x + C_2$
11. $y = x^4/8 + C_1 x + C_2$
13. (a) $y = Ce^{-.05x}$
    (b) $y = 90e^{-.05x}$ (c) about 55 grams

Section 8.3 *(pages 269–271)*

1. (a) 156 (b) 198 (c) 206 (d) 6
3. (a) 1080 (b) 1125 (c) 1000 (d) 625
5. about 13.5 grams
7. about 332,000
9. (a) $y = 3000e^{.19x}$
   (b) about 133,000
11. (a) about 243 (b) about 388 (c) about 1369

Chapter 9 Pretest *(page 275)*

1. -3 2. -23 3. -33
4.

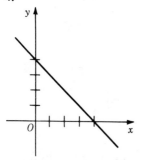

5.

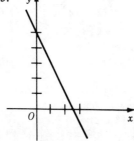

6.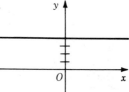

7.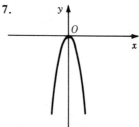

8.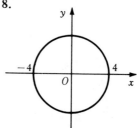

9.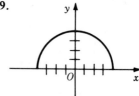

10. $f'(x) = 12\pi x$
11. $f'(x) = -\pi/(\pi x + 2\pi) = -1/(x+2)$
12. $f''(x) = 16 + 3x^{-5/2}$
13. $f''(x) = 6kx + 10k$
14. minimum at $x = 2$
15. minimum at $x = 3$; maximum at $x = -2$
16. (a) -7 (b) 161/5 or 32.2

Section 9.1 *(pages 278–279)*

1. 6   9. -17 17. 1300
3. -8 11. 47   19. 1996
5. 3   13. $\sqrt{43}$ 21. 136
7. 92   15. $\sqrt{19}$ 23. 304

Section **9.2** *(page 287)*

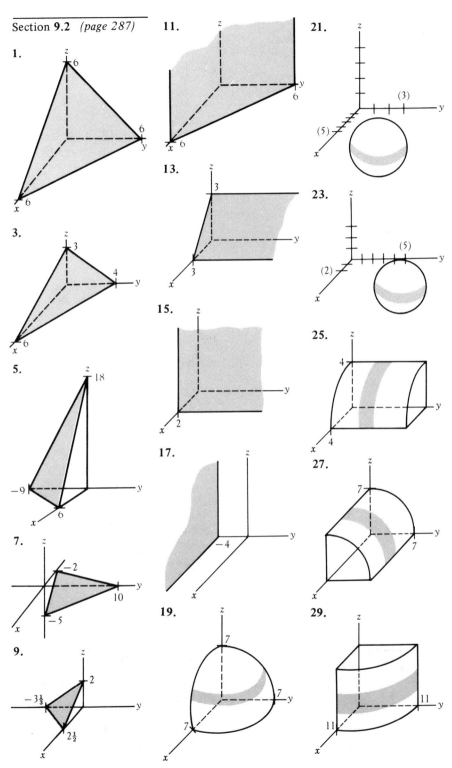

340

31.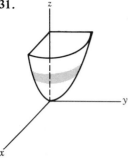

33.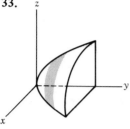

35.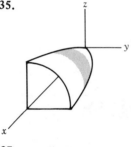

37.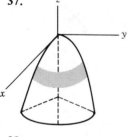

39.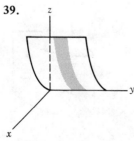

41.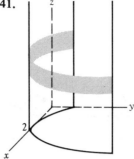

43.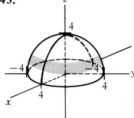

45.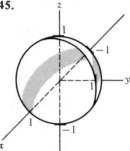

47.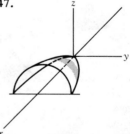

Section **9.3** *(page 292)*

In this section the answers are given in the order $f_x, f_y, f_{xx}, f_{yy}, f_{xy}$.

1. 8; 18y; 0; 18; 0
3. $8xy - 12y^2$; $4x^2 - 24xy$; $8y$; $-24x$; $8x - 24y$

5. $30x + 18y^3$; $36y^2 + 54xy^2$; 30; $72y + 108xy$; $54y^2$
7. $8x - 5y^3 + 24y^2x$; $-15xy^2 + 24yx^2$; $8 + 24y^2$; $-30xy + 24x^2$; $-15y^2 + 48yx$
9. $18x^2y^2 - 4y^4 - 16y^3$; $12x^3y - 16xy^3 - 48xy^2$; $36xy^2$; $12x^3 - 48xy^2 - 96xy$; $36x^2y - 16y^3 - 48y^2$
11. $e^xe^y + e^{2y} + 2y^3e^x$; $e^xe^y + 2xe^{2y} + 6y^2e^x$; $e^xe^y + 2y^3e^x$; $e^xe^y + 4xe^{2y} + 12ye^x$; $e^xe^y + 2e^{2y} + 6y^2e^x$
13. $x/\sqrt{x^2+y^2}$; $y/\sqrt{x^2+y^2}$; $y^2/(x^2+y^2)^{3/2}$; $x^2/(x^2+y^2)^{3/2}$; $-xy/(x^2+y^2)^{3/2}$
15. $-2y/(x-y)^2$; $2x/(x-y)^2$; $-4y/(x-y)^3$; $4x/(x-y)^3$; $-2(x+y)/(x-y)^3$
17. $2x/(x^2+y^2)$; $2y/(x^2+y^2)$; $(2y^2-2x^2)/(x^2+y^2)^2$; $(2x^2-2y^2)/(x^2+y^2)^2$; $-4xy/(x^2+y^2)^2$
19. $2/x$; $3/y$; $-2/x^2$; $-3/y^2$; 0
21. e^y; xe^y; 0; xe^y; e^y
23. 15; 0; 0
25. 28; 36; 150
27. $-1/\sqrt{5}$; 0; $6/13^{3/2}$
29. -2; does not exist; 0
31. e^2; 2; e^{-2}
33. (a) 80 (b) 180

Section 9.4 *(pages 299–300)*

1. relative minimum at $(-1,-1/2)$
3. relative minimum at $(-2,-2)$
5. relative minimum at $(15,-8)$
7. relative maximum at $(2/3, 4/3)$
9. saddle point at $(-2,2)$
11. saddle point at $(1,-1)$
13. relative maximum at $(10/3, 8/3)$
15. minimum cost of 59 at $x = 4, y = 5$

Section 9.5 *(pages 303–304)*

1. $f(6,6) = 72$
3. $f(4/3, 4/3) = 64/27$
5. $f(5,3) = 28$
7. $f(20,2) = 360$
9. 50 by 50
11. 30, 30, 30
13. 20 Doggie, 0 Petunia kits

Section 9.6 *(pages 308–309)*

1. $y = 4.0x + 5.3$
3. $y = 3.125x - 8.875$
5. (a)

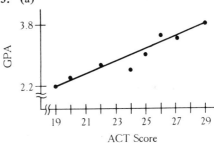

(b) $y = .16x - .89$ (c) 3.6

7. (a)

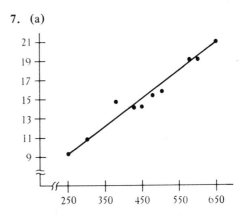

(b) $y = .03x + 1.56$
(c) 22.6; 24.1

Bibliography

Books for Algebra Review

Lial, Margaret, and Charles D. Miller, *Beginning Algebra*, 2nd ed. (Glenview, Illinois: Scott, Foresman and Company, 1976).

⸺, *Intermediate Algebra*, 2nd ed. (Glenview, Illinois: Scott, Foresman and Company, 1976). Study Guides for each of the books are available. These Study Guides alone might suffice for review purposes.

Calculus Books

Books for Reference

Ayers, Frank, *Calculus*, 2nd ed. (New York: McGraw-Hill Book Company, 1964). Ayers' book, one of the Schaum's Outline Series, can be used as a reasonable reference for this book. It is hard to understand, but features many worked-out examples. Most of the examples come from the physical sciences.

Nelson, Kaj, *Calculus* (New York: Barnes & Noble Books, 1958). Contains fewer worked problems than the Ayers book, but is a little easier to understand.

Books for Proofs

Each of the books listed below contains just about every proof omitted in this text. (Listed alphabetically.)

Clarke, Phillip, *Calculus and Analytic Geometry* (Lexington, Mass.: D. C. Heath & Company, 1974). Very similar in content and treatment to the Leithold book listed below.

Goodman, A. W., *Analytic Geometry and Calculus*, 3rd ed. (New York: Macmillan Publishing Co., Inc., 1974). Goodman's book has just about the most attractive illustrations of any of the books listed.

Johnson, R. E., and F. L. Kiokmesiter, *Calculus and Analytic Geometry*, 5th ed. (Boston: Allyn & Bacon, Inc., 1974).

Leithold, Louis, *Calculus and Analytic Geometry*, 2nd ed. (New York: Harper & Row, Publishers, 1972). This book is usually clear, direct, and to the point.

Riddle, Douglas, *Calculus*, 2nd ed. (Belmont, California: Wadsworth Publishing Co., Inc., 1974). A very good book, with many interesting treatments.

Salas, Saturnino, and Einar Hille, *Calculus, One and Several Variables*, 2nd ed. (Stamford, Conn.: Xerox Corporation, 1974). A nice, informal style of writing.

Spitzbart, Abraham, *Calculus* (Glenview, Illinois: Scott, Foresman and Company, 1975).

Stein, Sherman, *Calculus and Analytic Geometry* (New York: McGraw-Hill Book Company, 1974). Stein's book is the largest of those listed.

Thomas, George B., *Calculus and Analytic Geometry*, alternate ed. (Reading, Mass.: Addison-Wesley Publishing Co., Inc., 1972). A reprint, with minor changes, of the third edition of 1968.

Table of Integrals

Selby, S. M., ed. *CRC Standard Mathematical Tables*, 20th ed. (Cleveland: CRC Press, 1972). This is the most commonly available table of integrals. The company seems to have a perpetual sale—this book is available for around $5 if you order ten or more. Ask your instructor.

Index

Absolute maximum, 139
Absolute minimum, 139
Absolute value, 6
　function, 43
Addition property of
　equality, 8
Addition property of inequality, 10
Antiderivative, 210, 211
Area, 214
　between two curves, 226
Asymptote, 55
Axis, 30

Boeing Company, 172
Booz, Allen and Hamilton, 129
Boundary condition, 259

Calculus, fundamental theorem, 218
Carbon-14 dating, 191
Cartesian coordinate system, 30
Center of a circle, 49
Chain rule, 120
Change in x, 37
Characteristic, 200
Circle, 49
Closed interval, 84
Common denominator, 9
Common logarithms, 185, 199
Common ratio, 91
Composite function, 60
Compound interest, 203
Concave, 157
Constraint, 300
Consumer's surplus, 233
Continuity, 80
Continuous, 82
Coordinate system, 30

Cost-benefit curve, 57
Counting numbers, 3
Critical point, 136
Cube root, 21
Curve sketching, 153
Cylinder, 283

Decreasing function, 154
Definite integral, 214, 217
Delta x, 37
Denominator, common, 9
Dependent variable, 30, 277
Depletion of minerals, 251
Derivative
　applications of, 108
　chain rule, 120
　definition of, 96, 100, 102
　exponential, 195
　natural logarithm, 197
　partial, 287
　product, 116
　quotients, 118
　second, 142
　second partial, 290
Difference between two squares, 12
Difference quotient, 98
Differential, 235
Differential calculus, 96
Differential equations, 258
　applications of, 266
　general solution of, 259
Differentiation, techniques of, 103
Discontinuous, 82
Distributive property, 7
Domain, 30

e, 178
Ecology, 271
Element, 2
Empty set, 3
Entropy, 70
Equal sets, 3
Equation
　linear, 7
　of lines, 39
　quadratic, 11
Equilibrium
　demand, 41
　price, 41
　supply, 41
Explicit function, 123
Exponential, 20, 23
　derivative of, 196
　functions, 176, 178
Expression, rational, 16

Factoring, 11
First component, 29
First derivative test, 137
First octant, 280
FMC Corporation, 24
Function
　absolute value, 43
　composite, 60
　decreasing, 154
　definition of, 30
　explicit, 123
　exponential, 176, 178
　implicit, 124
　increasing, 154
　inverse, 183
　Lagrange, 300, 303
　limit of, 65
　linear, 35
　logarithmic, 182
　of two variables, 277
　operations on, 59
　polynomial, 52
　quadratic, 44

344

rational, 54
root, 47
Fundamental theorem of calculus, 218

General solution, 259
Genotype, 90
Geometric sequence, 91
   infinite, 92

Half-open interval, 85
Hares, 272
Horizontal asymptote, 55

Implicit differentiation, 124
Implicit function, 124
Increasing function, 154
Independent variable(s), 30, 277
Indeterminate form, 165
Inequalities, quadratic, 14
Infinite geometric sequence, 92
Inflection point, 159
Initial condition, 259
Instantaneous rate of change, 99
Integers, 4
Integral calculus, 96
Integral(s)
   definite, 214, 217
   sign, 217
   table of, 241
Integration
   by parts, 244, 245
   numerical, 248
   techniques of, 235
Intercept, 36
Interest, 203
Interval, 84
Inverse function, 183
Irrational numbers, 4

Lagrange function, 300
Lagrange multipliers, 300, 303
Learning curve, 191

Least squares regression line, 306
L'Hopital's rule, 165
   extended, 167
Limit(s)
   definition of, 67
   of a function, 65
   of a sequence, 90
   of integration, 217
   properties, 75
   to infinity, 86
Linear equation, 7
Linear function, 35
Logarithmic functions, 182
Logarithms
   common, 185, 199
   natural, 185
Lynx, 272

Mantissa, 200
Marginal cost, 113
Mathematical model, 7
Member, 2
Mineral depletion, 251
Model, mathematical, 7
Montgomery Ward, 206
Multiplication property of equality, 8
Multiplication property of inequality, 10
Multiplier, 92
Mutation, 92

Natural logarithm, 185
   derivative of, 197
Natural numbers, 3
Negative x integer exponents, 20
Neighborhood, 133, 293
Newton's method, 161, 163
n-th root, 21
Null set, 3
Number line, 3
Number(s)
   counting, 3
   integers, 4
   irrational, 4
   natural, 3

rational, 4
real, 3
whole, 4
Numerical integration, 248

Octant, 280
Oil tanker case, 62
Open interval, 84
Optimization, 133
Ordered pair, 29, 279
Ordered triple, 279
Origin, 30

Parabola, 44
Paraboloid, 285
Partial derivative, 287
Particular solution, 259
Plane, 280
Point of inflection, 159
Point-slope form, 39
Polynomial function, 52
Producer's surplus, 232
Product rule, 116
Properties of exponents, 20, 23

Quadrant, 31
Quadratic equation, 11
Quadratic formula, 13
Quadratic function, 44
Quadratic inequalities, 14
Quotient rule, 118

Radius of a circle, 49
Range, 30
Rational exponent, 21
Rational expression, 16
Rational function, 54
Rational numbers, 4
Real numbers, 3
Regression line, 306
Relation, 29
   domain, 30
   range, 30
Relative extrema, 134
Relative maximum, 134, 293
Relative minimum, 134, 293

Retinoblastoma, 92
Right circular cylinder, 283
Root, 21
Root function, 47

Saddle point, 295
Scatter diagram, 304
Scrap value, 201
Secant lines, 109
Second component, 29
Second derivative, 142
  test, 143
Second partial derivative, 290
Separation of variables, 263
Sequences, 89
  geometric, 91
Set, 2
  braces, 2
  -builder notation, 3
  element, 2

empty, 3
equal, 3
member, 2
null, 3
subset, 3
Σ-notation, 305
Simpson's rule, 249, 250
Slope, 37
Slope-intercept form, 39
Slope of a tangent line, 108, 111
Solving an equation, 7
Species interaction, 271
Sphere, 282
Square root, 21
Subset, 3

Table of integrals, 241
Tangent line, 108
Techniques of differentiation, 103
Techniques of integration, 235

Terms of a sequence, 90
Trapezoidal rule, 249

Variable
  dependent, 30
  independent, 30
Variables separable, 263
Vertex of a parabola, 44
Vertical asymptote, 55
Vertical line test, 31

Warranty costs, 254
Whole numbers, 4

x-axis, 30
x-coordinate, 30
x-intercept, 36

y-axis, 30
y-coordinate, 30
y-intercept, 36

Zero factor property, 12